E&M Endocrinology and Metabolism

Progress in Research and Clinical Practice

Margo Panush Cohen   Piero P. Foà

Series Editors

Endocrinology and Metabolism
Progress in Research and Clinical Practice

Margo Panush Cohen   Piero P. Foà
Series Editors

Cohen and Foà (eds.): Hormone Resistance and Other Endocrine Paradoxes (Vol. 1)
Jovanovic (ed.): Controversies in Diabetes and Pregnancy (Vol. 2)
Cohen and Foà (eds.): The Brain as an Endocrine Organ (Vol. 3)
Ginsberg-Fellner and McEvoy (eds.): Autoimmunity and the Pathogenesis of Diabetes (Vol. 4)

Forthcoming volumes:

Cohen and Foà (eds.): Endocrinology of Bone, Joints, and Skeletal Muscle (Vol. 5)
Cohen and Foà (eds.): Growth Factors and Hormones in Angiogenesis and Hemopoiesis (Vol. 6)

Fredda Ginsberg-Fellner   Robert C. McEvoy
Editors

Autoimmunity and the Pathogenesis of Diabetes

With 71 Figures in 110 Parts, 5 in Full Color

Springer-Verlag
New York Berlin Heidelberg
London Paris Tokyo Hong Kong

Fredda Ginsberg-Fellner, M.D.
Professor of Pediatrics
Director, Division of Pediatric Endocrinology
and Metabolism
Mount Sinai School of Medicine
New York, New York 10029-6574
USA

Robert C. McEvoy, M.D.
Professor of Pediatrics
(Endocrinology) and Cell Biology
Mount Sinai School of Medicine
New York, New York 10029-6574
USA

Series Editors
Margo Panush Cohen, M.D., Ph.D.
Director, Institute for Metabolic Research
University City Science Center
Philadelphia, Pennsylvania 19104
USA

Piero P. Foà, M.D., Sc.D.
Professor Emeritus of Physiology
Wayne State University
Chairman Emeritus
Department of Research
Sinai Hospital
Detroit, MI
Mailing Address:
2104 Rhine Road
West Bloomfield, Michigan 48033
USA

Library of Congress Cataloging-in-Publication Data
Autoimmunity and the pathogenesis of diabetes / Fredda Ginsberg
-Fellner, Robert C. McEvoy, editors.
p. cm. — (Endocrinology and metabolism ; vol. 4)

1. Diabetes—Pathogenesis. 2. Diabetes—Immunological aspects.
3. Autoimmune diseases. I. Ginsberg-Fellner, Fredda. II. McEvoy,
Robert C. III. Series: Endocrinology and metabolism (New York,
N.Y.) ; 4.
[DNLM: 1. Autoimmune Diseases. 2. Diabetes Mellitus—etiology.
3. Diabetes Mellitus—immunology. W1 EB396SN v. 4 / WK 810 A939]
RC660.A93 1990
616.4'62071—dc20
DNLM/DLC
for Library of Congress 89-21968

Softcover reprint of the hardcover 1st edition 1990

Typeset by Publishers Service, Bozeman, Montana.

9 8 7 6 5 4 3 2 1

ISBN-13: 978-1-4612-7920-4 e-ISBN-13: 978-1-4612-3218-6
DOI: 10.1007/978-1-4612-3218-6

Preface

It has been a challenge for us to edit this volume of *Endocrinology and Metabolism: Progress in Research and Clinical Practice*. The topic of the pathogenesis of insulin-dependent, type I diabetes mellitus is particularly appropriate for this series, since advances in this area have been made, to a large extent, by applying state-of-the-art laboratory techniques to clinical samples.

Over the last several years, a number of lines of evidence have been gathered, suggesting that classic type I diabetes mellitus results from the autoimmune destruction of pancreatic beta-cells in genetically susceptible individuals. This hypothesis is particularly appealing because it offers a rational approach to the prevention of diabetes by immunosuppression.

We have tried to present a balanced, authoritative summary of the information currently available to support the autoimmune hypothesis for the pathogenesis of human type I diabetes, to place this information in historical perspective, to include relevant information from animal models of type I diabetes in which more invasive experimentation is ethical, and, finally, to update the reader on the current status of attempts to intervene in the progression of diabetes with immunosuppressive drugs.

New York, New York

Fredda Ginsberg-Fellner
Robert C. McEvoy

Contents

Contributors

HEIKE BÄRMEIER, M.D.
Robert H. Williams Laboratory, Department of Medicine, RG-20, University of Washington, Seattle, Washington 98195, USA

GIAN FRANCO BOTTAZZO, M.D., M.R.C.P., M.R.C.PATH.
Reader in Clinical Immunology, Department of Immunology, Middlesex Hospital, Medical School, London, United Kingdom

MICHAEL CHRISTIE, PH.D.
C. H. Best Institute, University of Toronto, Toronto, Ontario, Canada

JOHN DUPRÉ, F.R.C.P.(C.), F.R.C.P.(LONDON)
Department of Medicine, University of Western Ontario, University Hospital; Robarts Research Institute, London, Ontario, Canada

G.S. EISENBARTH, M.D., PH.D.
Senior Investigator, Joslin Diabetes Center; Associate Professor of Medicine, Brigham and Women's Hospital, Harvard Medical School, Boston, Massachusetts 02115, USA

HELGA GLEICHMANN, M.D., PH.D.
Professor of Immunology, Clinical Department, Diabetes-Research-Institute at the University of Düsseldorf, Düsseldorf, Federal Republic of Germany

LIMING HAO, M.D.
Barbara Davis Center for Childhood Diabetes, Department of Microbiology and Immunology, Pediatrics, University of Colorado Health Science Center, Denver, Colorado 80262, USA

BETSY HEROLD, M.D.
Children's Memorial Hospital, Chicago, Illinois 60614, USA

KEVAN HEROLD, M.D.
Department of Medicine, University of Chicago, Chicago, Illinois 60637, USA

KEVIN J. LAFFERTY, PH.D.
Research Director; Professor of Microbiology/Immunology and Pediatrics, Barbara Davis Center for Childhood Diabetes, University of Colorado, Health Sciences Center, Denver, Colorado 80262, USA

ÅKE LERNMARK, PH.D.
Robert H. Williams Professor of Medicine, Department of Medicine, University of Washington, Seattle, Washington 98195, USA

THOMAS MANDRUP-POULSEN, M.D., PH.D.
Steno Memorial Hospital, Gentofte, Denmark

R.C. NAYAK, B.SC., PH.D.
Investigator, Joslin Diabetes Center; Instructor in Medicine (Biochemistry), Harvard Medical School, Boston, Massachusetts 02215, USA

JØRN NERUP, M.D.
Chief Physician, Head of Research, Steno Memorial Hospital, Gentofte, Denmark

JERRY P. PALMER, M.D.
Head, Division of Endocrinology and Metabolism, Seattle V.A. Medical Center; Professor of Medicine, University of Washington, Seattle, Washington 98195, USA

PABLO RUBINSTEIN, M.D.
Senior Investigator, and Head, Immunogenetics Laboratory, The Lindsley F. Kimball Research Institute of the New York Blood Center, New York, New York 10021, USA

C. STILLER, M.D., F.R.C.P.(C.)
Department of Medicine, University of Western Ontario, University Hospital; Robarts Research Institute, London, Ontario, Canada

NANCY M. THOMAS, B.S.
Research Coordinator, Department of Anatomy, Mount Sinai School of Medicine of the City University of New York, New York, New York 10029-6574, USA

YI WANG, M.D.
Barbara Davis Center for Childhood Diabetes, Department of Microbiology/ Immunology and Pediatrics, University of Colorado Health Science Center, Denver, Colorado 80262, USA

JI-WON YOON, PH.D.
Professor and Chief, Division of Virology, Department of Microbiology and Infectious Diseases; Director, Laboratory of Viral and Immunopathogenesis of Diabetes, Julia McFarlane Diabetes Research Center, Health Science Center, The University of Calgary, Calgary, Alberta, Canada

1
The Autoimmune Hypothesis of Insulin-Dependent Diabetes: 1965 to the Present

THOMAS MANDRUP-POULSEN AND JØRN NERUP

Introduction

When considering hypotheses of autoimmunity in insulin-dependent diabetes mellitus (IDDM) the selection of 1965 as a starting point for discussion is not arbitrary. In his 1965 paper,[1] Willy Gepts cautiously interpreted histopathologic findings in pancreases of recent-onset juvenile diabetic patients as supportive of the view pointing "to a possible immunologic derangement . . . of the greatest interest in view of the high frequency with which we have found inflammatory infiltrates in our recent-onset juvenile diabetics." In other words Gepts not only stressed the common occurrence of inflammatory lesions in the islets of recent-onset young diabetics, but also linked the histopathologic findings with pathogenetic considerations. No doubt this early hypothesis inspired diabetologists to search for further evidence to support an immunologic basis for the development of IDDM, a search that has grown ever since, making diabetes immunology a rapidly expanding field. This volume is but one of several specialized issues that have been devoted to the etiopathogenesis of IDDM.

Since the individual chapters in this book cover in detail important aspects of the autoimmune hypothesis in IDDM, the aim of the following review is partly to provide a general introduction to the immune system, autoimmunity, and autoimmune disease and partly to focus on areas of controversy in IDDM pathogenesis, wherein recent experimental evidence necessitates a reformulation of currently accepted hypotheses.

Autoimmunity

Autoimmunity refers to the apparent termination of a natural unresponsive state to self.[2] It is important to realize that the manifestation of autoimmunity by the presence of lymphoid cells and of circulating antibodies with reactivity to self is by no means equivalent to disease. In fact it has become increasingly evident that autoreactivity is an integral part of the normal immune system. Thus circulating autoantibodies can be detected in most normal healthy individuals. Controlled

TABLE 1.1. Four criteria for autoimmune diseases.

Circulating or cell-bound antibodies reactive with the autoantigen must be detectable
The autoantigen(s) must be identified
A mononuclear cell infiltration must be present in the target tissue
Passive transfer of the disease should be possible in experimental models with lymphoid cells or serum

suppression of autoreactivity in fetal life is an important part of the development of tolerance, and autoantibodies may even play a role in preventing autoimmune disease.[3] How immunologic tolerance is broken to initiate an autoimmune response is not known, but several hypotheses have been suggested, including lack of deletion of forbidden clones, liberation of antigens previously inaccessible to the immune system, altered self, cross-reactivity between nonself-antigens and self-antigens, or withdrawal of the normal suppression of autoreactive clones. Autoimmune disease results when autoimmunity leads to structural or functional damage to cells or tissues to the extent that clinical symptoms and signs appear. What makes autoimmune reactivity progress to autoimmune disease is only incompletely understood, but it is conceivable that autoimmune activity encompasses a continuous spectrum, ranging from the presence of harmless autoantibodies to severe autoimmune tissue destruction.

The classic criteria[4] for autoimmune diseases are summarized in Table 1.1. However many diseases now considered to be autoimmune in nature do not fulfill all these criteria, and the heterogeneity of histopathologic and immunologic manifestations of autoimmune diseases makes it difficult to formulate exhaustive, unifying definitions. Often the beneficial clinical effects of *ex juvantibus* immunosuppressive treatment on the course of a disease of suspected autoimmune origin is taken as sufficient evidence to prove a causal relationship between autoimmune phenomena and the pathogenesis of the disease in question.

There is little reason to believe that the autoimmune response differs from the normal immune response in terms of the fundamental processes of recognizing the antigen, triggering immunologic help, or mounting the inflammatory infiltrate. In other words the immune response may not function in an abnormal way in autoimmunity, but rather the specificity of reactivity and perhaps the control of the response may be abnormal, leading to functional or structural tissue or cell disturbances. Thus before we consider autoimmune hypotheses in IDDM pathogenesis, we will briefly highlight some important basic features of the immune response (Fig. 1.1).

The first step in the immune response is usually the *uptake* and *processing* of antigen by an antigen-presenting cell (APC), which is often the macrophage (MO). The processed antigen is *presented* on the surface of the APC in the context of class II molecules of the major histocompatibility complex (MHC), in man, termed *HLA-D molecules*.[5] The presented antigen is *recognized* in the context of self HLA-D molecules by T- (thymus-derived) helper (Th) cells expressing specific receptors for the antigen.[6] *Activitation* of the APC by antigen processing

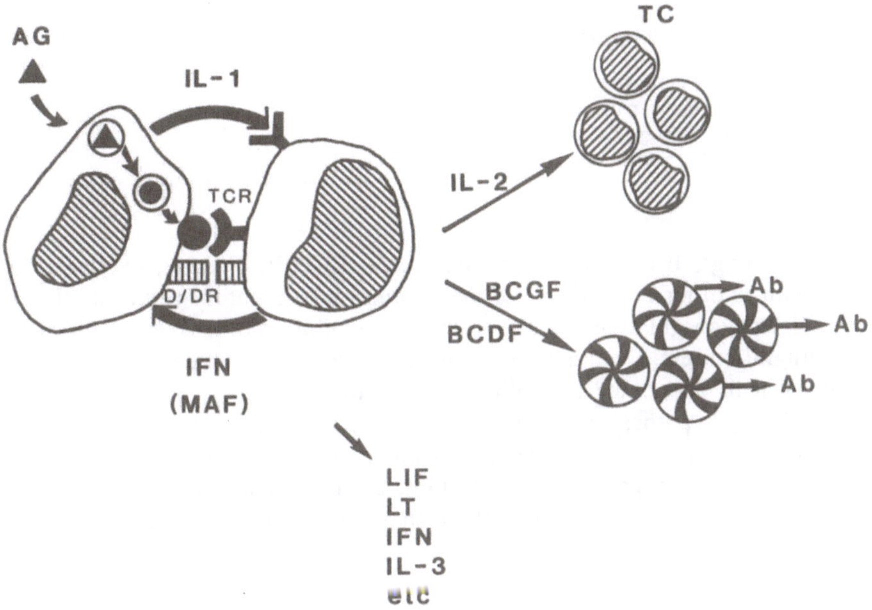

FIGURE 1.1. Scheme of some basic elements of immune response initiation. See text for an explanation. **Left:** The macrophage/antigen-presenting cell (to the left) and the helper T-cell (to the right) are interacting. Ag, soluble antigen; D/DR, MHC class II antigens; TCR, T-cell receptor complex; IL-1, interleukin-1; IFN, interferon; MAF, macrophage-activating factors. **Right**: Above, Tc, cytotoxic T-cells; IL-2, interleukin-2. Below, Ab, antibody-producing plasma cells; BCGF, B-cell growth-promoting factor; BCDF, B-cell differentiation factor; LIF, leukocyte migration inhibitory factor; LT, lymphotoxin; IL-3, interleukin-3.

and by interaction with the Th-cell during the recognition process[7] leads to the secretion of a soluble antigen-nonspecific factor from the APC, the monokine interleukin-1 (IL-1), which binds to surface receptors on the Th-cell.[8] Recognition of antigen and binding of IL-1 activate the Th cell to secrete a variety of other soluble, antigen-nonspecific factors, lymphokines. Monokines and lymphokines, together called cytokines, are important mediators of immunity.[9] Thus macrophage-activating factors (MAF), for example, interferon-gamma (IFN-gamma), stimulate the MO to produce more IL-1,[10] thereby *amplifying* the immune response by establishing a positive feedback loop between MO- and MAF-producing Th-cells. *Recruitment* of T- and B-lymphocytes, MO, natural killer- (NK) and killer- (K) cells to the target tissue is achieved by the release of chemotactic cytokines and cell constituents. The secretion of the T- and B-cell, growth-promoting lymphokines, interleukin-2 (IL-2), and B-cell growth factor (BCGF) in the presence of antigen leads to a *clonal expansion* of precursor T- and B-lymphocytes. These differentiate into cytotoxic (Tc) or suppressor (Ts) T-cells, or antibody-producing plasma cells, respectively, under the influence of other

TABLE 1.2. Features of IDDM in man and experimental models that should be addressed in hypotheses of IDDM pathogenesis.

Beta-cell selectivity of the disease process
Slow, progressive course in many cases of IDDM
Heterogeneous islet histopathology; desynchronized in time and space
HLA-association, but not HLA-restriction of the disease

cytokines (e.g., B-cell differentiation factor, BCDF). Further, cytokines are important in other components of the immuno-inflammatory response, that is, vasodilation, coagulation and clot formation, inhibition of leukocyte and monocyte migration, etc. Eventually the immune response is *limited* and *terminated*, events usually associated with elimination of the antigen.

Ideally any hypothesis concerning the beta-cell destructive immune process in IDDM should address the general fundamental issues of how this process is initiated, amplified, perpetuated, and limited; and in addition, it should be able to explain certain features of IDDM that are becoming clear from recent experimental evidence (Table 1.2).

The Concept of IDDM as an Autoimmune Disease

Until recently the concept of IDDM as an autoimmune disease was based primarily on circumstantial evidence, consisting of the following autoimmune phenomena: the association of IDDM susceptibility with immune response genes in the HLA-DR region[11]; the observation of islet-related autoimmunity, that is, insulitis,[1] cell-mediated, antipancreatic autoimmunity,[12] circulating antibodies reacting with cytoplasmic (ICA),[13] or surface structures (ICSA)[14-16] of the normal islet cells; the association of IDDM with other organ-specific autoimmune diseases[17]; and the presence in IDDM of other organ-specific autoantibodies.[18] As in other diseases of suspected autoimmune nature, intervention studies with specific immunosuppressive drugs have provided additional direct evidence for the involvement of autoimmune mechanisms in beta-cell destruction in IDDM. The much higher than expected frequency of remission in newly diagnosed IDDM patients treated with cyclosporin (Cs) in open pilot studies[19,20] and the significantly higher frequency of remission in the Cs-treated IDDM patients in double-blind, clinically controlled trials[21,22] confirm the results of studies in animal models[23,24] that have shown that immunosuppression can arrest the beta-cell destructive process. These and other aspects of IDDM autoimmunity will be dealt with in detail elsewhere in this volume. The following discussion briefly reviews the hypotheses of autoimmunity in IDDM from 1965 to the present by considering the single earliest event in the activation of the immune system dealt with in general terms above, that is, genetic and etiologic factors with implications for the pathogenesis of IDDM, triggers and genetic control of the autoimmune response, antigen presentation and recognition, triggering of the T-helper cell and the effector-phases of the autoimmune response to the beta-cells.

The Association Between IDDM and HLA

In 1974, the genetic susceptibility to IDDM was linked to the A8 and w15 haplotypes of the MHC system (HLA in man).[11,25] This important observation led to the classification of IDDM and non-IDDM as two genetically different diseases and increased the possibility of identifying the mode of inheritance of IDDM and individuals at genetic risk for development of IDDM. It also suggested that IDDM might be pathogenetically related to the autoimmune endocrinopathies Graves' disease and idiopathic Addison's disease, both of which were associated with HLA-A8. It was implied that "one or more immune-response genes associated with HL-A8 and/or W15 might be responsible for an altered T-lymphocyte response" in IDDM.[11] Later it was shown that the association of IDDM with A8 and w15 was due to linkage disequilibrium with loci (particularly the DQ locus) in the HLA-DR region,[26,27] which is considered to govern the immune response.[28] More than 90% of IDDM patients carry the phenotypes HLA-DR3 or HLA-DR4 or both, in contrast to approximately 50% of the nondiabetic population.[29] The highest relative risk is conferred by DR 3/4 heterozygosity, but the positive association between HLA phenotypes and IDDM in fact represents a continuum,[25] in which the relative risk of IDDM development of DR4 > DR3 > DRw9, and 8% of IDDM cases are non-DR 3, non-DR 4 or both. The diabetogenic HLA phenotypes are also highly associated with other autoimmune diseases.

It is possible that HLA-DR3 and HLA-DR4 are not identical in IDDM patients and controls or, in other words, that minor differences in the hypervariable regions of the HLA-D antigens may exist between IDDM patients and controls. Studies in molecular genetics, particularly in restriction fragment length polymorphism (RFLP), have identified RFLP patterns that confer a much higher relative risk (about 10-fold) for IDDM than conventional serological HLA-typing,[30] but genetic markers, even at the nucleotide sequence level specific for IDDM, have not been identified in coding regions and may not exist. Indeed the IDDM-associated mutations that have been identified are located in intervening sequences,[31] which are spliced out during transcription.

The mode of inheritance of IDDM remains an enigma and it is puzzling that the penetrance of IDDM in familial cases is higher than in sporadic cases of IDDM,[25] a fact that may indicate the involvement of additional non-HLA-linked susceptibility genes (e.g., polymorphisms flanking the insulin gene?)[32] or environmental triggering factors.

In the NOD mouse, susceptibility of IDDM seems to be associated with a recessive MHC class II gene and with at least one non-MHC gene.[33] In the BB rat, an MHC class II-related gene seems essential for IDDM development, whereas a non-MHC-related gene for lymphopenia is permissive, although not necessary for IDDM development.[33,34] Polymorphisms flanking the rat insulin II gene were not associated with IDDM in the BB rat.[35] In IDDM patients and healthy control subjects, the T-cell response against mumps and Coxsackie B4 virus was stronger in HLA-DR4– than in HLA-DR3–positive individuals but was identical between IDDM and controls.[36,37] One way HLA-D/DR genes could quantitatively influence the immune response is by determining the amount of immune mediators produced by interacting immune cells in response to a given antigenic stimulus. In fact the production of IL-1,[38,39] and certain lymphokines[40,41] may be controlled by the HLA-D/DR molecules. Further the genes for the MO mediator tumor necrosis factor (TNF) and the lymphokine lymphotoxin (LT) were recently mapped to the HLA region; in man they were either centromeric to DP or between DR and B,[42] and in the mouse, in the class III(D) region.[43] These chromosomal assignments indicate that TNF and LT genes may be closely linked to the HLA-B gene in man in linkage disequilibrium with DR and that certain polymorphisms of these cytokine genes could exist in certain HLA haplotypes. Such assignments would have implications for the level of TNF/LT secretion by interacting mononuclear cells (see below).

In summary the evidence suggests a quantitative rather than a qualitative role of the immune response genes associated with IDDM susceptibility. In other words, specific diabetogenic genes may not exist, but differences in host immune response against particular antigens related to certain HLA-types or subtypes may account for the HLA-IDDM linkage.[25] Such HLA-DR–related differential patterns of immune responses are well known for a number of antigens (for a survey, see Ref. 44).

Etiologic Factors with Implications for the Pathogenesis of IDDM

As discussed above, the mechanisms underlying the breakage of tolerance and the processes that make autoantigen accessible to the immune system are not well understood. A role for environmental triggering agents in the etiology of IDDM has been claimed since 1970, when it was demonstrated that the concordance rate of IDDM in identical twin pairs is less than 15%[45]; this is supported by the lower penetrance of IDDM in familial cases than in sporadic cases of IDDM.[25] However identical twins may not have the same immunologic repertoire, since rearrangement of immunoglobulin genes and T-cell receptor genes occurs,[46,47] and additional HLA-nonrelated genes may be involved in explaining the differences in IDDM penetrance. Although viruses (see Chapter 10) and chemicals (for a review, see Ref. 48) can undoubtedly cause IDDM-like syndromes in experimental animals, the role of specific, single environmental factors in the etiology of human IDDM is debatable. The seasonal variation in IDDM incidence, the association of IDDM to virus epidemics, and the finding of higher antibody titers against certain virus strains in newly diagnosed IDDM cases[49] may reflect precipitating rather than causative factors in IDDM development. In one case of human IDDM associated with virus infection,[50] histopathologic examination of the child's pancreas indicated that the excessive beta-cell damage observed may have preceded the viral infection.[51] Congenital rubella infections are associated with a high risk of developing IDDM later in life,[52] but whether this increased risk is caused by a direct effect of the virus on the antigenicity of the endocrine pancreas, or through effects on the immune system, is not known. In the BB rat, IDDM develops in spite of gnotobiotic breeding.[53] Genetic viral transmission cannot be excluded, however, as a possible trigger unaffected by gnotobiotic breeding. N-nitroso-compounds found in smoked food have been causally incriminated in IDDM,[54] but an epidemiologic association between the intake of smoked food and IDDM was not confirmed by others.[55,56] Dietary protein may play a role as a trigger for IDDM development in the BB rat.[57]

In summary, at present, it is not possible to prove or disprove the role of specific environmental factors as etiologic triggers in IDDM. Multiple subdiabetogenic, beta-cell noxious environmental events might induce cumulative beta-cell damage and trigger IDDM development.[58] In addition the possible protective role of environmental factors cannot be discarded.[59,60] However perhaps the time has come to question the need for specific exogenous triggers in IDDM, since two decades of intensive research in this area have not led to the identification of common environmental factors of substantive importance. Recently it was suggested that breakage of tolerance could be related to delayed tissue-specific expression of beta-cell autoantigen.[61] As discussed below, there may also be reasons to look for endogenous triggers (e.g., interleukin-1).

The Autoantigen in IDDM

The autoantigen(s) that is of primary importance in the activation of the autoimmune reaction against the beta-cells is unknown. The approach that has been used to identify putative autoantigens in IDDM has been to search for antigens that bind to the circulating autoantibodies. Thus ICA may bind to sialic acid containing glycolipids (gangliosides) in human islet tissue (Chapter 6).[62] However such gangliosides are not islet-specific, since monoclonal antibodies raised against islet tissue can react with glycoprotein and glyco-lipids in several neuroendocrine tissues.[63]

ICSA-positive sera from children with recent-onset IDDM[16] or from diabetic BB rats[64] immunoprecipitate a 64-kD protein from normal human or rat islet tissue. It has not been conclusively demonstrated that this islet-cell protein is beta-cell–specific or that it is a cell surface protein. Further characterization of this antigen is obviously of the utmost importance, not only to provide better assays for islet-cell autoantibodies but also to determine whether the 64-kD antigen has any pathogenetic relevance. Theoretically the antibody response to the 64-kD protein may be secondary to a polyclonal B-cell activation, with antibodies being generated in response to the liberation of a number of different islet-cell proteins following beta-cell destruction by an antibody-unrelated mechanism (see below).

Presentation of Autoantigen

Normally the presentation of antigen depends on specialized antigen-presenting cells, that is, cells with the ability to phagocytize antigen, to transform it into a form that can be recognized by T-helper cells, to present the antigen in the context of HLA-DR molecules, and to provide the necessary costimulatory signal for the T-helper cell in the form of interleukin-1 (Fig. 1.1). The APCs that possess all these properties are ordinarily tissue macrophages (MO) but include a number of other cells of the monocytic/phagocytic lineage (i.e., endothelial cells, astroglial cells, Langerhans cells of the skin, Kupffer cells in the liver sinusoids). Alternatively it has been proposed that endocrine cells that do not ordinarily express HLA-DR antigens can act as antigen-presenting cells if they are induced to so-called "aberrant" HLA-DR expression, thereby committing a sort of immunologic suicide.[65]

According to this hypothesis, IFN-gamma, possibly in combination with TNF[66] produced as a result of a local immune response to a viral infection, induces the beta-cells to express HLA-DR, enabling the Th-cell to recognize beta-cell–specific surface antigens. Activation of the Th-cell would then depend on IL-1 derived from sources other than the antigen-presenting cell, since beta-cells may not be able to produce IL-1.[67] For example IL-1 secretion could not be detected from cultured thyroid cells.[68] Although DR expression has been detected on beta-cells at the onset of IDDM[69] and on thyroid epithelium in autoimmune thyroid disease,[70] the importance of aberrant HLA-DR expression in the induction of endocrine autoimmunity is controversial. Thus aberrant HLA-DR–expressing cells may be poor antigen-presenting cells[71] and only seem to be efficient in presenting antigen to presensitized, activated T-cells.[72] In addition there is evidence to suggest that HLA-DR expression on the endocrine target cell is a late, secondary phenomenon to the specific autoimmune assault[73] and that HLA-DR expression may, in fact, protect the target cell from the immunologic attack,[73] for example, from NK-cells.[74]

Autoantigen Recognition in IDDM

Recognition of beta-cell autoantigen depends on T-helper lymphocytes. Much interest has focused on the establishment of T-cell lines or even T-cell clones with specific reactivity for islet-cell antigens. Such T-cell lines have been established in animal models of, for example, experimental allergic encephalitis. Injection of T-cells specific for basic myelin protein passively transferred the disease, and immunization with such attenuated T-cells prevented the disease.[75] Islet-antigen–specific, T-helper lymphocyte lines that proliferated and secreted IL-2 in an MHC-restricted manner have been established[76] from spleen and from inflammed islets of diabetic BB rats, but such T-cell lines rapidly lost specificity in culture. Obviously if such T-cell lines could be maintained and cloned, they would be invaluable tools for studying the spectrum of islet antigen epitopes that are recognized in experimental or human IDDM and for delineating which epitopes are of primary pathogenetic importance. Ultimately one could envision preventing IDDM by immunization with the relevant clonotypes.

The Effector Phase in IDDM

The effector phase of the immune response, that is, the mechanisms that lead to functional or structural cell damage depend on (1) antibodies, (2) leukocytes (T-cells, B-cells, natural killer- (NK) cells, macrophages (MO), and granulocytes), and (3) inflammatory mediators and other soluble nonantibody products of activated leukocytes.

Islet-Cell Autoantibodies As Effector Molecules

Islet-cell autoantibodies (reviewed in detail elsewhere in this volume) may be markers of ongoing beta-cell destruction,[77-79] although there is conflicting evidence as to whether persistence of ICA is related to better or poorer beta-cell function. Thus ICA may precede the onset of overt IDDM by several years,[80] but their predictive value is unclear in view of the report that only about one-third of ICA-positive unaffected individuals develop IDDM when followed for more than five years.[81] Also ICSA (reviewed in this volume) and 64-kD antibodies may be additional markers for IDDM,[82,83] but prospective studies of the relationship between these antibodies and beta-cell function have not been published.

Pathogenetic importance has been ascribed to islet-cell autoantibodies through complement-dependent, antibody-mediated cytotoxicity (CAMC) or antibody-dependent, cellular cytotoxicity (ADCC) to beta-cells. Although ICA may fix complement when bound to cytoplasmic antigens on cryostat sections of human pancreas,[84] and ICSA may induce CAMC to xenogeneic islet cells in vitro, a number of observations have raised questions regarding a primary, in vivo pathogenetic role of circulating islet-cell autoantibodies. Islet-cell antibodies (1) bind to cytoplasmic rather than surface antigens; (2) bind nonspecifically to all islet cells; (3) are unrelated to the beta-cell function during cyclosporin-induced remission (Fig. 1.2);[85] and (4) when transplacentally transferred from ICA-positive pregnant mothers to their fetuses, do not transfer IDDM.[86] Islet-cell surface antibodies (1) in the sera from 25% of unaffected first-degree relatives cause in vitro CAMC, that is, in vitro CAMC to islet cells is present in five times the number of first-degree relatives that develop IDDM[87]; (2) are found in 30% of nondiabetic patients with thyroid disease[15]; (3) are unrelated to insulitis, development of IDDM, or glucose intolerance in the BB-rat[88]; and (4) do not transfer IDDM or cause beta-cell destruction when injected repeatedly into mice.[89] Thus islet-cell autoantibodies are most probably immunologic epiphenomena that are caused by polyclonal B-cell activation following liberation of antigen from destroyed beta-cells. In support of polyclonal B-cell activation in IDDM is the diversity of circulating islet-nonrelated autoantibodies

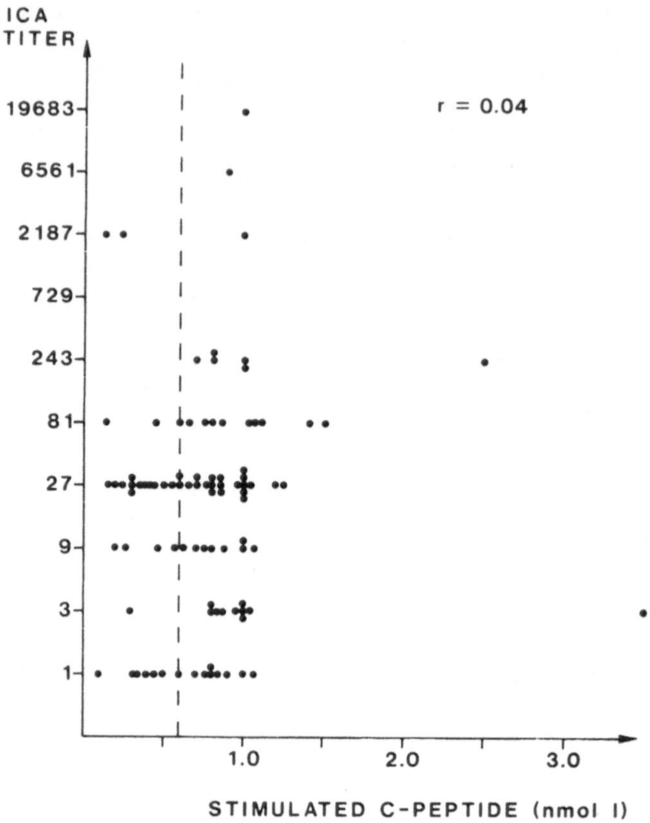

FIGURE 1.2. Lack of correlation between ICA titer and amount of glucagon-stimulated C-peptide in cyclosporin-treated IDDM patients of recent onset. Data from (85).

detectable in recent-onset IDDM patients.[90] Whether islet-cell autoantibodies participate in later stages in beta-cell destruction has not been demonstrated. It is difficult to imagine, however, that circulating beta-cell cytotoxic antibodies constitute the only effector mechanism that could explain the slow, progressive course and heterogeneous islet histopathology that characterize IDDM (Table 1.2).

Mononuclear Cells

Mononuclear cells (MNC) infiltrate the islets of Langerhans and this lesion (insulitis) resembles a typical type IV hypersensitivity reaction.[1] The cells present in late-stage insulitis, that is, after most of the beta-cells have been eradicated, in man are T-cytotoxic/suppressor cells, T-helper/inducer cells, NK-cells, K-cells, MO, and B-cells; in the rat they are activated T-cells, T-helper/inducer cells, and T-cytotoxic/suppressor cells.[91-93] In early insulitis in the BB rat, infiltration with a few macrophages, observed by electron microscopy, precedes light microscopic insulitis.[94,95] The insulitis infiltrate observed in segmental isografts transplanted from healthy,

TABLE 1.3. Summary of studies of allogeneic endocrine pancreatic transplantation. For details and references, see text.

Graft	IDDM model	
	Nonspontaneous	Spontaneous
Untreated		
Islet	Destroyed	Destroyed
Non-islet	Destroyed	Destroyed
Culture conditioned		
Islet	Accepted	Destroyed
Non-islet	Accepted	Accepted

nondiabetic donors to their identical twin recipients with longstanding IDDM contains activated T-cells, T-cytotoxic/suppressor cells, MO, and monocytes as the characteristic cells in this secondary immune response to beta-cells.[96] The functional role and specificity of these infiltrating cell types are uncertain. However, it has been reported that peripheral blood mononuclear cells (PBMNC) from recent-onset IDDM patients recognize pancreatic antigens[12] and may kill xenogeneic islet-cell targets in vitro in a non-MHC–restricted fashion.[97] The cytotoxic subset was claimed to belong to the OKT3+, OKT4− subpopulation, that is, T-cells of the cytotoxic subtype,[98] but adherent cells (MO and NK-cells) may have contaminated these cell preparations. Further recent evidence has suggested that Tc may not be the effector cell in animal models of spontaneous IDDM. When allogenic islets were precultured at room temperature, in 95% O_2, or treated with antibodies to MHC class II molecules or antilymphocyte antiserum, they were accepted and remained functional for prolonged periods of time even when transplanted across MHC barriers to animal recipients that do not spontaneously develop IDDM[99,100] (Table 1.3). The reduced immunogenicity of such pretreated islet allografts is thought to be due to deletion of MHC class II molecule-bearing passenger cells present in the isolated islets, for example MO, lymphocytes, endothelial cells, or dendritic cells.[99] If allogeneic islets pretreated as described were transplanted with similarly pretreated allogeneic adrenal or pituitary tissue to BB rats[101,102] or to NOD mice,[103] the islets but not the adrenal or pituitary allografts were destroyed by an inflammatory process similar to and concomitant with the spontaneous insulitis in the recipients' own islets (Table 1.3). These studies indicate that the effector mechanism in animal models of IDDM is not MHC restricted by either class I or class II molecules. In addition PBMNC-mediated, islet-cell killing in vitro is not MHC restricted, since PBMNC from recent onset IDDM patients or BB rats are cytotoxic to xenogeneic or allogeneic islet-cell targets.[97,104] Since Tc must see foreign antigen in context with self-MHC class I (HLA-ABC) molecules to be able to lyse target cells,[105] these cells cannot be the beta-cell killing cells in the BB rat, NOD mouse, or in man if the BB rat and NOD mouse are to be models for human IDDM. A T-cell line expressing the T-cell receptor complex has been reported to lyse tumor cells in an MHC non-restricted fashion.[106] These so-called "non-MHC restricted Tc" preferentially lyse NK-cell targets and may be identical to a subpopulation of NK-cells that express the beta-chain of the T-cell receptor.[107,108] Faced with the lack of MHC restriction of the islet-destructive mechanism in vitro and in vivo in the BB rat and NOD mouse, we offer the provocative view that (1) the MHC class I restricted T-cytotoxic cells are not the beta-cell killers in IDDM or (2) there is no animal or in vitro model of human IDDM. These studies indicate that T-cells do not themselves destroy beta-cells, but they do not exclude the fact that the T-cell system may be necessary for IDDM to develop. Thus measures that inhibit the T-cell-dependent immune response, for example, neonatal[109] or adult[110] thymectomy, cyclosporin (Cs) administration,[23] or injection of monoclonal antibodies against T-cell phenotypic markers,[111] can prevent diabetes development in BB rats. Cyclosporin prevented IDDM development in the NOD mouse[24] and induced and maintained clinical remission in recent onset IDDM patients.[19,21] Therefore the role of the T-cell system may be to focus the immune reaction to the islet organ by recognizing APC-presented, beta-cell autoantigen, recruiting other inflammatory cells to the islets, and providing helper signals (lymphokines) for the recruited cells. That cytotoxic T-cells participate in later stages of the beta-cell destructive process cannot be excluded, but they do not appear to be necessary for the efficient beta-cell destruction that is observed in culture-conditioned islet allografts.

Recent studies in vitro and in animal models of IDDM have attracted attention to non-MHC–restricted, effector cells, that is, MO and the NK-cells. Macrophages can recognize and kill certain tumor cells and are able to discriminate accurately between such tumor cells and normal cells. The recognition structure has not been identified, but MO do not need to see self-MHC class I molecules on the target cell.[112] The lytic process is not fully understood, but the generation of free radical species (hydroxyl or oxygen radicals) or the liberation of proteases or cytotoxic cytokines are thought to be involved.[112] Activated MOs from healthy individuals were cytotoxic to or inhibited insulin release from isolated rodent islets.[113,114] If MO function in diabetes-prone BB rats was blocked by intraperitoneal injection of silica, IDDM did not develop,[115] and islet allograft rejection in nondiabetic recipients could be blocked by both silica and carrageenan, another MO-suppressive agent.[116]

Natural killer-cells are large granular lymphocytes characterized by the ability to lyse a broad range of tumor cells and virus-infected cells, but NK-cell recognition involves a narrow spectrum of antigenic epitopes and does not require MHC class I sharing. Natural killer-cell activity against irrelevant targets was depressed in IDDM patients[117] and NOD mice,[118] but NK-cell number and activity toward irrelevant targets on a per cell basis was enhanced in diabetes-prone BB rats. Also NK-cells may be the most important cytotoxic cell in the lymphopenic BB rat.[119] It has been suggested that the beta-cell cytotoxic activity of acutely diabetic or diabetes-prone, BB-rat spleen cells is conferred by NK-cells,[104] since administration of antibodies against NK-cells prevented IDDM development in BB rats.[111] Thus at present, non-MHC–restricted effector cells such as the MO or the NK-cell or both appear to be the best candidates for beta-cell killers in IDDM.

How do MO and NK-cells kill beta-cells? One of the similarities between MO and NK-cells is their ability to elaborate the monokine interleukin-1 (IL-1).[120,121]

Recent studies have shown that IL-1 by itself is beta-cell cytotoxic in vitro,[122] as reviewed in the following paragraph.

A Role for Lymphokines in Beta-Cell Destruction?

Crude, cell-free supernatants of activated MNC can markedly affect islet function and morphology.[67,123] Crude cytokine preparations (CCP) were obtained by in vitro stimulation of PBMNC with antigen (purified protein derivative of tuberculin, PPD) or a plant lectin (phytohemagglutinin, PHA). When human or rat islets isolated by collagenase digestion were cultured in the presence of 50% CCP, the glucose-stimulated, accumulated insulin release was markedly inhibited after 24 hours and the insulin and glucagon contents in rat islets were reduced after seven days of culture.[123] The decrease in rat islet insulin release, insulin content, and glucagon content after seven days in culture with CCP was dose-dependent, with a half-maximal effect at 1 to 2% supernatant dilution.[67] When newborn rat, islet-cell monolayer cultures were exposed to as little as 0.3 percent CCP for six days, a subsequent acute challenge with 20 mmol/l glucose and 0.1 mM phosphodiesterase-inhibitor 3-isobutyl-l-methylxanthine (IBMX) failed to elicit first- and second-phase insulin responses.[124] Recent studies have shown that (pro)insulin biosynthesis, total protein biosynthesis, and preproinsulin mRNA are reduced after exposure to CCP for 24 hours.[125]

Rat islets cultured in the presence of CCP have a lowered DNA content and, when viewed by light microscopy, appeared to have disintegrated.[123] In addition electron microscopy showed severe ultrastructural changes of organelles and nuclear chromatin in all rat islet cells after a seven-day-exposure to CCP.[67] The islet cytotoxic effects could not be ascribed to the presence of stimulants (PPD/PHA) or endotoxin in the CCP or to the dialysis and lyophilization of CCP, and CCP did not degrade insulin. Preculture of rat islets under conditions that do not support the viability of passenger nonendocrine cells in the isolated islets did not preclude CCP-mediated islet cytotoxicity.[126] Based on these functional and morphologic studies, it was concluded that CCP has a direct, dose-dependent cytotoxic effect on isolated islets of Langerhans.

Preliminary studies suggested that the cytotoxic activity in CCP was conferred by a molecule above 10,000 M_r which was labile even when stored at −80°C, stable in lyophilized form, and sensitive to heating at 56°C for 60 minutes.[67] When highly purified human lymphokines and cytokines were screened for cytotoxic activity with isolated rat islets, only preparations with interleukin-1 (IL-1) activity reproduced the effects of CCP, that is, they reduced rat islet insulin release, they reduced insulin and glucagon content, and they disrupted normal islet morphology.[122,127] A half-maximal reduction in insulin release from rat islets occurred at a concentration of affinity-purified human monocyte IL-1 (native IL-1, nIL-1) of approximately 5 lymphocyte-activating factor (LAF) units/ml, where 1 LAF unit is the IL-1 concentration necessary to double the background proliferation of murine thymocytes in the costimulator assay.[128] Thus the

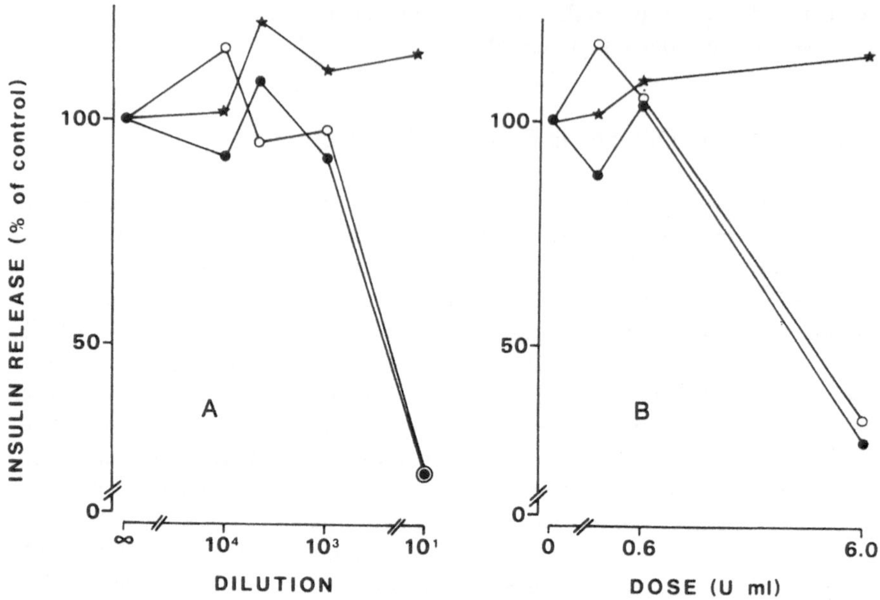

FIGURE 1.3. Blocking of the beta-cell cytotoxic effect of human crude IL-1-containing MNC supernatants (**A**) or affinity-purified human monocyte IL-1 (**B**) with anti-IL-1 -antibody. Key to **A**: •, crude supernatant; ○, crude supernatant + 1% v/v preimmune rabbit serum; *, crude supernatant + 1% v/v rabbit anti-human IL-1–anti serum. Key to **B**: •, native IL-1: ○, native IL-1 + 1% v/v preimmune rabbit serum; *, native IL-1 + 1% v/v rabbit-anti-human IL-1-anti-serum.

cytotoxic effect of IL-1 on isolated islets was observed at concentrations that were optimal for T-cell proliferation in vitro.[127,129]

Three recent studies in other laboratories have confirmed the inhibitory action of IL-1 in isolated rat islets.[130-132]

Screening studies with purified cytokines did not exclude the possibility that additional cytokines could be cytotoxic to isolated islets.[122] Therefore islet-cytotoxic CCP (a pool of dialyzed and lyophilized, PHA-stimulated MNC supernatants) were separated by affinity chromatography on Sepharose columns with or without a polyclonal, antimonocyte, IL-1 antibody with an IL-1–binding capacity of approximately 800 U.[127] Eluates from the anti-IL-1 column had no significant effect on insulin release when tested on rat islets, whereas eluates from the control column were as active as the prechromatographic material. Acid elution of the anti-IL-1 column, but not of the control column, resulted in a 100% recovery of the islet cytotoxic activity, which coeluted with LAF activity. The interleukin-2 (IL-2) or IFN-gamma activities in the CCP were unaffected by he anti-IL-1 immunoadsorbent. These experiments strongly suggested that IL-1 was the only islet cytotoxic cytokine in human PHA-stimulated CCP. Further addition of anti-IL-1-antibody to the culture medium of rat islets blocked the beta-cell cytotoxic effect of CCP and affinity-purified IL-1 (Fig. 1.3). In addition recombinant IL-1 (rIL-1), but not human rIFN-gamma or rTNF, reproduced the dose-dependent effects of CCP and nIL-1 on rat islet insulin release, insulin/glucagon content, and morphologic changes.[127]

Very low (< 100 U/ml) concentrations of TNF were found in PHA-stimulated CCP.[126] rTNF reduced insulin release 30 to 50% only at extremely high concentrations (225,000 U/ml), but rTNF in these concentrations increased the islet insulin content, in contrast to rIL-1.[127] The light and electron microscopic morphology were unaffected.[133] Tumor necrosis factor is resistant to heating at 56 °C, whereas the islet cytotoxic activities of CCP[67] and nIL-1 were completely abolished by heating to 56 °C for 60 minutes. Tumor necrosis factor has no obvious amino acid sequence homology to IL-1, but it shares most biological properties with IL-1.[134] Although it is ineffective in causing beta-cell cytotoxicity alone, 500 to 5000 U/ml of rTNF, but not 5000 U/ml of rLT, or 50 U/ml of rIFN, markedly potentiated the beta-cell cytotoxic effect of rIL-1.[135]

Interleukin-I is a family of polypeptides encoded for by at least two genes in man and in the mouse.[136] The two mature gene products have isoelectric pHs (pI) of 7 (beta) and 5 (alpha), respectively. In man the pI 7 form of IL-1 predominates whereas in the mouse, the pI 5 form seems to be the most predominant IL-1. The amino acid homology between the pI 7 and 5 forms is only 26%, but there is a 62 percent homology between human and mouse pI 5 IL-1.[136] When chromatofocused fractions of human monocyte IL-1 were tested for LAF activity and rat islet cytotoxicity, only the pI 7 form was cytotoxic to islet cells. Experiments with rIL-1 alpha and rIL-1 beta have shown that the pI 5 form is toxic to islet cells, but the half-maximal effect on insulin release is obtained with 10-fold higher molar concentrations than that with the pI 7 form.[135]

A sequential study of islet ultrastructure was made to determine when morphologic signs of cell damage could be seen at a given IL-1 concentration. Rat or human islets were incubated with or without the lowest concentration of IL-1 sufficient to inhibit the glucose-induced insulin release maximally, that is, 25 U/ml.[133] Portions of 20 IL-1-treated or control islets were sampled in parallel and examined by electron microscopy by an investigator who was "blinded" as to whether the islets were IL-1–treated or –untreated. The earliest ultrastructural abnormality observed in IL-1–treated rat and human islets was an accumulation of spherical inclusions lacking a distinct surrounding membrane. The opaque bodies, the contents of which are unknown, were seen in rat islets cultured with IL-1 for only 30 minutes and in human islets after 24 hours; they were specific for IL-1–treated islets, since they were not observed in any control islet cells.[133] Autophagic vacuoles were also prominent in IL-1–treated islets. No signs of damage to organelles or nuclear chromatin in rat islets were detectable after a few hours of IL-1 exposure. The pathophysiologic importance of these very early findings is unclear; they may be reversible and may not indicate cellular damage per se.

The first signs of damage to organelles or nuclear chromatin were observed in rat islets after 12 hours of exposure to IL-1.[133] Islets sampled after more than 12 hours of exposure to IL-1 revealed a heterogeneous morphologic pattern with marked degenerative changes in most (> 80%) beta-cells. Affected cells were not confined to the islet periphery exposed directly to the culture medium, but were scattered randomly throughout the islets. Specific for IL-1 induced, beta-cell damage was the accumulation of osmiophilic bodies, which were not observed in control islet cells, even in the 1 to 2% spontaneously degenerating cells found in control islets. Such bodies may represent a later stage of the opaque bodies.

These morphologic changes seem to agree with the results of the functional studies, in that reduction of rat islet function occurred almost synchronously with morphologic evidence of cell damage.[133]

In order to investigate a possible selective effect of IL-1–mediated, islet cell cytotoxicity, rat and human islets were examined by electron microscopy after exposure to IL-1 for up to 7 days. Within 24 hours of exposure to 25 U/ml of nIL-1, semiquantitative analyses of rat and human islet cells revealed that only the beta-cells had morphologic abnormalities.[133] Further, in human islets, the ultrastructural degenerative changes after seven days of exposure to IL-1 were observed only in beta-cells. This study indicated that, within a defined dose and time interval, the islet cytotoxic effect of IL-1 was beta-cell selective. A similar, in vitro selective activity of two diabetogenic compounds, streptozotocin and alloxan, which specifically destroy beta-cells in vivo, have been reported.[137] That IL-1 selectively, but not exclusively, kills beta-cells in vitro may be compatible with observations in the BB rat, in which alpha- and delta-cells exhibit morphologic injury in the insulitis process,[138] and with clinical studies that show a decreased alpha-cell function (peak response to intravenous arginine) in prediabetic patients who have impaired beta-cell response and who subsequently developed IDDM.[139]

THE CELLULAR MECHANISM OF IL-1–MEDIATED ISLET CYTOTOXICITY

The cellular mechanism of IL-1 action is unknown. It has been proposed that IL-1 may act as a calcium ionophore or that it activates cell membrane-associated phospholipase A_2 or both. High (20 mM) glucose decreased the islet inhibitory action of 1 and 10% CCP,[126] and a similar inhibitory effect of 20 and 40 mM glucose on the co-mitogenic effect of IL-1 on murine thymocytes has been observed (Dinarello, personal communication). High glucose also reduced alloxan- and streptozotocin-induced beta-cell cytotoxicity.[137] It has been proposed that alloxan is selectively beta-cell cytotoxic because it induces the formation of intracellular free hydroxyl anions.[140] Beta-cells are particularly susceptible to the toxic action of free radicals because of their low content of the scavenging enzyme superoxide dismutase.[140] Glucose is thought to protect against alloxan toxicity by increasing the amount of NADPH in beta-cells. In turn, (NADPH), reduces the glutathione scavenging substrate that enables the beta-cell to deal with free radical species.[140] Interleukin-1 has been shown to induce free radical formation in human eosinophils[141] and in endothelial cells.[142] Thus there is evidence to suggest that IL-1 may be selectively beta-cell cytotoxic because of its ability to induce free radical formation.

In summary soluble products of human PBMNC are cytotoxic to isolated rat and human islets, and this effect is conferred by IL-1, since (1) CCP depleted of

IL-1 by column affinity chromatography has no cytotoxic activity, (2) affinity-purified and recombinant IL-1 mimic all the effects of CCP, and (3) rIFN-gamma, rTNF, purified LT, IL-2, MIF, and LIF are not cytotoxic for isolated islets. When rat islets are exposed to IL-1 for 0 to 24 hours, only the beta-cells show ultrastructural changes. The earliest IL-1–induced changes in rat beta-cell ultrastructure are detectable after 30 minutes of exposure, and signs of cellular damage are evident after 16 hours of exposure to IL-1. The cellular mechanism of IL-1–induced, beta-cell damage is unknown, but may relate to the generation of an excess of intracellular metabolic products (e.g., free radicals). Thus, IL-1 selectively damages islet beta-cells in vitro, and this dose- and time-dependent effect is markedly potentiated by another macrophage product, TNF. Although in vivo studies to prove a role of IL-1 and TNF as effector molecules in IDDM are still lacking, IL-1 secreted in situ in the islets by MO, NK-cells, B-lymphocytes, and/or endothelial cells may explain certain features of the pathogenesis of IDDM: namely, (1) the beta-cell selectivity of the disease process and (2) the lack of MHC restriction of the beta-cell destructive mechanism in vitro and in vivo in animal models of IDDM.[25] We have suggested that the strong association of IDDM with the HLA-DR3 and/or DR4 phenotype(s), and the protective effect of other HLA-DR types, for example, DR2, may be related to the quantitative control these genes exert on the immune response, by determining the magnitude of cytokine production of activated MNC.[25] Indeed preliminary data indicate that IL-1 secretion by MO in response to endotoxin is much lower in healthy donors carrying HLA-DR2,x or 2,2 than in donors with other non-2 phenotypes.[143] And as mentioned above, the TNF and LT genes have been assigned to the HLA-region on the short arm of chromosome 6, probably in linkage disequilibrium with the HLA-B and DR locus in man.[42,43] This chromosomal location may imply that linkage disequilibrium exists between IDDM-associated, HLA-D region genes and the TNF gene. If TNF gene polymorphisms occur, certain HLA-haplotypes may be associated with high or low TNF-secretor phenotypes. This leads to the intriguing possibility that diabetes susceptibility is conferred by the HLA-DR3 and/or DR4 phenotype(s) because these haplotypes are associated with a high TNF-secretor phenotype, which determines the degree of potentiation of IL-1–mediated, beta-cell destruction.

Thus IL-1 may be directly involved in tumor cell killing and tissue destruction in autoimmune diseases, for example, rheumatoid arthritis.[144] The high local concentration of immune mediators obtained in mononuclear cell infiltrates may have effects on the target tissues that differ from those mediated by low circulating levels of cytokines. Our studies suggest that IL-1 may be involved in beta-cell destruction in IDDM. Interesting preliminary studies suggest that a similar toxic effect of IL-1 can be observed on isolated human thyroid cells,[133,145] indicating that IL-1 may play a role in the pathogenesis of other autoimmune endocrinopathies.

A recent study supports the in vivo importance of cytokines in IDDM development. Intraperitoneal injection of CCP [supernatants of concanavalin A (Con A)-activated spleen cells] not only accelerated the appearance of IDDM in diabetes-prone BB rats but also induced IDDM in diabetes-resistant BB rats.[146] The Con

A-activated spleen cells from acutely diabetic BB rats only transferred IDDM, whereas CCP prepared from both acutely diabetic rats and normal WF as well as Buffalo rats induced IDDM, although IDDM could not be induced in normal Wistar–Furth (WF) rats. The CCP-injected diabetic rats all showed peritonitis, severe pancreatitis, and insulitis. The active molecule(s) was not IL-2 but rather was found in the molecular weight range above 50 kD. Interpretation of this study is not totally clear. Thus it is not known whether intraperitoneally injected cytokines (1) contain IL-1 that directly damages the endocrine pancreas and/or triggers an immune response against the beta-cells, (2) unmasks a latent antipancreatic autoimmunity through costimulator activity (IL-1, IL-2) for T-cell clones reactive against beta-cells, or (3) causes beta-cell destruction secondary to a nonspecific local inflammation, for example, peritonitis and pancreatitis. The fact that CCP did not induce IDDM in WF rats indicates that these animals may not carry beta-cell, autoreactive T-cell clones that could be activated by the CCP, or alternatively, that genetic differences between rat strains determine the in vivo susceptibility of beta-cells to immune effector molecules like IL-1 or determine how much mediator is produced in the islet inflammatory response, as suggested above. Nevertheless it is possible that IL-1 in the injected CCP plays a role in any or all of these interpretative models. Indeed aggregates of IL-1 or IL-1 bound to carrier proteins could reach molecular weights above 50 kD. Further IL-1 release from acutely diabetic BB rat spleen cells[147] or human PBMNC from recent-onset, IDDM patients[148] (Schernthaner, personal communication) is normal.

If the beta-cell cytotoxic effect of IL-1 in vitro can be confirmed in vivo, how could IL-1 be envisaged as an effector molecule in the integrated immune reaction to the beta-cells in IDDM? Figure 1.4 shows a recently proposed model.[25] Any beta-cell cytotoxic agent (e.g., multiple subdiabetogenic hits by a chemical or a virus or both) that can kill a limited number of beta-cells could trigger an autoimmune response by liberating sequestered antigen from damaged beta-cells. The beta-cell antigen will be taken up, processed, and presented by a macrophage or a dendritic cell to a Th-cell with specificity for the antigen, if such a cell is present in the islet when the triggering event occurs. It should be noted that IL-1 produced in regional inflammatory responses unrelated to the islets (a local vasculitis or pancreatitis) could theoretically reach toxic concentrations in the adjacent islets, thereby acting as an endogeneous trigger. The activated Th-cell will produce MAF (e.g., IFN-gamma), which induces IL-1 production from the antigen presenting as well as by the bystander MO, leading to further IL-1 production. Natural killer-cell activity can be boosted by IL-1, IL-2, and IFN-gamma, and B-lymphocytes, and endothelial cells can be stimulated to produce more IL-1. The beta-cell cytotoxic action of IL-1 is potentiated by TNF. Eventually the IL-1 and TNF concentration in the inflammed islet microenvironment may reach beta-cell cytotoxic levels, and beta-cell destruction could take place through the induction of intracellular free radical formation in beta-cells by IL-1. Because of the beta-cell–selective action of IL-1 and possibly because of a low regenerative potential of the adult beta-cell, most non-beta–cells survive the insulitis process, which burns out as the beta-cell antigenic mass is

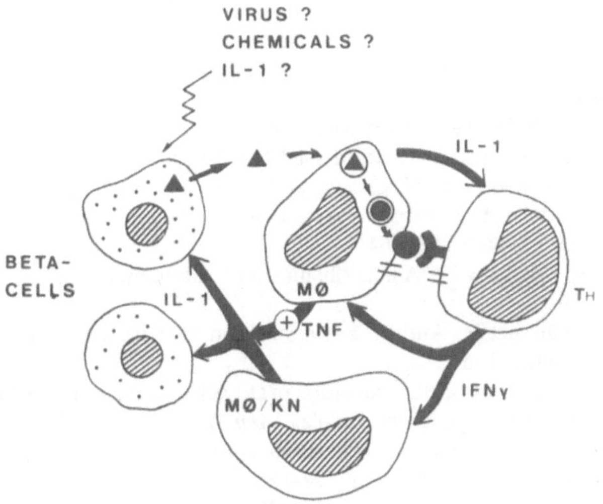

FIGURE 1.4. Model for IL-1-mediated, beta-cell destruction; the macrophage is the antigen-presenting cell. For explanation, see text and the legend to Figure 1.1. From Nerup et al.[25] Copyright © 1987, John Wiley and Sons. Reprinted by permission of John Wiley and Sons and Nørhaven Bogtrykkeri A/S.

eliminated. This model may explain the requirements for the pathogenetic process listed in Table 1.2 and how beta-cell destruction is initiated, perpetuated, amplified, and terminated.

Conclusion

The model proposed herein suggests that IL-1, secreted by macrophages, natural killer-cells, B-lymphocytes and endothelial cells, and potentiated by tumor necrosis factor, can act as effector molecules of beta-cell destruction in the pathogenesis of insulitis. The model does not exclude the role of other effector molecules secreted by MO (free radicals, hydrogen peroxide, proteases, etc.) or by natural killer-cells (NK-cell cytotoxic factor, NKCF). Nor does the model exclude the later involvement of cytotoxic T-cell or autoantibody-related effector mechanisms, but it suggests that IL-1 could be an early, primary, and sufficient mechanism by which "autoimmune" beta-cell destruction in IDDM is effected.

Acknowledgments. This work was supported by the Michaelsen Foundation, the Danish Diabetes Association, and the Nordisk Insulinlaboratorium. We thank Ms. Anne Rafn for preparing the manuscript. We also gratefully acknowledge contributions by the following collaborators: K. Bendtzen, G. Bendixen, L. Bæk, C.A. Dinarello, J. Dupré, J. Egeberg, B.S. Hansen, D.W. Jørgensen, J.H. Nielsen, S.J. Prowse, G.A. Spinas, C.R. Stiller, and M. Svenson.

References

1. Gepts W. Pathologic anatomy of the pancreas in juvenile diabetes mellitus. *Diabetes* 1965;14:619–633.
2. Weigle WO. Cellular events in experimental autoimmune thyroiditis, allergic encephalomyelitis and tolerance to self, in Talal N (ed): *Autoimmunity. Genetic, Immunologic, Virologic and Clinical Aspects*. New York: Academic Press, 1977, 141–170.
3. Cohen IR, Cooke A. Natural autoantibodies might prevent autoimmune disease. *Immunol Today* 1986;7:363–364.
4. Milgrom F, Witebsky E. Autoantibodies and autoimmune disease. *JAMA* 1962; 181:706–716.
5. Grey HM, Chesnut R. Antigen processing and presentation to T cells. *Immunol Today* 1985;6:101–106.
6. Schwartz RH. T-lymphocyte recognition of antigen in association with gene products of the major histocompatibility complex. *Ann Rev Immunol* 1985;3:237–261.
7. Unanue ER. Cooperation between mononuclear phagocytes and lymphocytes in immunity. *N Engl J Med* 1980;303:977–985.
8. Oppenheim JJ, Stadler BM, Siraganian RP, Mage M, Mathieson B. Lymphokines: Their role in lymphocyte responses. Properties of interleukin-1. *Fed Proc* 1982; 41: 257–262.
9. Bendtzen K. Lymphokines in inflammation, in Venge P, Lindbom A, (eds): *Inflammation. Basic Mechanisms, Tissue-Injuring Principles, and Clinical Models*. Stockholm: Almqvist & Wiksell International, 1985, 187–217.
10. Arenzana-Seisdedos F, Virelizier JL, Fiers W. Interferons as macrophage-activating factors. III. Preferential effects of interferon-gamma on the interleukin-l secretory potential of fresh or aged human monocytes. *J Immunol* 1985;134:2444–2448.
11. Nerup J, Platz P, Andersen OO, Christy M, Lyngsøe J, Poulsen JE, Ryder L, Nielsen LS, Thomsen M, Svejgaard A. HL-A antigens and diabetes mellitus. *Lancet* 1974;i:864–866.
12. Nerup J, Andersen OO, Bendixen G, Egeberg J, Gunnarsson R, Kromann H, Poulsen JE. Cell-mediated immunity in diabetes mellitus. *Proc Roy Soc Med* 1974;67:506–513.
13. Bottazzo GF, Florin-Christensen A, Doniach D. Islet-cell antibodies in diabetes mellitus with autoimmune polyendocrine deficiencies. *Lancet* 1974;ii:1279–1282.
14. Maclaren NK, Huang S-W. Antibody to cultured human insulinoma cells in insulin-dependent diabetes. *Lancet* 1975;i:997–1000.
15. Lernmark Å, Freedman ZR, Hofmann C, Rubinstein AH, Steiner DF, Jackson RL, Winter RJ, Traisman HS. Islet-cell-surface antibodies in juvenile diabetes mellitus. *N Engl J Med* 1978;229:375–380.
16. Bækkeskov S, Nielsen JH, Marner B, Bilde T, Ludvigsson J, Lernmark Å. Autoantibodies in newly diagnosed diabetic children immunoprecipitate human pancreatic islet cell proteins. *Nature* 1982;298:167–169.
17. Carpenter CCJ, Solomon N, Silverberg SG, Bledsoe T, Northcutt RC, Klinenberg JR, Bennett IL, McGehee HA. Schmidt's syndrome (thyroid and adrenal insufficiency): A review of the literature and a report of fifteen new cases including ten instances of co-existent diabetes mellitus. *Medicine* 1964;43:153–180.
18. Irvine WJ, Clarke BF, Scarth L, Cullen DR, Duncan LJP. Thyroid and gastric autoimmunity in patients with diabetes mellitus. *Lancet* 1970;ii:163–168.
19. Stiller CR, Dupré J, Gent M, Jenner MR, Keown PA, Laupacis A, Martell R, Rodger

NW, Graffenried BV, Wolfe BMJ. Effects of cyclosporine immunosuppression in insulin-dependent diabetes mellitus of recent onset. *Science* 1984;223:1362–1367.

20. Assan R, Feutren G, Debray-Sachs M, Quiniou-Debrie MC, Laborie C, Thomas G, Chatenoud L, Bach JF. Metabolic and immunological effects of cyclosporin in recently diagnosed type 1 diabetes mellitus. *Lancet* 1985;i:67–71.
21. Feutren G, Papoz L, Assan R, Vialettes B, Karsenty G, Vexiau P, Rostu HD, Rodier M, Sirmai J, Lallemand A, Bach J-F. Cyclosporin increases the rate and length of remissions in insulin-dependent diabetes of recent onset. Results of a multicentre double-blind trial. *Lancet* 1986;ii:119–123.
22. The Canadian-European Randomized Control Trial Group (1988). Cyclosporin-induced remission of IDDM after early intervention. Association of 1 yr of cyclosporin treatment with enhanced insulin secretion. *Diabetes* 37:1574–1582.
23. Laupacis A, Stiller CR, Gardell C, Keown P, Dupré J, Wallace AC, Thibert P. Cyclosporin prevents diabetes in BB Wistar rats. *Lancet* 1983;i:10–12.
24. Mori Y, Suko M, Okudaira H, Matsuba I, Tsuruoka A, Sasaki A, Yokoyama H, Tanase T, Shida T, Nishimura M, Terada E, Ikeda Y. Preventive effects of Cyclosporin on diabetes in NOD mice. *Diabetologia* 1986;29:244–247.
25. Nerup J, Mandrup-Poulsen T, Mølvig J. The HLA-IDDM association: Implications for etiology and pathogenesis of IDDM. *Diab Metab Rev* 1987;3:779–802.
26. Thomsen M, Platz P, Andersen OO, Christy M, Lyngsøe J, Nerup J, Rasmussen K, Ryder LP, Nielsen LS, Svejgaard A. MLC typing in juvenile diabetes mellitus and idiopathic Addison's disease. *Transplant Rev* 1975;22:125–147.
27. Platz P, Jakobsen BK, Morling N, Ryder LP, Svejgaard A, Thomsen M, Christy M, Kromann H, Benn J, Green A, Hauge M. HLA-D and -DR antigens in genetic analysis of insulin-dependent diabetes mellitus. *Diabetologia* 1981;21:108–115.
28. Gonwa TA, Peterlin MB, Stobo JD. Human I genes: Structure and function. *Adv Immunol* 1983;34:71–96.
29. Nerup J. Etiology and pathogenesis of insulin-dependent diabetes mellitus: Present views and future developments, in Martin JM, Ehrlich RM, Holland FJ, (eds): *Etiology and Pathogenesis of Insulin-Dependent Diabetes Mellitus.* New York, Raven Press, 1981, 275–288.
30. Dausset J, Cohen D. HLA at the gene level, in Albert ED, Baur MP, Mayr WR, (eds): *Histocompatibility Testing 1984.* Berlin, Heidelberg, New York, Toronto: Springer-Verlag, 1984, 22–27.
31. Michelsen B, Lernmark Å. Molecular cloning of a polymorphic DNA endonuclease fragment associates insulin-dependent diabetes mellitus with HLA-DQ. *J Clin Invest* 1987;79:1144–1152.
32. Bell GI, Horita S, Karam JH. A polymorphic locus near the human insulin gene is associated with insulin-dependent diabetes mellitus. *Diabetes* 1984;33:176–183.
33. Eisenbarth GS. Type I diabetes mellitus. A chronic autoimmune disease. *N Engl J Med* 1986;314:1360–1368.
34. Like AA, Guberski DL, Butler L. Diabetic BioBreeding/Worcester (BB/Wor) rats need not be lymphopenic. *J Immunol* 1986;136:3254–3258.
35. Winter WL, Beppu H, Maclaren NK, Cooper DL, Bell GI, Wakeland EK. Restriction-fragment-length polymorphisms of 5′-flanking region of insulin I gene in BB and other rat strains. *Diabetes* 1987;36:193–198.
36. Bruserud Ø, Jervell J, Thorsby E. HLA-DR3 and -DR4 control T-lymphocyte responses to mumps and Coxsackie B4 virus: Studies on patients with type 1 (insulin-dependent) diabetes and healthy subjects. *Diabetologia* 1985;28:420–426.

37. Bruserud O, Stenersen M, Thorsby E. T-lymphocyte responses to Coxsackie B4 and mumps virus. II. Immunoregulation by HLA-DR3 and -DR4 associated restriction elements. *Tissue Antigens* 1985;26:179–192.
38. Durum SK, Higuchi C, Ron Y. Accessory cells and T-cell activation. The relationship between two components of macrophage accessory cell function: I-A and IL-1. *Immunobiology* 1984;168:213–231.
39. Palacios R. Monoclonal antibodies against human Ia antigens stimulate monocytes to secrete interleukin-1. *Proc Natl Acad Sci USA* 1985;82:6652–6656.
40. Caruso C, Bellina L. B8, DR3 antigens and production of human leucocyte migration inhibitory factor (LIF) by mononuclear cells stimulated with concanavalin A (Con A). *Tissue Antigens* 1983;22:167–169.
41. MacSween JM, Eastwood SL, Cohen AD, Rajaraman K, Fox RA. Lymphokine responses to concanavalin A stimulation: Association with HLA DR antigens. *Clin Immunol Immunopathol* 1984;32:368–377.
42. Spies T, Morton CC, Nedospasov SA, Friers W, Pious D, Strominger JL. Genes for the tumor necrosis factor α and β are linked to the human major histocompatibility complex. *Proc Natl Acad Sci USA* 1986;83:8699–8702.
43. Müller U, Jongeneel CV, Nedospasov SA, Lindahl KF, Steinmetz M. Tumour necrosis factor and lymphotoxin genes map close to H-2D in the mouse major histocompatibility complex. *Nature* 1987;325:265–267.
44. Papadopoulos G, Lernmark Å. Islet cell antibodies and cellular immunity in human diabetes. *Behring Inst Mitt* 1984;75:50–57.
45. Tattersall RB, Pyke DA. Diabetes in identical twins. *Lancet* 1972;ii:1120–1125.
46. Sakano H, Huppi K, Heinrich G, Tonegawa S. Sequences at the somatic recombination sites of immunoglobulin light-chain genes. *Nature* 1979;280:288–294.
47. Siu G, Kronenberg M, Strauss E, Haars R, Mak TW, Hood L. The structure, rearrangement and expression of Dp gene segments of the murine T-cell antigen receptor. *Nature* 1984;311:344–350.
48. Kromann H. Aspects of the aetiology and pathogenesis of insulin-dependent diabetes mellitus. *Dan Med Bull* 1982;29:257–265.
49. Graighead JE. Viral diabetes mellitus in man and experimental animals. *Am J Med* 1981;70:127–134.
50. Yoon J-W, Austin M, Onodera T, Notkins AL. Virus-induced diabetes mellitus: Isolation of a virus from the pancreas of a child with diabetic ketoacidosis. *N Engl J Med* 1979;300:1173–1179.
51. Gepts W. The pathology of the pancreas in human diabetes, in Andreani D, DiMario U, Federlin KF, Heding LG (eds): *Immunology in Diabetes*. London: Kimpton, 1984, 21–34.
52. Rubenstein P, Walker ME, Fedun B, Witt ME, Cooper LZ, Ginsberg-Fellner F. The HLA-system in congenital rubella patients with and without diabetes. *Diabetes* 1982;31:1088–1091.
53. Rossini AA, Williams RM, Mordes JP, Appel MC, Like AA. Spontaneous diabetes in the gnotobiotic BB/W rat. *Diabetes* 1979;28:1031–1032.
54. Helgason T, Jonasson MR. Evidence for a food additive as a cause of ketosis-prone diabetes. *Lancet* 1981;ii:716–720.
55. Christy M, Christau B, Mølbak AG, Nerup J. Diabetes and month of birth. *Lancet* 1982;ii:216.
56. Symon DK, Hennessy ER, Smail PJ. Smoked foods in the diets of mothers of diabetic children. *Lancet* 1984;ii:55.

57. Elliott RB, Martin JM. Dietary protein: A trigger of insulin-dependent diabetes in the BB rat? *Diabetologia* 1984;26:297–299.
58. Tonioli A, Onodera T, Yoon J-W, Notkins AL. Induction of diabetes by cumulative environmental insults from viruses and chemicals. *Nature* 1980;288:383–385.
59. Borch-Johnsen K, Mandrup-Poulsen T, Zachau-Christensen B, Joner G, Christy M, Kastrup K, Nerup J. Relation between breast-feeding and incidence rates of insulin-dependent diabetes mellitus. *Lancet* 1984;ii:1083–1086.
60. Scott FW, Mongeau R, Kardish M, Hatina G, Trick KD, Wojcinski Z. Diet can prevent diabetes in the BB rat. *Diabetes* 1985;34:1059–1062.
61. Adams TE, Alpert S, Hanahan D. Nontolerance and autoantibodies to a transgenic self antigen expressed in pancreatic beta cells. *Nature* 1987;325:223–228.
62. Nayak RC, Omar MAK, Rabizadeh A, Srikanta S, Eisenbarth GS. "Cytoplasmic" islet cell antibodies. Evidence that the target antigen is a sialoglycoconjugate. *Diabetes* 1985;34:617–619.
63. Srikanta S, Eisenbarth GS. Islet cell antigens. Initial studies of their biology and function. *Mol Biol Med* 1986;3:113–127.
64. Bækkeskov S, Dyrberg T, Lernmark Å. Autoantibodies to a 64-kilodalton islet cell protein precede the onset of spontaneous diabetes in the BB rat. *Science* 1984;224:1348–1350.
65. Bottazzo GF, Todd I, Mirakian R, Belfiore A, Pujol-Borrell R. Organ-specific autoimmunity: A 1986 overview. *Immunol Rev* 1986;94:136–169.
66. Pujol-Borell R, Todd I, Doshi M, Sutton R, Gray D, Feldmann M, Bottazzo GF. In vitro induction of HLA class II expression in human B cells: Requirement for a two mediator signal. *Diabetologia* 1986;29:585A.
67. Mandrup-Poulsen T, Bendtzen K, Nerup J, Egeberg J, Nielsen JH. Mechanisms of pancreatic islet cell destruction. Dose-dependent cytotoxic effect of soluble blood mononuclear cell mediators on isolated islets of Langerhans. *Allergy* 1986;41:250–259.
68. Weetman AP, Volkman DJ, Burman KD, Gerrard TL, Fauci AS. The *in vitro* regulation of human thyrocyte HLA-DR antigen expression. *J Clin Endocrinol Metab* 1985;61:817–824.
69. Foulis AK, Farquharson MA. Aberrant expression of HLA-DR antigens by insulin-containing beta-cells in recent-onset type 1 diabetes mellitus. *Diabetes* 1986;35:1215–1224.
70. Hanafusa T, Pujol-Borell R, Chiovato L, Russell RCG, Doniach D, Bottazzo GF. Aberrant expression of HLA-DR antigen on thyrocytes in Graves' disease: Relevance for autoimmunity. *Lancet* 1983;ii:1111–1114.
71. Rubinstein D, Roska AK, Lipsky PE. Liver sinusoidal lining cells express class II major histocompatibility antigens but are poor stimulators of fresh allogeneic T lymphocytes. *J Immunol* 1986;137:1803–1810.
72. Londei M, Lamb JR, Bottazzo GF, Feldmann M. Epithelial cells expressing aberrant MHC class II determinants can present antigen to cloned human T cells. *Nature* 1984;312:639–641.
73. Iwatani Y, Gerstein HC, Litaki M, Row VV, Volpé R. Thyrocyte HLA-DR expression and interferon-gamma production in autoimmune thyroid disease. *J Clin Endocrinol Metab* 1986;63:695–708.
74. Harel-Bellan A, Quillet A, Marchiol C, DeMars R, Tursz T, Fradelizi D. Natural killer susceptibility of human cells may be regulated by genes in the HLA region on chromosome 6. *Proc Natl Acad Sci USA* 1986;83:5688–5692.

75. Ben-Nun A, Wekerle H, Cohen IR. Vaccination against autoimmune encephalomyelitis with T-lymphocyte line cells reactive against myelin basic protein. *Nature* 1981;292:60–61.
76. Prud'Homme GJ, Fuks A, Colle E, Guttmann RD. Isolation of T-lymphocyte lines with specificity for islet cell antigens from spontaneously diabetic (insulin-dependent) rats. *Diabetes* 1984;33:801–803.
77. Crossley JR, James AG, Elliott RB, Berryman CC, Edgar BW. Residual beta-cell function and islet cell antibodies in diabetic children. *Pediatr Res* 1981;15:62–65.
78. Mustonen A, Knip M, Huttunen N-P, Puukka R, Käär M-L, Åkerblom HK. Evidence of delayed beta-cell destruction in type 1 (insulin-dependent) diabetic patients with persisting complement-fixing cytoplasmic islet cell antibodies. *Diabetologia* 1984;27:421–426.
79. Marner B, Agner T, Binder C, Lernmark Å, Nerup J, Mandrup-Poulsen T, Waldorff S. Increased reduction in fasting C-peptide is associated with islet cell antibodies in type 1 (insulin-dependent) diabetic patients. *Diabetologia* 1985;28:875–880.
80. Srikanta S, Ganda OP, Eisenbarth GS, Soeldner JS. Islet-cell antibodies and beta-cell function in monozygotic triplets and twins initially discordant for type 1 diabetes mellitus. *N Engl J Med* 1983;308:322–325.
81. Spencer KM, Tarn A, Dean BM, Lister J, Bottazzo GF. Fluctuating islet-cell autoimmunity in unaffected relatives of patients with insulin-dependent diabetes. *Lancet* 1984;i:764–766.
82. Pujol-Borell R, Hanafusa T, Cudworth AG, Bottazzo GF. Islet cell surface antibodies and the natural history of type 1 (insulin-dependent) diabetes. *Diabetologia* 1982; 23:194A.
83. Bækkeskov S, Kristensen JK, Srikanta S, Bruining GJ, Mandrup-Poulsen T, de Beaufort C, Soeldner JS, Eisenbarth G, Lindgren F, Lernmark Å. Antibodies to a M_r 64,000 human islet cell antigen precede the clinical onset of insulin dependent diabetes. *J Clin Invest* 1987;79:926–934.
84. Bottazzo GF, Dean BM, Gorsuch AN, Cudworth AG, Doniach D. Complement-fixing islet-cell antibodies in type-I diabetes: Possible monitors of active beta-cell damage. *Lancet* 1980;i:668–672.
85. Mandrup-Poulsen T, Nerup J, Stiller CR, Marner B, Bille G, Heinrichs D, Martell R, Dupre J, Keown PA, Jenner MR, Rodger NW, Wolfe B, Graffenried BV, Binder C. Disappearance and reappearance of islet cell cytoplasmic antibodies in cyclosporin-treated insulin-dependent diabetics. *Lancet* 1985;i:599–602.
86. Tingle AJ, Lim G, Wright VJ, Dimmick JE, Hunt JA. Transplacental passage of islet cell antibodies in infants of diabetic mothers. *Pediatr Res* 1979;13:1323–1325.
87. Dobersen MJ, Scharff JE, Ginsberg-Fellner F, Notkins AL. Cytotoxic autoantibodies to beta cells in the serum of patients with insulin-dependent diabetes mellitus. *N Engl J Med* 1980;303:1493–1498.
88. Dyrberg T. Humoral autoimmunity in the pathogenesis of insulin-dependent diabetes mellitus. Studies in the spontaneously diabetic BB rat. *Acta Endocrinol* 1986; 113 (suppl. 280):1–29.
89. Svenningsen A, Dyrberg T, Gerling I, Lernmark Å, MacKay P, Rabinovitch A. Inhibition of insulin release after passive transfer of immunoglobulin from insulin-dependent diabetic children to mice. *J Clin Endocrinol Metab* 1983;57:1301–1304.
90. Drell DW, Notkins AL. Multiple immunological abnormalities in patients with type 1 (insulin-dependent) diabetes mellitus. *Diabetologia* 1987;30:132–143.
91. Bottazzo GF, Dean BM, McNally JM, MacKay EH, Swift PGF, Gamble DR. In situ

characterization of autoimmune phenomena and expression of HLA molecules in the pancreas in diabetic insulitis. *New Engl J Med* 1985;313:353–360.
92. Like AA, Forster RM, Woda BA, Rossini AA. T-cell subsets in islets and lymph nodes of Bio Breeding/Worcester (BB/W) rats. *Diabetes* 1983;32 (suppl. 1):51A.
93. Dean BM, Walker R, Bone AJ, Baird JD, Cooke A. Pre-diabetes in the spontaneously diabetic BB/E rat: Lymphocyte subpopulations in the pancreatic infiltrate and expression of rat MHC class II molecules in endocrine cells. *Diabetologia* 1985;28:464–466.
94. Kolb-Bachofen V, Kolb H. New concept of insulitis and B-islet cell destruction. *Diabetes* 1983;32 (suppl. 1):22A.
95. Kolb H, Kantwerk G, Treichel U, Kürner T, Kiesel U, Hoppe T, Kolb-Bachofen V. Prospective analysis of islet lesions in BB rats. *Diabetologia* 1986;29:559A.
96. Sibley RK, Sutherland DER, Goetz F, Michael AF. Recurrent diabetes mellitus in the pancreas iso- and allograft. A light and electron microscopic and immunohistochemical analysis of four cases. *Lab Invest* 1985;53:132–144.
97. Charles MA, Suzuki M, Waldeck N, Dodson LE, Slater L, Ong K, Kershnar A, Buckingham B, Golden M. Immune islet killing mechanisms associated with insulin-dependent diabetes: *In vitro* expression of cellular and antibody-mediated islet cell cytotoxicity in humans. *J Immunol* 1983;130:1189–1194.
98. Boitard C, Chatenoud L-M, Debray-Sachs M. *In vitro* inhibition of pancreatic B cell function by lymphocytes from diabetics with associated autoimmune diseases: A T cell phenomenon. *J Immunol* 1982;129:2529–2531.
99. Faustman D, Hauptfeld V, Lacy P, Davie J. Prolongation of murine islet allograft survival by pretreatment of islets with antibody directed to Ia determinants. *Proc Natl Acad Sci USA* 1981;78:5156–5159.
100. Prowse SJ, Simeonovic CJ, Lafferty KJ, Bond BC, Magi CE, Mackie D. Allogeneic islet transplantation without recipient immunosuppression, in Larner J, Pohl SL (eds): *Methods in Diabetes Research, Vol. 1. Laboratory Methods. Part B.* New York: Wiley, 1984, 253–269.
101. Weringer EJ, Like AA. Immune attack on pancreatic islet transplants in the spontaneously diabetic BioBreeding/Worcester (BB/W) rat is not MHC restricted. *J Immunol* 1985;134:2383–2386.
102. Prowse SJ, Bellgrau D, Lafferty KJ. Islet allografts are destroyed by disease occurrence in the spontaneously diabetic BB rat. *Diabetes* 1986;35:110–114.
103. Nomikos IN, Prowse SJ, Carotenuto P, Lafferty KJ. Combined treatment with nicotinamide and desferrioxamine prevents islet allograft destruction in NOD mice. *Diabetes* 1986;35:1302–1304.
104. MacKay P, Jacobson J, Rabinovitch A. Spontaneous diabetes mellitus in the Bio-Breeding/Worcester rat. Evidence in vitro for natural killer cell lysis of islet cells. *J Clin Invest* 1986;77:916–924.
105. Zinkernagel RM, Doherty PC. H-2 compatibility requirement for T-cell–mediated lysis of target cells infected with lymphocytic choriomeningitis virus. Different cytotoxic T-cell specificities are associated with structures coded for in H-2K or H-2D. *J Exp Med* 1975;141:1427–1436.
106. Lanier LL, Phillips JH. Evidence for three types of human cytotoxic lymphocyte. *Immunol Today* 1986;7:132–134.
107. Yanagi Y, Caccia N, Kronenberg M, Chin B, Roder J, Rohel D, Kiyohara T, Lauzon R, Toyonaga B, Rosenthal K, Dennert G, Acha-Orbea H, Hengartner H, Hood L, Mak TW. Gene rearrangement in cells with natural killer activity and expression of the beta-chain of the T-cell antigen receptor. *Nature* 1985;314:631–633.

108. Moingeon P, Ythier A, Goubin G, Faure F, Nowill A, Delmon L, Rainaud M, Forestier F, Daffos F, Bohuon C, Hercend T. A unique T-cell receptor complex expressed on human fetal lymphocytes displaying natural-killer-like activity. *Nature* 1986; 323:638–640.
109. Like AA, Kislauskis E, Williams RM, Rossini AA. Neonatal thymectomy prevents spontaneous diabetes mellitus in the BB/W rat. *Science* 1982;216:644–646.
110. Like AA, McGill PD, Sroczynski E. Adult thymectomy prevents diabetes mellitus in BB/W rats. *Diabetologia* 1986;29:565A.
111. Like AA, Biron CA, Weringer EJ, Byman K, Sroczynski E, Guberski DL. Prevention of diabetes in BioBreeding/Worcester rats with monoclonal antibodies that recognize T lymphocytes or natural killer cells. *J Exp Med* 1986;164:1145–1159.
112. Herberman RB, Reynolds CW, Ortaldo JR. Mechanism of cytotoxicity by natural killer (NK) cells. *Ann Rev Immunol* 1986;4:651–680.
113. Debray-Sachs M, Boitard C, Assan R, Hamburger J. Are "activated" macrophages cytotoxic against normal cells? *Transplant Proc* 1981;13:1111–1113.
114. Schwizer RW, Leiter EH, Evans R. Macrophage-mediated cytotoxicity against cultured pancreatic islet cells. *Transplantation* 1984;37:539–544.
115. Oschilewski U, Kiesel U, Kolb H. Administration of silica prevents diabetes in BB-rats. *Diabetes* 1985;34:197–199.
116. Nash JR, Everson NW, Wood RFM, Bell PRF. Effect of silica and carrageenan on the survival of islet allografts. *Transplantation* 1980;29:206–208.
117. Negishi K, Waldeck N, Chandy G, Buckingham B, Kershnar A, Fisher L, Gupta S, Charles MA. Natural killer cell and islet killer cell activities in type 1 (insulin-dependent) diabetes. *Diabetologia* 1986;29:352–357.
118. Kataoka S, Satoh J, Fujiya H, Toyota T, Suzuki R, Itoh K, Kumagai K. Immunologic aspects of the nonobese diabetic (NOD) mouse. Abnormalities of cellular immunity. *Diabetes* 1983;32:247–253.
119. Woda BA, Biron CA. Natural killer cell number and function in the spontaneously diabetic BB/W rat. *J Immunol* 1986;137:1860–1866.
120. Dinarello CA. Interleukin-1. *Rev Infect Dis* 1984;6:51–95.
121. Scala G, Allavena P, Djeu JY, Kasahara T, Ortaldo JR, Herberman RB, Oppenheim JJ. Human large granular lymphocytes are potent producers of interleukin-1. *Nature* 1984;309:56–59.
122. Mandrup-Poulsen T, Bendtzen K, Nerup J, Dinarello CA, Svenson M, Nielsen JH. Affinity-purified human Interleukin 1 is cytotoxic to isolated islets of Langerhans. *Diabetologia* 1986;29:63–67.
123. Mandrup-Poulsen T, Bendtzen K, Nielsen JH, Bendixen G, Nerup J. Cytokines cause functional and structural damage to isolated islets of Langerhans. *Allergy* 1985;40:424–429.
124. Nielsen JH, Mandrup-Poulsen T, Spinas GA, Hansen BS, Mølvig J, Nerup J, Bendtzen K. Possible role of interleukin-1 (IL-1) in the pathogenesis of insulin-dependent diabetes mellitus (IDDM), in Jaworsky M (ed): *The Immunology of Diabetes Mellitus*. Excerpta Medica Int Congr Ser. Amsterdam: Elsevier Press 1986;717:97–103.
125. Spinas GA, Hansen BS, Linde S, Kastern W, Mølvig J, Mandrup-Poulsen T, Dinarello CA, Nielsen JH, Nerup J. Interleukin 1 dose-dependently affects the biosynthesis of (pro)insulin in isolated rat islets of Langerhans. *Diabetologia* 1987;30:474–480.
126. Mandrup-Poulsen T, Spinas GA, Prowse SJ, Hansen BS, Jørgensen DW, Bendtzen K, Nielsen JH, Nerup J. Islet cytotoxicity of interleukin 1: Influence of culture conditions and islet donor characteristics. *Diabetes* 1987;36:641–647.

127. Bendtzen K, Mandrup-Poulsen T, Nerup J, Nielsen JH, Dinarello CA, Svenson M. Cytotoxicity of human pI 7 interleukin-1 for pancreatic islets of Langerhans. *Science* 1986;232:1545–1547.
128. Rosenwasser LJ, Dinarello CA. Ability of human leukocytic pyrogen to enhance phytohaemagglutinin induced murine thymocyte proliferation. *Cell Immunol* 1981;63:134–142.
129. Spinas GA, Mandrup-Poulsen T, Mølvig J, Bæk L, Bendtzen K, Dinarello CA, Nerup J. Low concentrations of interleukin-1 stimulate and high concentrations inhibit insulin release from isolated rat islets of langerhans. *Acta Endocrinol* 1986;113:551–558.
130. Zawalich WS, Diaz VA. Interleukin 1 inhibits insulin secretion from isolated perifused rat islets. *Diabetes* 1986;35:1119–1123.
131. Sandler S, Andersson A, Hellerström C. Inhibitory effects of interleukin-1 on insulin secretion, insulin biosynthesis and oxidative metabolism of isolated rat pancreatic islet. *Endocrinology* 1987; 121:1727–1731.
132. Comens PG, Wolf BA, Unanue ER, Lacy PE, McDaniel ML. Interleukin-1 is potent modulator of insulin secretion from isolated rat islets of Langerhans. *Diabetes* 1987;36:963–970.
133. Mandrup-Poulsen T, Egeberg J, Nerup J, Bendtzen K, Nielsen JH, Dinarello CA. Ultrastructural study of time-course and cellular specificity of interleukin-1 mediated islet cytotoxicity. *Acta Pathol Microbiol Immunol Scand (C)* 1987;95:55–63.
134. Dinarello CA, Cannon JG, Wolff SM, Bernheim HA, Beutler B, Cerami A, Figari IS, Palladino MA, O'Connor JV. Tumor necrosis factor (cachectin) is an endogeneous pyrogen and induces production of interleukin 1. *J Exp Med* 1986;163:1433–1450.
135. Mandrup-Poulsen T, Bendtzen K, Dinarello CA, Nerup J. Human tumor necrosis factor potentiates human interleukin-1 mediated rat pancreatic β-cell cytotoxicity. *J Immunol* 1987;139:4077–4082.
136. Oppenheim JJ, Kovacs EJ, Matsushima K, Durum SK. There is more than one interleukin 1. *Immunol Today* 1986;7:45–56.
137. Pipeleers D, Van De Winkel M. Pancreatic B cells possess defense mechanisms against cell-specific toxicity. *Proc Natl Acad Sci USA* 1986;83:5267–5271.
138. Seemayer TA, Tannenbaum GS, Goldman H, Colle E. Dynamic time course studies of the spontaneously diabetic BB Wistar rat. III. Light-microscopic and ultrastructural observations of pancreatic islets of Langerhans. *Am J Pathol* 1982;106:237- 249.
139. Ganda OP, Srikanta S, Gleason RE, Soeldner JS, Eisenbarth GS. Diminished A-cell secretion in the early phase of type 1 diabetes mellitus. *Metabolism* 1986;35:1074–1077.
140. Malaisse WJ. Alloxan toxicity to the pancreatic B-cell. A new hypothesis. *Biochem Pharmacol* 1982;31:3527–3534.
141. Pincus SH, Whitcomb EA, Dinarello CA. Interaction of IL 1 and TPA in modulation of eosinophil function. *J Immunol* 1986;137:3509–3514.
142. Matsubara T, Ziff M. Increased superoxide anion release from human endothelial cells in response to cytokines. *J Immunol* 1986;137:3295–3298.
143. Mølvig J, Bæk L, Christensen P, Platz P, Nerup J. Low-responders of interleukin-1 secretion in *E. coli* lipopolysaccharide stimulated human monocyte cultures are HLA-DR2 positive. *Acta Endocrinol* 115 (suppl. 282):43A.
144. Dinarello CA. Interleukin-1 and the pathogenesis of the acute-phase response. *N Engl J Med* 1984;311:1413–1418.
145. Rasmussen ÅK, Bech K, Feldt-Rasmussen U, Poulsen S, Siersbaek-Nielsen K, Friis T, Bendtzen K. The influence of interleukin-1 on the function of in vitro cultured

human thyroid cells in monolayer. *Acta Endocrinol* (Copenh) 1987; (suppl. 281): 93–95.

146. Handler ES, Mordes JP, Seals J, Koevary S, Like AA, Nakano K, Rossini AA. Diabetes in the Bio-Breeding/Worcester rat. Induction and acceleration by spleen cell-conditioned media. *J Clin Invest* 1985;76:1692–1694.
147. Prud'Homme GJ, Fuks A, Colle E, Seemayer TA, Guttmann RD. Immune dysfunction in diabetes-prone BB rats. Interleukin 2 production and other mitogen-induced responses are suppressed by activated macrophages. *J Exp Med* 1984;159:463–478.
148. Luger A, Schernthaner G, Urbanski A, Luger TA. Cytokine production in patients with newly diagnosed insulin dependent (type I) diabetes mellitus. *Eur J Clin Invest* 1988;18:233–236.

2
Immunogenetics and the Pathogenesis of Insulin-Dependent Diabetes Mellitus

PABLO RUBINSTEIN

Clinicians have long suspected that diabetes includes different disease entities. In fact Nancy Simpson quotes Hindu physicians, writing perhaps some 2000 years ago, who clearly identified two types of diabetes, one inherited (probably type I or insulin-dependent diabetes mellitus, IDDM), the other caused by eating unwholesome foods (type II or non-insulin-dependent diabetes mellitus).[1] These comments, as well as more modern clinical observations, were substantiated in the early 1970s when unambiguously distinct genetic backgrounds for type I and type II diabetic patients were reported. The pertinent evidence derived from studies of identical twins[2] [concordance close to 100% in those with onset after age 40 (type II?) but 50% or lower[3] in those with onset before age 40 (type 1?)][2]; from the study of HLA antigens (associations with B8 and B15 present in type I but not in type II disease),[4,5] and from the finding of genetic linkage between susceptibility to type I diabetes and the HLA loci.[6,7] In addition it was discovered almost simultaneously that many type I patients have autoantibodies reactive with the islets of Langerhans,[8] whereas patients with type II disease were essentially always negative, as were healthy individuals. These findings allowed the separation of the two major types of human diabetes into genetically, and perhaps, mechanistically, distinct nosologic conditions, which made it possible to explore their respective modes of inheritance and their pathophysiologic mechanisms. In this context a set of clinical, virologic, and epidemiologic studies that implicate diverse environmental agents in the pathogenesis of IDDM[9] became important. These environmental agents may affect a given individual with a probability that is likely to vary with the family history of disease. Thus their contribution to the expression of genetic susceptibility is not random, which obviously limits the strict applicability of simple mathematical models of inheritance.

We must, therefore, understand the pathogenesis of type I diabetes in terms of a genetic millieu, a triggering process, and an effector mechanism that eventually results in overt insulin-dependence. These three aspects of IDDM pathogenesis are most likely interdependent in that the susceptibility genotype in distinct genetic backgrounds may be more or less easily triggered by different environmental agents to produce the required amount of autoimmune, beta-cell destruction. In order to develop preventive and therapeutic strategies for IDDM, there must be an understanding not only of the nature of these elements per se but also of their interrelations in the pathogenesis of diabetes.

Genetic Background

As already stated, a major milestone in the unraveling of the genetics of IDDM was the demonstration that the mode of inheritance differed from that of the adult-onset form of glucose intolerance.

Given the importance of the HLA-linked genetic component in the association of IDDM with specific HLA antigens and haplotypes and in the risk for sibs of patients, it is likely that this component is a primary determinant of susceptibility.[10] The evidence is reviewed in the following section. Multiple unlinked genes, however, might modify the expression of the HLA-linked susceptibility even if they are not diabetogenic per se. In addition to several popular candidates (see below) these genes might act by, for example, such diverse and apparently exotic mechanisms as the regulation of the total number of pancreatic beta-cells, the determination of physiologic responsiveness to stress, and the efficiency of insulin production by beta-cells.

Twin Studies

In an early study, the empirical risk to the twin of a proband was almost 100% for non-insulin-dependent (NIDDM) patients (onset at or after the age of 40) and not more than about 50% for IDDM (onset before age 40). In addition concordant pairs usually had onsets separated by not more than some three years, indicating that both members of each pair had only one of the two types of diabetes.[2] The significantly incomplete penetrance of IDDM in most identical twin pairs establishes that environmental factors must contribute to this liability, since the genomes of identical twins are identical except for the possible occurrence of somatic mutations and, perhaps, some rearrangements. It is possible that the 50% estimate for the concordance rate may be high because juvenile diabetes (IDDM?) was defined on the basis of age-at-onset below 40 in this study, and some of those twin pairs may have had NIDDM of earlier than usual onset, and not IDDM. In addition because the ascertainment was based on the presence of disease, concordantly affected twin pairs had a higher chance of being recruited into the study. Data from American identical twin pairs reported by the Joslin Clinic group,[3] support the view that the "true" figure may be lower. In this rather small series, concordance was only about 30%. Since the period of discordance was not specifically considered, however, it might increase with time.

Twin studies therefore have revealed not only that there is a specific genetic background to IDDM that is different from that of NIDDM but also that noninherited, environmental influences determine whether the IDDM-susceptible genotype will be expressed in a given individual. Some heterogeneity of the penetrance also has been revealed by twin studies; in at least one study, those pairs with the HLA DR*3,4 genotype were concordant more often than the group as a whole.[11] Data from twin studies also have been used, together with data from HLA-identical sibs, to suggest a role of non-HLA genes in susceptibility to IDDM[12] (see below).

HLA as a Genetic Marker of Susceptibility

The association between IDDM and some antigens of the HLA system was formally demonstrated in 1974.[5,6] The term *association* is used to indicate that the frequency of the associated antigen is significantly higher or, in the case of negative associations, lower in patients than in the general population when calculated by the relative risk (RR) statistic.[13] That this association is the result of linkage between the marker gene HLA and a putative IDDM-susceptibility gene was discovered very shortly thereafter by an analysis of segregation patterns in complete families with multiple affected members (multiplex families)[7,10] and of HLA-haplotype sharing between affected sibs.[11]

The HLA System

Within the segment of chromosome 6 that carries HLA a large number of genes are organized in three major classes. (The structure of the HLA system is summarized in Fig. 2.1.)

Class I Genes

Class I genes encode several distinct types of molecules that form heterodimers with beta-2-microglobulin (itself encoded by a gene in chromosome 15)[15] for expression on the cell membrane. The best characterized class I molecules are those determined by the HLA A, B, and C loci, but at least two more such gene products, thought to be the human equivalents of the murine T1 and Qa antigens, are now known.[16] These differ from standard class I antigens in that they are restricted in their expression to a subset of T-cells, instead of being present on all nucleated cells. In addition to the expressed genes, there is evidence of the existence of a number of pseudogenes which are thought to contribute to the generation of genetic variability in the region.[17] An extra class I gene, designated HLA-E, has been found between the A and C loci, but no products have been identified as yet (Fig. 2.1).

The function of class I antigens appears to be related to their capacity to present cellular antigens to effector T-lymphocytes[18] and thus to mediate the phenomenon of effector cell "restriction." Because of MHC restriction, effector T-cells can recognize cell-bound antigens only when they are present in a cell that carries

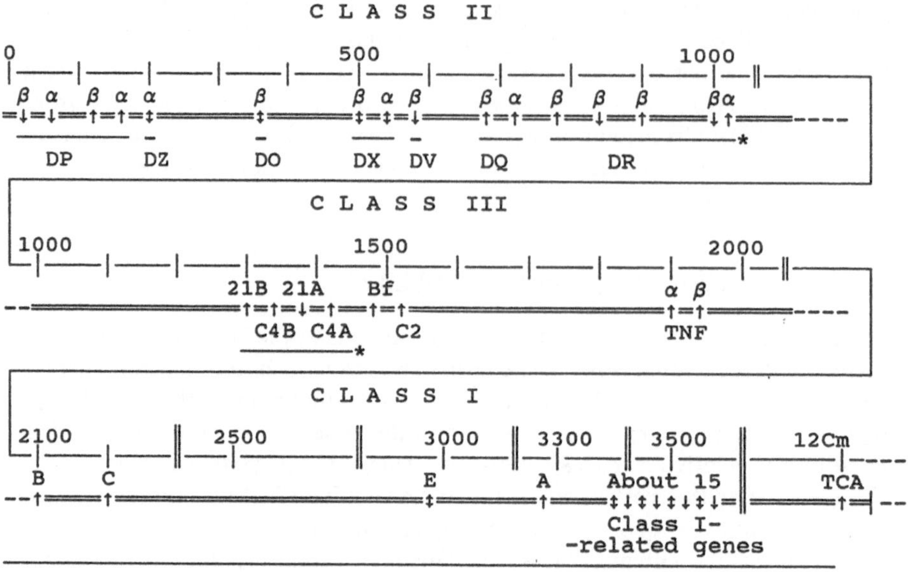

FIGURE 2.1. Structure of the HLA system. Notes: Data from review in reference 120 and (TCA) from ref. 12. Symbols: ↑ = expressed gene; ↓ = pseudogene; ↕ = expression status unknown; α,β = heavy and light chain genes, respectively; * = variable number of genes in these regions. Scale shown in Kbp (kylobase pairs) beginning (0 Kpb) at the centromeric end. Beyond ≈ 3600 Kpb, distances are given in cM (centimorgans). Interruptions of the continuous scale are indicated with a " ‖ " symbol.

class I antigens identical to those in the cells that originally stimulated the antigen-specific response, which are usually the responder's own cells.

CLASS III

Although class III genes are located between the class I and class II regions, their structure and function is unrelated to both. Four complement components (C2, C4A, C4B, and B) and an enzyme (21-hydroxylase) are encoded in this central region of MHC.[19,20] The genes for the tumor necrosis factor (TNF) alpha and beta molecules also map in this region.[21] We do not know why class III genes occupy such a strategic position within the MHC of both humans and mice. The function of C2 and C4 is to form the C3 proactivator molecule of both classical and alternative pathways, but C3 itself, although linked to the MHC in the mouse,[22] maps in a different chromosome in the human.[23] Both 21-hydroxylase (21-OH) and TNF have functions unrelated (as far as is known) to those of the other MHC gene products. An interesting feature of some class III genes is the existence of duplications and "null" alleles. Thus in each haplotype, C4, 21-OH and the TNF have two expressed genes each and at least the first two produce significantly variable, quantitative phenotypes that result from the expression of two genes, one gene, or no genes in each haplotype.[24]

Class II

The most centromerically-situated region of the MHC, class II includes three gene clusters, DR, DQ and DP. Each cluster has at least two expressed genes encoding, respectively, an alpha and a beta-chain and may have several other genes and/or pseudogenes. Thus the DR cluster has an alpha gene; one or usually two expressed beta genes; and one beta pseudogene. The DP cluster includes an expressed and a pseudogene for each chain and there are two *DQ-alpha* (*DQA1* and *DQA2*, formerly *DQ-alpha* and *DX-alpha*, respectively) and two *DQ-beta*-chain genes (*DQB1* and *DQB2*, corresponding to those previously designated *DQ-beta* and *DX-beta*). There are two other subregions, integrated by a single gene each, *DO-beta* and *DN-alpha* (formerly *DZ-alpha*) (Fig. 2.1).

Class II antigen structure includes two different (alpha and beta) noncovalently linked, glycoprotein molecules, of 32- to 34- and 29-kD, respectively, both transmembrane molecules organized into similar regions. They have two extracellular domains, a connection region, a transmembrane portion and an intracytoplasmic piece. Serologic variants are based mainly on the structural variations of the two outer domains of both alpha and beta chains, that is, of their NH_2-terminal regions.[25] Although the beta chains of DR, DQ and DP are all polymorphic, only DQ and DP alpha chains display genetic variation even though there are restriction-fragment length polymorphisms (RFLPs) of the *DR-alpha* gene as well.[26] It is interesting that, except for some haplotypes, two *DR-beta* gene products are expressed with DR-alpha by most HLA haplotypes.[27] The main beta chain is the product of the *DR-beta-1* gene; the second is encoded by either *DR-beta-3* or by *DR-beta-4*, in *DRw52* and *DRw53* haplotypes, respectively.[28] The *DQ* and *DP* regions of all haplotypes, however, appear to express singular alpha- and beta-chains. It has been recently discovered, moreover, that class II alpha–beta dimers may form spontaneously between the alpha chains of one region (e.g., DR-alpha) and the beta chains of another (DQ-beta) in lymphoblastoid cell lines (LCLs).[29] In mice the homologues of these mismatched molecules are expressed by L cells after transfection of non-isotypic *IA-beta* and *IE-alpha* genes[30]; the efficiency of the combination is, however, strikingly different for different IA-beta alleles and, in at least that case, no expression of the reverse IE-beta:IA-alpha mismatched pair could be detected serologically. Transfection of mouse IA-beta-d into a human LCL resulted in its expression paired to DR-alpha, that is, in a cross-isotypic hybrid isotypically analogous to DQ-beta:DR-alpha. Although this evidence suggests that such mixed-isotype molecules might be of functional importance, particularly with regard to HLA-disease associations, T-lymphocytes restricted by mixed-isotype class II molecules thus far have not been unequivocally demonstrated. There has been speculation that just such non-isotypic molecules may underlie the excessive risk for IDDM in DR*3,4 heterozygotes by mediating a novel function responsible for this increased susceptibility. However, since the DR-alpha molecules of both haplotypes are identical, this phenomenon would result in mismatched pairs identical to those that would be produced in the corresponding homozygotes, making this suggestion difficult to sustain. If this type of mechanism were involved in susceptibility, it might be mediated by differences in the regulatory sequences of these genes resulting in non-stoichiometric expression of the alpha and beta genes of one haplotype, which would favor the formation of mismatched pairs, as suggested by Lechler.[31] There is no reported evidence of this type of polymorphic difference in the regulation of class II gene expression, however. The immunologic functions of the class II molecules include the presentation of antigens to T-lymphocytes in the afferent phase of immune responses. Because this presentation involves physical binding of antigenic peptides to variable regions of class II molecules,[32] class II alleles may be structurally responsible for the genetic control of immune responsiveness. It is thus possible that this property of class II gene products may mediate the different susceptibility to IDDM and perhaps to all MHC-associated autoimmune diseases.

Linkage Disequilibrium

The existence of association and linkage between HLA and IDDM susceptibility indicates that "linkage disequilibrium" mediates the observed associations, that is, disease susceptibility alleles occur with significantly different frequency on chromosomes bearing different HLA genes. Linkage disequilibrium is defined as the absence of random association (equilibrium) between the alleles of closely linked genes. The persistence of linkage disequilibrium suggests that selective forces maintain the association in opposition to the effects of genetic recombination. The "preferred" allelic combinations would, thus, provide some advantage to their carriers, compared to individuals who have some but not all the alleles normally included in such haplotypes.

Linkage disequilibrium in the MHC underlies our understanding of the association of IDDM with class II alleles.[33] The first associations of IDDM with HLA antigens involved the HLA-B (class I) antigens B8 and Bw15,[4,5] but soon thereafter it was realized that these were the result of linkage disequilibrium between those class I antigens and those for the class II specificities Dw3 and Dw4, respectively, which have an even stronger association with IDDM susceptibility.[34] (Dw specificities reflect T-cell recognition of allotypic variants of class II alleles and are identified by cellular techniques using "homozygous typing cells" as typing reagents.) It was also found that B7 and Dw2 are in a significantly negative linkage disequilibrium that IDDM and, again, with the negative association of Dw2 is stronger than that of B7. Interestingly the presence of both B8 and B15

TABLE 2.1. Affected and unaffected HLA haplotypes: The HRR (New York[a] data) on affected sib pairs.

	Affected[b]				HRR[c]	
DR	1st	2nd	1+2	Unaffected	1st	2nd
1	46	4	50	49	0.75	0.74
2	22	0	22	69	0.23	0.21
3	148	7	155	40	3.94	3.65
4	183	10	193	75	2.62	2.43
5	29	1	30	7	0.29	0.27
w6	31	9	40	38	0.65	0.77
7	34	4	38	52	0.50	0.51
w8	1	0	1	5	0.16	0.15
9	10	0	10	2	4.18	3.78
w10	4	0	4	1	3.31	3.00
	488	35	539	402		

[a] New York data refers to our family study (Rubinstein, Ginsberg-Fellner et al., in progress).
[b] First and second refer to the HLA haplotypes inherited by the first and second affected sibs in multiplex sibships.
[c] HRR, haplotype relative risk.

(or of both Dw3 and Dw4) in a patient seemed to occur with an even higher frequency, and the corresponding heterozygotes had the highest relative risks for the development of IDDM, a phenomenon akin to "overdominance."[35] More recently serologic typing of HLA-D–related (HLA-D) antigens has demonstrated the same associations, that is, positive with DR3 and DR4 and negative with DR2, as expected, and, additionally, with DR5.[36]

Table 2.1 shows the haplotype relative risks (HRRs)[37] of HLA-DR alleles for IDDM in New York City. The precise risk estimates vary somewhat in different populations, but are, in general, highest for class II antigens, DR3 and DR4 in Caucasoids and DR8 in Japanese,[38] DR3 in Chinese,[39] and possibly DR1 in Caucasoids.[40] The other DR antigens may not be neutral with regard to IDDM: The conferred risks are not random but rather display a significant, consistent ranking.[41] In addition to the higher HRRs associated with DR antigens, the HLA class II region was directly implicated in IDDM susceptibility by the study of affected sib pairs in which one sib had inherited an intra-HLA–recombinant haplotype. In these cases crossovers that separate class I from class II from GLO resulted in the sharing of the class II region.[42,43] As mentioned earlier, family data indicated that the susceptibility not only is associated with certain HLA antigens, but also is inherited in block with the HLA loci,[10] regardless of whether they carry specificities associated with IDDM in the population. Thus affected sib pairs heterozygous for DR*3-or-DR*4 and a non-DR*3-or-DR*4 allele are as likely to share one as they are the other affected haplotype.[10,38] This suggests that genetic determinants distinct from, but closely linked to those responsible for serologic specificities underlie genetic susceptibility. These putative

TABLE 2.2. Common extended HLA haplotypes (supratypes) in IDDM.[a]

Class I			Class III				Class II			Risk[b]
A	Cw	B	C2	C4A	C4B	Bf	DR	DQ	DP	
1	7	8	C	Q*0	1	S	3,w52	2.1	C	High
X	3	62	C	3	2.9	S	4,w53	3.2	X	High
25	5	18	C	3	Q*0	F1	3,w52	2.1	C	High
26	–	38	C	2	1	S	4,w53	3.2	X	High
3	7	7	C	3	1	S	2,–	1	X	Low
2	–	44	C	3	Q*0	S	4,w53	3.1	X	Low

[a] Data from Refs. 46 and 47 and R. de Cordoba et al., submitted.
[b] The values of the respective relative risks vary in different populations.

determinants must, therefore, maintain high but not complete linkage disequilibrium with DR*3 and DR*4.

Linkage disequilibrium, in general, keeps specific alleles of the HLA class I, II and III loci together, which makes it difficult to determine which of these loci has the primary role. In fact the RR values may not be completely independent of the population frequency of the marker allele.[44] It is possible that an infrequent allele at a closely linked locus yields a higher RR than even the "disease" allele at the main locus with which it is in linkage disequilibrium, if the latter is substantially more common. For instance the class III *Bf*F1* allele (phenotype frequency about 1%), which is in very high linkage disequilibrium with the *HLA-B18,DR3* haplotype, generated substantial interest because its RRs were higher than either B18 or DR3 (phenotype frequency between 10 and 20%).[45] (Given their position in the haplotype, between the class I and II gene clusters, specific combinations of class III genes, or "complotypes," are also involved when class I and II alleles are in linkage disequilibrium in the same haplotypes, giving rise to "supratypes" or "extended haplotypes."[46,47]) The most common supratypes in caucasoid IDDM are shown in Table 2.2.

In summary, because of the existence of linkage disequilibrium, it is very difficult to attribute susceptibility to a distinct locus rather than to the HLA haplotype unit that encodes the various associated markers.[47]

Mode of Inheritance of IDDM

The inheritance of IDDM susceptibility, together with the HLA haplotype, provides a genetic marker of enormous practical importance. It allows investigation of the mode of inheritance of IDDM susceptibility, which is necessary to estimate risks for the relatives of probands, and it may be useful in elucidating the nature of the genetic contribution to the individual variance in IDDM susceptibility and risk.

TABLE 2.3. Segregation of HLA haplotypes and disease status of sibs of probands (New York data).[a]

		Sibs of probands [N (%)]		
		IDDM	Unaffected	Total
Haplotypes shared	2	36(25)	110(75)	146(26)
with proband	1	22(8)	253(92)	275(50)
(N)	0	4(3)	131(97)	135(24)
		62	494	556
		$\chi^2 = 38.8$, $p < .0001$.		

[a] Index cases have been excluded.

The mode of inheritance is most often inferred from the segregation of a trait in the relatives of a carrier. If a disease trait is inherited by the sibs of patients in a Mendelian proportion, that proportion directly indicates the mode of inheritance. For this to happen, the presence of its genetic determinant(s) must be a sufficient condition for the expression of the trait, that is, for the presence of disease. When noninherited factors are also required, as in IDMM, the Mendelian ratios are almost inevitably distorted. Generally distortions may result from reduced penetrance or the variable expressivity of a single susceptible genotype, from epistatic effects of genes not directly producing the susceptibility, from genetic heterogeneity of the condition (i.e., more than one genetic background producing the same clinical syndrome), or from several other conditions. For example our own data (see Table 2.3) indicate that susceptibility is expressed as disease by some 25% of sibs who share both HLA haplotypes with patients, but only by about 8% of sibs who share one and 3% of those that share none ($P < .001$).

Assuming homogeneity of the penetrance, a dominant susceptibility gene would be expressed by half as many one-haplotype–sharing as two-haplotype–sharing sibs, since only half of the former but all of the latter would carry the patient's one necessary gene. A recessive gene, if not extremely frequent, should be expressed by substantially less than half as many sibs who carry only one of the patient's HLA haplotypes (since an extra, susceptibility-carrying haplotype would then be required) as by those carrying both. This indicates that an overall dominant mode of inheritance of the susceptibility genotype is extremely unlikely, while recessivity may be compatible with the above proportions and with the segregation data. Although similar results have been obtained by different mathematical analyses on diverse family studies,[38,48,49] simple recessivity does not account for all the features of the inheritance of the disease trait.

On the basis of a largely recessive mode of inheritance for the major component of IDDM susceptibility, two important statistics can be estimated: the average penetrance, from the recurrence rate for HLA-identical sibs; and the disease gene frequency, from the proportion of affected sibs who do not share HLA haplotypes.[38,49,50] Using this approach, susceptibility gene frequency estimates for various Caucasoid populations have been remarkably similar, about 30 to 40%, resulting in an expected number of homozygotes (the square of the gene frequency), of about 0.12, or 12%. From these estimates, a prevalence of about 2 to 3% would be anticipated if the overall penetrance were the same as in the sibs of patients (0.5–0.25). The actual prevalence is much lower (about 0.2% in the United States),[51] which suggests that the HLA-linked gene may not be strictly recessive,[50] that the genetic background of IDDM may be heterogeneous,[52] or that the numerical inconsistency may be the result of nongenetic influences.[42,53] These reasons for the reduced penetrance of the HLA-linked susceptibility genes are briefly discussed below, and some are treated more fully elsewhere in this article.

Environmental Effects

Evidence of penetrance-determining environmental effects has been obtained from the study of congenital rubella and gestational diabetes, which increase the prevalence of IDDM about 50-fold[54] (see below). In addition although IDDM patients with these diseases have the same gene frequencies of DR*3 and DR*4 as do conventional IDDM patients, they do not have excessive frequencies of the DR*3,4 heterozygous genotype.[55,56] The excess of DR*3,4 heterozygotes,[57] therefore, may reflect a facilitating interaction between these DR alleles and environmental factors that would enhance the expression of the homozygous IDDM-susceptibility genotype under some, but not all conditions.

There are also epidemiologic indications of associations between exposure to viral diseases and the subsequent onset of IDDM (for reviews, see Refs. 58–60). A role for infectious agents in pathogenesis also is supported by the finding that certain experimental viral infections induce glucose intolerance in animals[60] and, possibly, by the isolation of Coxsackie virus from the pancreas of a child who died in the early stages of IDDM.[61] Other indirect evidence also has been obtained; the most persuasive evidence is the increased incidence of IDDM in several European countries (for review, see Ref. 62), with time spans short enough to reject genetic change as an explanation. The role of viral diseases in IDDM is discussed more fully below, and elsewhere in this volume.

Genetic Heterogeneity

The first evidence for possible heterogeneity of the HLA-linked, IDDM-susceptibility gene(s) was the higher than expected frequency of the DR*3,4 genotype in IDDM patients.[25,50,57] This high frequency suggests that the susceptibility gene(s) associated with DR*3 differ from those linked to DR*4, so that the DR*3,3 and DR*4,4 diabetic genotypes are qualitatively different from each other and from the DR*3,4 heterozygotes.[52,57] This model, however, implies that the higher than expected penetrance of the DR*3,4 genotype, compared to the two homozygotes, should be present regardless of how the patients were selected. This requirement is not always met. In addition to the data presented above from the studies of congenital rubella and gestational diabetes, quantitatively variable

TABLE 2.4. Testing for genetic heterogeneity in IDDM. (Is the DR*3,4 genotype more likely than other DR types to recur in affected sib pairs?) (New York data).

DR type of first affected sib	Haplotypes shared by second affected sib (N)			
	2	1	0	Total
DR3,4	7(54)[a]	5	1	13
Other	18(60)[a]	9	3	30
All pairs	25	14	4	43

p = NS (no significant increase in sharing by sibs of DR3 DR4 probands).

[a] Percent of pairs in this genotype class.

penetrances have been encountered in multiplex siblings.[53,63] In multiplex families that may have DR*3,4 offspring, that is, in families in which one parent is heterozygous for DR*3 and the other is heterozygous for DR*4, the first-affected sibs had a significantly higher frequency of the DR 3,4 phenotype than the corresponding second-affected sibs[53,63] (Table 2.4). Thus the excess of DR*3,4, which is not consistent with the expectations of a simple recessive model, may itself be nonrandomly associated with environmental variables in the pathogenesis of IDDM.

A second type of evidence offered in support of the existence of heterogeneity suggests that, contrary to segregation data, dominance of the IDDM-genetic determinant may account for the higher than expected frequency of IDDM in Black Americans.[64] If it is assumed that genetic susceptibility to IDDM was extremely rare in the African ancestors of Black American patients, the overall admixture of Caucasoid genes, that is, the proportion of Caucasoid and Negroid genes, would determine population prevalence—directly if the mode of inheritance were dominant and as its square if the mode of inheritance were recessive. Because the prevalence in Black Americans is one-third to one-half that in Caucasians, roughly the same as the gene admixture, a dominant gene seems to be implicated.[64] Because this interesting idea depends on the lack of IDDM susceptibility in the original Black immigrants, however, its appeal is limited. The hypothesis requires the demonstration that the HLA genes of Black American IDDM patients are identical to those present in Caucasoid patients and also that they were not present in their West African Black ancestors. Contrary to this requirement, the proponents of the hypothesis reported an increase in the frequency of DR3 in a small group of Nigerian IDDM patients.[65] The frequency of DR4 was higher in IDDM than in NIDDM patients and controls, but it was low overall and the sample was too small to permit the conclusion that these patients differed from Black American IDDM patients. Also although Black American IDDM patients have increased frequencies of DR3 and DR4, there is considerable heterogeneity in the frequency of the HLA-B antigens that maintain linkage disequilibrium with these DR specificities in Caucasoids and in the respective relative risks (see, for example, Ref. 66). Although it would be premature to conclude that the IDDM-associated genes in Black Americans are the same as in the Nigerian patients, taken together these data do not support the hypothesis that those genes result exclusively from Caucasoid admixture. The samples are small, and the populations are most certainly not genetically homogeneous. The subject, however, deserves further study.

The hypothesis of dominant inheritance has been recently updated. In a study of affected father–child pairs, DR4 was shared more frequently and DR3 less frequently than expected.[67] This was taken to indicate that the DR4-associated susceptibility is dominantly inherited, whereas the DR3-associated susceptibility could be recessive. A pooled set of father–offspring patient pairs displayed a similar trend (G. Thomson, personal communication). More data are needed to establish the generality of the phenomenon and to determine whether it depends on the dominance of DR4-associated IDDM.

A third indication of genetic heterogeneity is suggested by the ranking of IDDM associations of haplotypes other than DR*2, DR*3 and DR*4. Our own data (Table 2.1) agree in general with the ranking suggested,[68] while the simplest recessive model would anticipate that these other haplotypes should be more or less equally associated with IDDM.

Fourth it has been suggested that the parental transmission ratios for the *B8,DR3* haplotypes may be higher than 50%. This hypothesis has been proposed several times;[69-71] it is occasionally restricted to the father's (but not the mother's)[70] segregation and, in at least one study, it also referred to DR*4.[71] Such distorted segregation ratios, however, most probably result from ignoring the selection bias so that both patients and healthy children are counted. Not surprisingly these distortions have not been confirmed in properly analyzed pooled data.[72]

If we accepted the evidence for genetic heterogeneity, the next step would be to determine the nature of that heterogeneity. The issue would then be whether deviations from simple genetic models derive from heterogeneity at the level of the susceptibility gene itself or from epistatic interactions of additional genes in the respective HLA haplotypes. The distinction is not merely academic, since it contrasts the type of genetic mechanism that produce the individual's susceptibility with those that influence the probability of expression of the susceptibility genotype.

Non-HLA Loci

Another explanation for the reduced penetrance, namely that susceptibility-determining genes that are not linked to HLA are required, has been inferred from several observations:

1. HLA-identical sibs of patients are most often discordant for diabetes.
2. The concordance for IDDM is higher in identical twins than in HLA-identical sib pairs. As stated earlier, this difference may be of a somewhat uncertain significance and comparisons with like-sexed and HLA-identical fraternal twins must also be made. (The study of fraternal twins would also allow us to deter-

mine analytically the respective contributions of other genetic factors and of environmental influences to the observed variance in penetrance statistics.) This study has yet to be done (G. Eisenbarth, personal communication, 1988).

3. Diabetes in the NOD (non-obese diabetic) mouse and in the BB (bio-breeding) rat models which, like human IDDM, have a major susceptibility determinant in the MHC, also involve non-MHC–linked genes.[73,74]

There have been several reports of the existence and linkage relationships of IDDM-susceptibility genes that segregate independently of HLA. The most startling involved the Kidd blood groups,[75] which indicated the genetic linkage between the gene encoding this blood group and a second susceptibility gene (the lod score was 2.83 for a recessive model at a recombination fraction = 0). The significance of this linkage in the original series decreased as the series became more numerous,[76] however, and it could not be confirmed by other investigators.[77]

A second potential candidate was suggested by a polymorphic DNA region next to the 5′ end of the insulin gene on chromosome 11. This polymorphism, of the insertion-deletion type,[78] has been explored in both Caucasians and Blacks by several investigators. It initially appeared that there might be an excess of homozygotes for the "short" alleles in Caucasoid though not in Black IDDM patients.[79] Detailed analyses of the interaction between this putative second susceptibility locus and HLA alleles in the determination of diabetes susceptibility have not been done. From work done at the Fifth Genetic Analysis Workshop, however, a significant effect of this genetic region on the HLA-linked susceptibility was discerned, although there was evidence for its independent segregation from (lack of linkage to) IDDM (R.S. Spielman, personal communication, 1988).

Third, findings suggestive of the association between IDDM susceptibility and polymorphic variants of both the T-cell receptor, C-beta-chain genes[80] and the IgG, CH-chain gene[81] have been reported. These candidate genes are particularly appealing for two reasons: first because they are involved in immune responsiveness, in general, and autoimmunity, in particular, and, second, because functional T-cell receptor genes as well as immunoglobulin heavy chain genes are organized by somatic events. Thus any two individuals, including identical twins, might have different gene rearrangements, and if these were associated with functionally different products, genetically identical people might conceivably be subject to different risks. Attractive though they may be, the original reports on these candidate genes have not received much support. In our own studies (Carrier et al., in preparation), affected sib pairs shared the T-cell receptor, C-beta alleles only as would be expected by chance, and there was no indication of linkage to the susceptibility gene. In addition as the numbers increased, the significance of the original series may have decreased (M. Hoover et al., presented at the Tenth International Histocompatibility Testing Conference, November 1987).

Finally there is a preliminary report on the possibility that the haptoglobin genotype may be important (in addition to the MHC) in families in which probands have parents with NIDDM.[82]

Molecular Approaches

A direct approach toward defining the mode of inheritance of IDDM has been recently initiated. It consists of the search for the HLA genetic variants that associate most closely, if not perfectly, with the disease, which may be present in all the haplotypes carrying the genes responsible for susceptibility. Such variants may be detectable through the use of RFLPs of HLA class II genes, or of biochemical polymorphisms of their protein products. The principal aims of this search are, first, to define the molecular difference between haplotypes present in patients ("affected" haplotypes) and serologically identical haplotypes present in healthy unrelated controls ("unaffected" haplotypes) and, second, to characterize the genetic similarities that may be consistently present in all serologically different, affected haplotypes. Additional expected benefits include the feasibility of identifying genetically at-risk individuals in the general population and a better chance of locating the HLA gene that is most contributory to pathogenesis.

Distinctive RFLPs have already been found in all HLA class II genes, and their association with IDDM has recently been the object of intense investigation (see, for example, Refs. 83–85). The results thus far indicate that, at least for DR*4-carrying haplotypes, molecular variations of the *DQ-beta* gene identify serologically *DR*-identical haplotypes with significantly different RRs. These RFLPs are accompanied by serologically and biochemically detectable variants (including amino acid sequence differences) of the DQw3 antigens[86] and it has been postulated that they identify *DQ-beta* as "the" HLA gene responsible for the IDDM association.[87,88] Such a conclusion is not entirely warranted, since all *DQ* alleles maintain high linkage disequilibrium with variants of other class II genes. Thus the molecular variant of *DQ-beta 3* (*DQ-beta 3.2*) that is positively associated with IDDM could simply mark a subset of haplotypes that carry susceptibility genes, irrespective of the relationship of the *DQ*-beta gene per se to the disease. In addition no *DQ-beta* variants that correlate with different risks for IDDM have been detected in *DR3*-carrying haplotypes, and very recent data from our own studies on class II protein polymorphisms indicate that DP-beta isoelectric-point variants may identify subsets of *DR3* haplotypes with very different risks (Rodriguez de Cordoba et al. 1988, submitted). Studies at the protein level have shown that some but not all *DQ*-heterozygous individuals present hybrid molecules, that is, class II molecules consisting of a DQ-alpha chain joined to the DQ-beta chain of the paired haplotype, a situation described as "transcomplementation." Individuals of the DR3,4 type (DQw2,w3) may present such hybrid molecules,[89] a provocative finding in view of the higher penetrance of this genotype and of the possibility that hybrid molecules might, under appropriate conditions, restrict antigen-specific, T-cell responses.[90]

In summary, (1) the genetic susceptibility to IDDM involves one or more loci within or very close to the HLA class II region; (2) HLA haplotypes encoding some class II alleles (especially DR*3 and DR*4) imply higher risks of disease than others, but all haplotypes found in patients are approximately equally likely to recur in the corresponding, second-affected sibs; (3) segregation data are consistent with an incompletely penetrant recessive mode of inheritance, but the population data, including the excessive frequency of the DR*3,4 genotype among patients and the inaccuracy of the predicted prevalence rates, suggest the influence of modifiers of genetic and/or environmental nature or the existence of genetic heterogeneity; (4) evidence of genetic heterogeneity has been taken to suggest that the mechanisms of the susceptibility and perhaps even the mode of inheritance may be different in haplotypes carrying different *DR* alleles; (5) penetrance-modifying secondary susceptibility loci have been proposed that either have not been confirmed in subsequent studies, or to date have been insufficiently studied; and (6) the effect of environmental factors in increasing the penetrance of the HLA-linked susceptibility has been disclosed by various epidemiologic, virologic, and clinical studies.

The Effector System: Autoimmunity

After IDDM onset, the endocrine pancreas is clearly deficient in the number of insulin-producing beta-cells. Lymphocytic infiltrates are present in the islets of Langerhans shortly after onset[91] and patients with IDDM produce islet cell-specific antibodies[8] and exhibit cell-mediated hypersensitivity toward pancreatic tissue-specific antigens.[92] Additionally IDDM patients frequently have other autoimmune manifestations.[93,94] These observations suggest that the susceptibility genotype might enable or enhance autoimmune reactivity, in general, which when directed toward islet-cell antigens, might eventually destroy beta-cells and, consequently, produce glucose intolerance.

Systematic reviews of this subject are given elsewhere in this volume and only a brief discussion of certain aspects selected for their relevance to the issue of individual risks will be included here.

Evidence of Autoimmune Activity in IDDM Patients

Technical improvements in immunofluorescence methodology[95] and the development of alternative techniques, including those that detect antibodies reactive with cell-surface (ICSA),[96,97] rather than cytoplasmic antigens (ICA), have confirmed the existence of anti-beta-cell humoral autoimmunity in most IDDM patients at onset.[98] Similar autoantibodies have been encountered in the BB rat and the NOD mouse models.[99,100] Recent progress in identifying beta-cell antigens targeted by these autoantibodies has demonstrated the existence of several distinct antigens.[101] Islet-cell–specific monoclonal antibodies, of both rodent and human origin, have been prepared and investigated[102] (see review, Ref. 103). Interestingly most if not all of these monoclonal antibodies also react with epitopes present in nonpancreatic tissues.

The Role of Beta-Cell Autoimmunity in IDDM Pathogenesis

In contrast to its recognized prevalence in IDDM patients, the role of autoimmunity to islet-cell antigens in the mechanism of the onset and, in general, in the pathogenesis of IDDM, is not clearly understood.

Cell-Mediated Autoimmune Mechanisms

Very little has been learned regarding cell-mediated autoimmunity in human IDDM, largely because of technical problems. Obviously recognition, identification, and characterization of effector cells would require special cellular targets that are not yet routinely available. Such targets should express both the beta-cell–specific epitopes and the restricting-HLA–specificities in order to support the (most likely) MHC-restricted interaction with the patient's putative, beta-cell–autoreactive T-lymphocytes. For this reason, although cellular hypersensitivity to beta-cells in human IDDM was recognized many years ago,[92] research on autoimmunity has been almost entirely limited to the study of humoral antibodies.

Humoral Autoimmunity

ICA are detectable in almost all partients shortly after onset, their prevalence is similar in patients of both sexes and decreases with time after onset. They are found in some healthy relatives of IDDM patients over a period of years.[93,96,104,105] In fact ICA and ICSA (along with HLA-identity to the proband and a decrease in insulin secretion) are part of a clinical triad that prospectively identifies children at high risk for IDDM.[107,108] Healthy relatives of IDDM probands, however, may have ICA and even complement-fixing ICSA for many years without clinically relevant deleterious effects on beta-cell function.[104] In our studies, the presence of ICA and ICSA in relatives of patients was largely limited to families with persistently ICA- and/or ICSA-positive patients.[96,104] Apparently therefore, these autoantibodies are not pathogenic in themselves and might represent responses to diverse types of beta-cell injuries, including those provoked by environmental triggers of IDDM. Their frequent presence in healthy relatives of ICA-positive, IDDM patients suggests that these relatives are more often exposed to the events that lead to their production than are individuals in the general population. It is still possible, however, that differences in the concentration, avidity, and/or specificity of various islet-cell antibodies exist between patients and their healthy relatives and that such differences account for the discordance in pathogenic significance. This possibility becomes more likely where the islet antibody prevalences observed with different methods are compared. Recognition of this problem led to the introduction of a standardized procedure under the aegis of the Juvenile Diabetes Foundation.[95] This new procedure appears to increase the consistence of findings in different laboratories and to decrease the prevalence of ICA among nonpatients, that is, it provides higher specificity. Its lower sensitivity, although useful in defining individuals at increased risk, may limit studies of the factors that trigger islet antibody formation.

HLA and the Prevalence of Islet-Cell Autoantibodies

Much excitement was initially generated by the report of a higher prevalence of anti-islet cell antibodies in DR3-positive patients, which seemed to support the concept of genetic heterogeneity between IDDM syndromes associated with

DR3 and DR4.[105] It is probably fair to state, however, that most investigators have not found clear differences in the proportion of positive islet-cell serologies between patients of different HLA types (see review, Ref. 53).

Nevertheless, there is some evidence that the production of islet-cell autoantibodies is associated with "affected" HLA haplotypes. For example healthy sibs who share one or two HLA haplotypes with their persistently ICSA-positive probands are equally likely to have ICSA, whereas sibs who share none do not differ from the controls, regardless of the HLA-DR phenotype.[96] These data suggest that patients could be homozygous for the ability to make islet-cell autoantibodies, which behaves as a dominantly inherited and HLA-linked trait. Either of their HLA haplotypes, separately or together, confers this ability with the same likelihood, in contrast to the risk of IDDM itself, which is much higher for sibs who share both of the proband's HLA haplotypes. It is also interesting that anti-beta-cell autoantibodies are found in a very high proportion of patients with congenital rubella syndrome,[55A] regardless of their DR types, and in women with gestational diabetes.[56] In the latter group, a significant HLA association was seen: 18 of 20 ICA-positive patients had DR3 or DR4.

Related Autoantibodies

In addition to ICA and ICSA, patients with IDDM have autoantibodies to insulin in the absence of exposure to exogenous insulin.[109] The presence of these anti-insulin autoantibodies may not be associated with the HLA phenotypes but there is a significant difference in treated patients: *DR3* homozygotes, in contrast to other patients, appear not to make anti-insulin antibodies or to make them in very low concentrations, whereas patients with other DR types appear not to differ significantly in this regard.[110] Since it is not clear how anti-insulin autoantibodies could participate in the destruction of beta-cells, their presence suggests that they are secondary to, rather than causative of, such destruction.

The negative association between *DR3* homozygosity and the production of anti-insulin antibodies is interesting in view of the generally positive association between the DR3 type and an enhanced capacity to produce diverse autoantibodies.[104] As stated earlier, diabetic patients and their immediate relatives frequently make antibodies to other endocrine organs, especially the thyroid, and to epitopes shared by many other issues,[93,94] and they manifest other conditions of a possibly autoimmune pathogenesis.[93,104,105] Most of these associated endocrine diseases, however, are not inherited with HLA and/or have HLA associations that are different from those of IDDM.

Thus a number of coincident but different functions of the HLA system may be involved in the many immunologic correlates of IDDM.

Other Autoimmune Effectors

Animal experiments indicate that an adoptive transfer of lymphoid cells can transfer insulin-dependent diabetes.[111] Despite the difficulty of their in vitro evaluation, there is strong evidence for the involvement of cellular effectors in the destruction of beta-cells in human IDDM. Thus immunosuppressive therapy instituted shortly after the onset of IDDM has led to a prolongation of the "honeymoon" period[112] and to the preservation of residual insulin-secreting capability.[113] Perhaps more strikingly, segmental grafts of human pancreas from a healthy to a diabetic twin resulted in the prompt reappearance of disease in the grafted pancreas.[114] Diabetes, in these cases, was an unambiguous consequence of the selective destruction of the beta-cells, apparently by a lymphocytic infiltration of islets.

Environmental Factors

The lack of concordance in at least 50% of the identical twins of IDDM probands is strong evidence to support the concept that noninherited environmental factors are necessary for the development of disease in those at genetic risk.

There have been many observations linking certain viral infections to the onset of IDDM. The publication in 1864 of a case in which mumps preceded the onset of diabetes (cited in Ref. 58), was followed by other early clinical observations involving the mumps virus (see review in Ref. 59). Other infectious agents subsequently have been implicated by epidemiologic animal model, and clinical case studies.[60] Direct and definitive demonstration of viral involvement in the pathogenesis of conventional IDDM, however, remains elusive. In a case of IDDM, in which onset was accompanied by a lethal diabetic coma, it was possible to isolate Coxsackie B4 virus from the pancreas at autopsy, which, after in vitro passage, induced hyperglycemia and other symptoms of diabetes upon inoculation into mice.[61] Histologic evidence of prior insulitis was later found, however, suggesting that the beta-cells had been damaged earlier, and that the onset of IDDM may have been unrelated to the virus isolated at autopsy. This somewhat reduces the strength of the evidence that this virus was causally associated with IDDM in this case. Coxsackie B4, however, also has been implicated in the triggering of type I diabetes by the finding of high titers of antibodies specific for this virus in IDDM patients of recent onset (see review, Ref. 115).

Very interesting data regarding viral involvement have emerged from studies of congenital rubella syndrome. Patients with this condition, which is caused by a rubella infection of the mother during the first trimester of pregnancy, have an approximately 50-fold increase in the prevalence of IDDM through the third decade of life.[54] The HLA associations of diabetes (increased frequencies of DR3 and DR4 and reduced frequency of DR2) are virtually the same for congenital rubella syndrome patients with diabetes as they are for conventional IDDM patients.[55] This suggests that even such devastating viral infections may have diabetogenic consequences only in individuals with the conventional IDDM-susceptibility genotype. Our ongoing studies have also shown that congenital rubella syndrome patients frequently produce ICSA ($>20\%$), conventional ICA ($>12\%$), and anti-insulin antibodies ($>10\%$) (McEvoy et al., 1988, in preparation). Antithyroid (microsomal and thyroglobulin) autoantibodies also are significantly more frequent in these patients than in age-matched controls.[55A] Thus congenital rubella syndrome patients in whom IDDM is unambiguously related to a specific virus have autoimmune and genetic correlates that do not differ from those in the conventional disease.

It is interesting that the onset of IDDM in children with congenital rubella syndrome usually occurs several years after birth: The viral infection does not appear to immediately destroy the beta-cells but rather it initiates a process that results in glucose intolerance many years later. We do not know whether a chronic rubella infection is slowly destroying the beta-cells, whether rubella virus initiates an effector of beta-cell-toxic autoimmunity, or whether the virus merely reduces the starting number of beta-cells and thus facilitates the development of diabetes in the presence of genetic risk.

The evidence for environmental factors in human IDDM also derives from other, apparently noninfectious, conditions, including gestational diabetes. In pregnant women with this syndrome, who appear to have a normal glucose tolerance after parturition, decompensation to overt IDDM is frequent. In ongoing studies of 130 women with gestational diabetes, Ginsberg-Fellner et al. found that, among a small sample of patients ($n = 50$), 20 had serum ICA (40%) and 18 of the 20 ICA-positive patients had the IDDM-associated HLA antigens DR3 or DR4.[56] Five years later, 11 of the 130 patients had become insulin-dependent; 9 of the 11 were DR3 and/or DR4 positive, and all but one were retrospectively shown to have been ICA-positive after parturition (unpublished observations). It appears, therefore, that women with gestational diabetes frequently produce anti-islet cell autoantibodies, just as do children with congenital rubella syndrome, and those who have the required genetic makeup are at an extremely high risk of developing IDDM. In this group, moreover, there may be a relationship between the presence of DR3 and DR4 and the formation of ICA and ICSA. Although the nature and mechanism of action of the specific risk factor are unknown, the relationship between gestational diabetes as an environmental risk factor and IDDM is clear.

These examples indicate that similar genetic and immunologic factors may operate in exceptional cases as well as in "conventional" IDDM and suggest that "triggering" environmental agents probably act by enhancing the penetrance of the IDDM-susceptibility genotype. It is possible that IDDM in some, and perhaps most conventional cases is precipitated by repeated islet-cell-tropic viral infections that may have no clinical signs and are thus usually not recognized.

Speculations on the Mechanism of IDDM Pathogenesis

The relationships among genetic susceptibility, autoimmunity, and environmental influences are not simple. Available data indicate that subsets of haplotypes of every DR specificity carry susceptibility but that the probability of this occurring depends on the respective DR antigens. Thus DR4 haplotypes as a group include a substantial proportion of "affected" haplotypes. The additional presence of DQw3.2 increases the relative risk, that is, the probability that these haplotypes carry susceptibility factors. If the *DQ* locus were, in fact, "the susceptibility" locus, given the importance of this locus in restricting suppressive immune responses, the data would suggest that the mechanism of IDDM pathogenesis involves diminished suppression of "normal" anti-beta-cell autoimmunity.

This hypotheses would agree with evidence for a mostly recessive segregation, since the presence of a single haplotype capable of allowing suppressive responses would be sufficient for suppression to take place. Haplotypes with a low probability of carrying the susceptibility gene, such as DR*2 and DR*5, would be most often capable of determining phenotypically normal suppression and would thus be considered "protective." A convincing example of this type of abnormal suppression has been described in the pathogenesis of liver damage that accompanies persistent autoimmunity in *Schistosoma japonicum* infections.[116] We also know that beta-cells are capable of de novo expression of class II antigens in the presence of interferon and tumor necrosis factor,[117] which is likely to occur with local inflammatory processes. One could therefore postulate that, in the evolution of certain viral diseases, pancreatic inflammation occurs and leads to class II antigen expression by beta-cells, making them capable of presenting tissue-specific autoantigens. This would result in the clonal stimulation and expansion of T-cells with specificity for the beta-cell and perhaps insulin-related autoantigens, and in the production of ICA and anti-insulin autoantibodies. It is clear, however, that this expansion does not by itself destroy sufficient beta-cells for the development of IDDM, since ICA may be present for many years in healthy individuals. Thus the regulatory influence of expanded suppressive T-cell clones in normal individuals must be considered in the scheme of expanded, autoaggressive, antibeta-cell lymphocyte populations. People homozygous for the genetic variant that confers IDDM susceptibility would be relatively inefficient in marshaling such a regulatory immune reaction. This inefficiency, which allows the beta-cell toxic phase to continue beyond that accompanying a viral inflammation, would produce more damage after each event. Epistatic effects of other MHC genes could modify the overall effects of a reduced suppressor-cell response, for example, by modulating the antiviral immune response itself. Thus, different haplotypes might affect the magnitude of pancreatic inflammation to significantly different degrees in response to similar infections, even if they carried the same IDDM-susceptibility alleles. Given the antigen specificity of some class II-nominal antigen interactions, it is possible that some viral infections may be epistatically enhanced by one DR type and not by another.[118] Moreover, given that many antigenically different viruses produce these pancreatic inflammations, such MHC-heterozygotes such as DR*3,4 or DR*1,4, in some populations, might be induced to excessive inflammation in response to more viruses than would the corresponding homozygotes, resulting in the appearance of overdominance. Overwhelming or especially severe chronic infections with a single virus, however, might be sufficient to produce IDDM even in *DR* homozygotes. An additional feature of IDDM, its slow development, becomes more understandable in this light, since the damage resulting from separate events would be cumulative.

An important weakness of this scheme is the lack of methods for objectively testing it in humans because the cellular targets are not available. Its main strengths are, first, that it accounts for the complex mode of inheritance and its interactions with the environmental factors in IDDM pathogenesis. Second, it is in general agreement with data that show decreased levels of nonspecific indices of suppressor-cell activity (see review, Ref. 94). Third, if proven correct, this hypothesis suggests an avenue for the prevention of IDDM in people at genetic risk; it might be possible to expand, selectively, the desirable suppressor-cell populations in vitro to an extent sufficient to prevent IDDM when they are returned to the donor. This would be the mirror image of a current approach to metastatic cancer therapy, in which in vitro expanded tumor-specific cytotoxic cell populations from the patient are used to treat that patient.[119]

In summary, we suggest that the central problem of type I diabetic is an HLA-linked, inherited deficiency in certain T-cell–mediated, suppressive immune responses, including those against self beta-cell antigens. This deficiency would be responsible for the inability to control normal T-cell responses triggered by inflammatory processes of the pancreatic islets and directed at viral and/or cell-specific antigenic targets. Different exposures to environmental triggers and allelic variants of non-HLA genes that control functions relevant to this multifaceted process of immunologic recognition, generation of effector and regulatory elements, local tissue responses, and number and regenerating capacity of specific targets probably determine whether a susceptible subject will develop IDDM.

References

1. Simpson NE. The genetics of diabetes mellitus—a review of family data, in Creutzfeldt W, Kobberling J, Neel J (eds): *The Genetics of Diabetes Mellitus*. Berlin, Heidelberg, New York: Springer-Verlag, 1976;12–20.
2. Tattersall RB, Pyke DA. Diabetes in identical twins. *Lancet* 1972;2:1120–1124.
3. Cahill GF. Current concepts of diabetic complications with emphasis on hereditary factors; a brief review, in Sing CF, Skolnick M (eds.): *Genetic Analysis of Common Diseases: Application to Predictive Factors in Coronary Heart Disease*. New York: Alan R. Liss, 1979;113–125.
4. Nerup J, Platz P, Ortved-Andersen O, et al. HLA antigens and diabetes mellitus. *Lancet* 1974;2:289–291.
5. Cudworth AG, Woodrow JC. HLA antigens and diabetes mellitus. *Lancet* 1974;2:1153–1155.
6. Rubinstein P, Suciu-Foca N, Nicholson JF, et al. The HLA system in the families of patients with juvenile diabetes mellitus. *J Exp Med* 1976;143:1277–1282.
7. Barbosa J, Chern MM, Noreen H, et al. Analysis of linkage between the major histocompatibility system and insulin-dependent diabetes in multiplex families. *J Clin Invest* 1978;62:492–495.

8. Bottazzo GF, Florin-Christensen A, Doniach D. Islet-cell antibodies in diabetes mellitus with autoimmune poly-endocrine deficiencies. *Lancet* 1974;2:1279–1283.
9. Craighead JE. The role of viruses in the pathogenesis of pancreatic disease and diabetes mellitus. *Prog Med Virol* 1975;19:161–214.
10. Rubinstein P, Suciu-Foca N, Nicholson JF: Genetics of juvenile diabetes mellitus. A recessive gene closely linked to HLA-D and with 50 percent penetrance. *N Engl J Med* 1977;297:1036–1040.
11. Cudworth AG, Spencer KM, Gorsuch AN, et al. Immunogenetic heterogeneity in insulin-dependent diabetes, in Kobberling J, Tattersall R (eds.): *The Genetics of Diabetes Mellitus*. London, New York: Academic Press 1982;63–78.
12. Thomson G. A two locus model for juvenile diabetes. *Ann Hum Genet* 1980;43: 383–398.
13. Svejgaard A, Jersild C, Staub-Nielsen L, Bodmer WF. HLA antigens and disease. Statistical and genetic considerations. *Tissue Antigens* 1974;4:95–104.
14. Cudworth AG, Woodrow JC. Evidence for HLA-linked genes in "juvenile" diabetes mellitus. *Br Med J* 1975;3:133–136.
15. Cox DR, Gedde-Dahl T Jr. Report of the committee on the genetic consultation of chromosomes 13, 14, 15, and 16. Eighth International Workshop on Human Gene Mapping. *Cytogenet Cell Genet* 1985;40:206–241.
16. van Leeuwen A, Festenstein H, van Rood JJ. Di-allelic alloantigenic systems on subsets of T cells. *Hum Immunol* 1982;4:109–121.
17. Barbosa JA, Kamarck ME, Biro PA, et al. Identification of human genomic clones coding the major histocompatibility antigens HLA-A2 and HLA-B7 by DNA-mediated gene transfer. *Proc Natl Acad Sci USA* 1982;79:6327–6331.
18. Benacerraf B. Role of MHC gene products in immune regulation. *Science* 1981; 212:1229–1238.
19. Carroll MC, Campbell RD, Porter RR. Mapping of steroid 21-hydroxylase genes adjacent to complement component C4 genes in HLA, the major histocompatibility complex in man. *Proc Natl Acad Sci USA* 1985;82:521–524.
20. White PC, Grossberger D, Onufer BJ, et al. Two genes encoding steroid 21-hydroxylase are located near the genes encoding the fourth component of complement in man. *Proc Natl Acad Sci USA* 1985;82:1089–1093.
21. Spies T, Morton CC, Nedospasov SA, et al. Genes for the tumor necrosis factor alpha and beta are linked to the human major histocompatibility complex. *Proc Natl Acad Sci USA* 1986;83:8699–8702.
22. Penalva da Silva F, Hoecker G, Day NK, et al. Murine complement component 3: Genetic variation and linkage to H-2. *Proc Natl Acad Sci USA* 1978;75:963–965.
23. Naylor S, Lalouel J-M, Show DJ. Report of the committee on the genetic constitution of chromosomes 17, 18 and 19. Eighth International Workshop on Human Gene Mapping. *Cytogenet Cell Genet* 1985;40:242–267.
24. Hauptmann A, Mauff G, Tongio MM, et al. Evidence for the presence of two structurally different C4A genes on one chromosome in a Caucasoid family (Abs). Fourth International Workshop on the Genetics of Complement. 1982.
25. Seyfried CE, Gregersen PK, Napom BS, et al. Functional polymorphisms among HLA-DR4+ DR-beta chains associated with limited peptide diversity. *Mol Immunol* 1987;24:471–481.
28. Rollini P, Mach B, Gorski J. Linkage maps of three HLA-DR-beta-chain genes: Evidence for a recent duplication event. *Proc Natl Acad Sci USA* 1985;82:7197–7282.
29. Lotteau V, Teyton L, Burroughs D, et al. A novel class II molecule (DR-alpha-DQ-beta) created by mismatched isotype pairing. *Nature* 1986;329:339–341.

30. Germain RN, Quill H. Unexpected expression of a unique mixed-isotype class II MHC molecule by transfected cells. *Nature* 1986;320:72–75.
31. Lechler RI. MHC molecular structure-permitted pairs? *Immunol Today* 1988;9:76–78.
32. Babbit BP, Matsueda G, Haber E, et al. Antigenic competition at the level of peptide-Ia binding. *Proc Natl Acad Sci USA* 1986;83:4509–4514.
33. Winchester RJ. The HLA system and susceptibility to diseases: An interpretation. *Clin Aspects Autoimmunol* 1986;1:9–26.
34. Thomsen M, Platz P, Ortved-Andersen O, et al. MLC typing in juvenile diabetes mellitus and idiopathic Addison's disease. *Transplant Rev* 1975;22:120–125.
35. Svejgaard A, Ryder JP. Associations between HLA and disease, in Dausset J, Svejgaard A (Eds.): *HLA and Disease.* Baltimore, Maryland: Williams & Wilkins, 1977;46–53.
36. Rubinstein P, Walker M, Carpenter C, et al. Genetics of HLA disease associations. *Hum Immunol* 1981;3:384 (Abs).
37. Falk CT, Rubinstein P. Haplotype relative risks: An easy reliable way to construct a proper control sample for risk calculations. *Ann Hum Genet* 1987;51:227–233.
38. Bertrams J, Baur MP. Disease reports: Insulin dependent diabetes mellitus, in Albert ED, Baur MP, Mayr WR (Eds.): *Histocompatibility Testing 1984.* Berlin: Springer-Verlag, 1984:348–358.
39. Lee DW, Chan SH, Tan SH, et al. HLA-system in chinese children with insulin-dependent diabetes mellitus: A strong association with DR3. *Metabolism* 1984; 33:1102–1105.
40. Winnearls BC, Bodmer JG, Bodmer WF, et al. A family study of the associations between insulin-dependent diabetes mellitus, autoantibodies and the HLA system. *Tissue Antigens* 1984;24:234–246.
41. Thomson G. HLA-DR antigens and susceptibility to insulin-dependent diabetes mellitus. *Am J Hum Genet* 1984;36:1309–1317.
42. Rubinstein P, Ginsberg-Fellner F, Falk CT. Genetics of Type 1 diabetes mellitus: A single recessive predisposition gene mapping between HLA-B and GLO. *Am J Hum Genet* 1981;865–882.
43. Bertrams J, Sodomann P, Grosse-Wilde H, et al. Two HLA-B/D,DR recombinants in diabetic families: The susceptibility gene travelled with HLA-D,DR fragment, in Terasaki PI (ed.): *Histocompatibility Testing, 1980.* Los Angeles: UCLA Press, 1980;949.
44. Rubinstein P. HLA-disease associations: In search of meaning, in Hackel E, Mallory D (eds.): *Theoretical Aspects of HLA.* New York: American Association of Blood Banks, 1982;103–128.
45. Raum D, Awdeh Z, Alper CA. Bf types and the mode of inheritance of insulin-dependent diabetes mellitus (IDDM). *Immunogenetics* 1981;12:59–74.
46. Raum D, Awdeh Z, Yunis E, et al. Extended Major Histocompatibility Complex haplotypes in Type 1 diabetes mellitus. *J Clin Invest* 1984;74:449–454.
47. Kelly H, McCann VJ, Kay PH, et al. Susceptibility to IDDM is marked by MHC supratypes rather than individual alleles. *Immunogenetics* 1985;22:643–651.
48. Rubinstein P, Suciu-Foca N. HLA and the genetic aspects of the predisposition to juvenile diabetes mellitus. A follow-up, in Camerini-Davalos RA, Hanover B (Eds.): *Treatment of Early Diabetes.* New York: Plenum Publishing, 1979,33–40.
49. Rich SS, Green A, Morton NE, Barbosa J. A combined segregation and linkage analysis of insulin-dependent diabetes mellitus. *Am J Hum Genet* 1987;40:237–249.

50. Svejgaard A, Platz P, Ryder LP. Insulin-dependent diabetes. Joint results of the 8th Workshop study, in Terasaki PI (ed.): *Histocompatibility Testing, 1980.* Los Angeles: UCLA Press, 1980;638–656.
51. Kyllo CJ, Nuttal FA. Prevalence of diabetes mellitus in school-age children in Minnesota. *Diabetes* 1978;27:57–60.
52. Anderson CE, Hodge SE, Rubin R, et al. A search for heterogeneity in insulin dependent diabetes mellitus (IDDM): HLA and autoimmune studies in simplex, multiplex and multigenerational families. *Metabolism* 1983;32:471–477.
53. Rubinstein P, Walker M, Ginsberg-Fellner F. The excess of DR3,4 in Type 1 diabetes: What does it portend? Diabetes 1986;35:985–989.
54. Forrest JM, Menser MA, Burgess JA. High frequency of diabetes mellitus in young adults with congenital rubella. *Lancet* 1971;2:332–334.
55. Rubinstein P, Walker ME, Fedun B, et al. The HLA system in congenital rubella patients with and without diabetes. *Diabetes* 1982;31:1088–1091.
55a. Ginsberg-Fellner F, Witt ME, Yagihashi S, et al. The congenital Rubella Syndrome as a model for Type I diabetes mellitus (IDDM); increased prevalence of islet-cell surface antibodies. *Diabetologia* 1984;27:87–89.
56. Rubinstein P, Walker M, Krassner J, et al. HLA antigens and islet-cell antibodies (ICA) in gestational diabetes (GD). *Hum Immunol* 1981;3:271–275.
57. Rotter JI. An HLA genotype study of IDDM: The excess of DR3/DR4 heterozygotes allows rejection of the recessive hypothesis. *Diabetes* 1983;32:169–174.
58. Maugh TH. Diabetes: Epidemiology suggests a viral connection. *Science* 1975; 188:347–351.
59. Gundersen E. Is diabetes of infectious origin? *J Infect Dis* 1927;41:197–202.
60. Rayfield EJ, Seto Y. Viruses and the pathogenesis of diabetes mellitus. *Diabetes* 1978;27:1126–1140.
61. Yoon JW, Austin M, Onodera T, Notkins AL. Virus-induced diabetes mellitus: Isolation of a virus from the pancreas of a child with diabetic ketoacidosis. *N Engl J Med* 1979;300:1173–1179.
62. Vaandrager GJ, Bruining GJ, Veenhof FJ, et al. Incidence of childhood diabetes in The Netherlands: A decrease from north to south in North-Western Europe? *Diabetologia* 1984;27:203–206.
63. Falk CT. HLA haplotype sharing in IDDM affected sib pairs as a function of HLA DR genotype. *Genet Epidemiol* 1986; supp.1:353–358.
64. MacDonald MJ. The frequencies of juvenile diabetes in American blacks and Caucasians are consistent with dominant inheritance. *Diabetes* 1980;29:110–114.
65. MacDonald MJ, Famuyiwa OO, Nwabuebo IA, et al. HLA-DR associations in Black Type I diabetics in Nigeria. Further support for models of inheritance. *Diabetes* 1986;35:583–589.
66. Neufeld MR, Maclaren NK, Riley WJ. HLA in American Blacks with juvenile diabetes. *N Engl J Med* 1980;303:111–112.
67. MacDonald MJ, Gottschall J, Hunter JB, et al. HLA-DR4 in insulin-dependent diabetic parents and their diabetic offspring: A clue to dominant inheritance. *Proc Natl Acad Sci USA* 1986;83:7049–7053.
68. Louis EJ, Thomson G. Three allelic synergistic model for insulin-dependent diabetes mellitus. *Diabetes* 1986;35:958–963.
69. Cudworth AG, Wolf E, Gorsuch AN, et al. A new look at HLA genetics with particular reference to type-1 diabetes. *Lancet* 1979;2:389–391.
70. Awdeh Z, Raum D, Yunis E, et al. Extended HLA/complement allele haplotypes: Evidence for T/t-like complex in man. *Proc Natl Acad Sci USA* 1983;80:259–263.

71. Vadheim CM, Rotter JI, MacLaren NK, et al. Preferential transmission of diabetic alleles within the HLA gene complex. *N Engl J Med* 1986;315:1314–1318.
72. Thomson G, Klitz W, Louis EJ, et al. HLA and IDDM predisposition: New aspects. *Genet Epidemiol* 1986; Suppl. 1:363–368.
73. Prochazka M, Leiter EH, Serreze DV, et al. Three recessive loci required for insulin-dependent diabetes in non-obese diabetic (NOD) mice. *Science* 1987;237:286–289.
74. Jackson RA, Buse JB, Rifai R, et al. Two genes required for diabetes in BB rats. Evidence from cyclical intercrosses and backcrosses. *J Exp Med* 1984;159:1269–1636.
75. Hodge SE, Anderson CE, Neiswanger K, et al. Close genetic linkage between diabetes mellitus and the Kidd blood group. *Lancet* 1981;2:893–895.
76. Hodge SE, Anderson CE, Neiswanger K, et al. The search for heterogeneity in insulin-dependent diabetes mellitus (IDDM): Linkage studies, two-locus models, and genetic heterogeneity. *Am J Hum Genet* 1983;35:1139–1155.
77. Dunsworth TS, Rich SS, Swanson HJ, et al. No evidence of linkage between diabetes and the Kidd marker. *Diabetes* 1982;31:991–993.
78. Bell GI, Selby KN, Rutter WJ. The highly polymorphic region near the human insulin gene is composed of simple tandemly repeating sequences. *Nature* 1982;295: 31–35.
79. Bell GI, Horita S, Karam JH. A highly polymorphic locus near the human insulin gene is associated with insulin-dependent diabetes mellitus. *Diabetes* 1984;33: 176–183.
80. Hoover ML, Angelini G, Ball E, et al. HLA-DQ and T-cell receptor genes in insulin-dependent diabetes mellitus. Cold Spring Harbor Symp Quant Biol 1986;51:803–809.
81. Field LL, Anderson CE, Neiswanger K, et al. Interaction of HLA and immunoglobulin antigens in Type 1 (insulin-dependent) diabetes. *Diabetologia* 1984;27: 504–508.
82. Rich SS, Reusch J, Panter S, Barbosa J. Is there a genetic connection between IDDM and NIDDM? *Diabetes* 1986;35(Suppl. 1):24A.
83. Cohen-Haguenauer O, Robbins E, Massart C, et al. A systematic study of the HLA-class II DNA restriction fragments in insulin dependent diabetes mellitus. *Proc Natl Acad Sci USA* 1985;83:3335–3339.
84. Hitman GA, Sachs J, Cassell P, et al. A DR-3-related DX gene polymorphism strongly associates with insulin dependent diabetes mellitus. *Immunogenetics* 1986;23:47–51.
85. Festenstein H, Awad J, Hitman GA, et al. New HLA-DQ associations with rheumatoid arthritis and insulin-dependent diabetes mellitus. *Nature* 1986;322:64–67.
86. Schreuder GMT, Tilanus MFJ, Bontrop RE, et al. HLA-DQ polymorphism associated with resistance to Type 1 diabetes detected with monoclonal antibodies, isoelectric point differences and restriction fragment length polymorphisms. *J Exp Med* 1986;164:938–943.
87. Todd JA, Bell JI, McDevitt HO. HLA-DQ-beta gene contributes to susceptibility and resistance to insulin-dependent diabetes mellitus. *Nature* 1987;329:599–604.
88. Holbeck SL, Nepom GT. Molecular analysis of DQ-beta-3.1 genes. *Hum Immunol* 1988;21:183–192.
89. Nepom BS, Schwartz D, Palmer J, Nepom GT. Transcomplementation of HLA genes in IDDM. HLA-DQ alpha- and -beta-chains produce hybrid molecules in DR3/4 heterozygotes. *Diabetes* 1987;36:114–117.
90. Gomard E, Henin Y, Sterkers G, et al. An influenza A virus-specific and DRw8-restricted T-cell clone cross-reacting with a transcomplementation product of the HLA-DR2 and DR4 haplotypes. *J Immunol* 1986;136:3961–3967.

91. Foulis AK, Liddle CN, Farquharson MA, et al. The histopathology of the pancreas in Type 1 (insulin-dependent) diabetes mellitus: A 25 year review of deaths in patients under 20 years of age in the United Kingdom. *Diabetologia* 1986;29:267–274.
92. Nerup J, Ortved-Andersen O, Bendixen G, et al. Anti-pancreatic cellular hypersensitivity in diabetes mellitus. *Diabetes* 1971;20:424–429.
93. Ginsberg-Fellner F, Witt ME, Toguchi Y, et al. Thyroid and islet-cell autoimmunity co-exist in families of Type 1 diabetics. *Pediatr Adolesc Endocrinol* 1986;15:105–110.
94. Drell DW, Notkins AL. Multiple immunological abnormalities in patients with Type 1 (insulin-dependent) diabetes mellitus. *Diabetologia* 1987;30:132–143.
95. Bottazzo GF, Gleichmann H (chairpersons) Workshop report. Immunology and diabetes workshops: Report of the First International Workshop on the Standardization of Cytoplasmic Islet-Cell Antibodies. *Diabetologia* 1986;29:125–126.
96. Toguchi Y, Ginsberg-Fellner F, Rubinstein P. Cytotoxic islet-cell surface antibodies (ICSA) in patients with Type 1 diabetes and their first degree relatives. *Diabetes* 1985;34:855–860.
97. Doberson MJ, Scharff JE, Ginsberg-Fellner F, et al. Cytotoxic autoantibodies to beta cells in the serum of patients with insulin-dependent diabetes mellitus. *N Engl J Med* 1980;303:1493–1498.
98. Lernmark Å, Freedman ZR, Hoffman C, et al. Islet-cell surface antibodies in juvenile diabetes mellitus. *N Engl J Med* 1978;299:375–380.
99. Kanazawa Y, Komeda K, Sato S, et al. Non-obese-diabetic mice: Immune mechanisms of pancreatic beta-cell destruction. *Diabetologia* 1984;27:113–115.
100. Dyrberg T, Nakhooda AF, Baekkeskov S, et al. Islet-cell surface and lymphocyte antibodies in the spontaneously diabetic BB Wistar rat. *Diabetes* 1982;31:278–281.
101. Baekkeskov S, Landin M, Kristensen JK, et al. Antibodies to a 64,000 M_r human islet cell antigen precede the clinical onset of insulin-dependent diabetes. *J Clin Invest* 1987;79:926–934.
102. Haspel MV, Onodera T, Prabhakar BS, et al. Virus induced autoimmunity, monoclonal antibodies that react with endocrine tissues. *Science* 1983;220:304–306.
103. Eisenbarth GS. Genes, generator of diversity, glycoconjugates and autoimmune beta-cell insufficiency in Type 1 diabetes. *Diabetes* 1987;36:355–364.
104. Ginsberg-Fellner F, Dobersen MJ, Witt ME, et al. HLA-antigens, cytoplasmic islet-cell antibodies and carbohydrate tolerance in families of children with insulin-dependent diabetes mellitus. *Diabetes* 1982;31:292–298.
105. Cudworth AG, Bottazzo GF, Doniach D. Genetic and immunological factors in Type 1 diabetes, in Irvine J (Ed.): *Immunology of Diabetes.* Edinburgh: Teviot Scientific Publications, 1980,67–100.
106. Ginsberg-Fellner F, Witt ME, Franklin BH, et al. Triad of markers for identifying children at high risk of developing insulin-dependent diabetes mellitus. *JAMA* 1985; 254:1469–1472.
107. Riley WJ, Maclaren NK. Islet cell antibodies are seldom transient. *Lancet* 1984; 1:1351–1352.
108. Srikanta S, Rabizadeh A, Ganda OP, et al. Prospective screening of first degree relatives of patients with type I diabetes: Islet cell antibodies (ICA) and abnormal insulin secretion. *Clin Res* 1985;33:445(A).
109. Palmer JP, Asplin CM, Dorman TL, et al. Insulin antibodies in insulin dependent diabetics before treatment. *Science* 1983;222:1337–1339.
110. McEvoy RC, Witt ME, Ginsberg-Fellner F, Rubinstein P. Anti-insulin antibodies in

children with Type 1 diabetes mellitus. Genetic regulation of production and presence at diagnosis before insulin replacement. *Diabetes* 1986;35:634–641.
111. Wicker LS, Miller BJ, Mullen Y. Transfer of autoimmune diabetes mellitus with splenocytes from nonobese diabetic (NOD) mice. *Diabetes* 1986;35:855–860.
112. Assan R, Feutren G, Debray-Sachs M, et al. Metabolic and immunological effects of cyclosporin in recently diagnosed Type 1 diabetes mellitus. *Lancet* 1985;1:67–71.
113. Stiller CR, Dupre J, Gent M, et al. Effects of cyclosporine immunosuppression in IDDM of recent onset. *Science* 1984;223:1362–1365.
114. Sibley RK, Sutherland DE, Goetz FC, Michael AF. Recurrent IDDM in pancreatic iso- and allografts. A light and electron microscopic and immunohistochemical analysis in 4 cases. *Lab Invest* 1985;53:132–144.
115. Barret-Connor E. Is insulin-dependent diabetes mellitus caused by Coxsackievirus B infection? A review of the epidemiologic evidence. *Rev Infect Dis* 1985;7:207–215.
116. Hirayama K, Matsushita S, Kikuchi I, et al. HLA-DQ is epistatic to HLA-DR in controlling the immune response to schistosomal antigens in humans. *Nature* 1987;327:426–430.
117. Pujol-Borrell R, Todd I, Doshi M, et al. HLA class II induction in human islet-cells by interferon-alpha plus tumor necrosis factor or lymphotoxin. *Nature* 1987;326:304–306.
118. Ilonen J, Salmi A. Comparison of HLA-Dw1 and -Dw2 positive adherent cells in antigen presentation to heterozygous T-cell lines: A low rubella antigen-specific response associated with HLA-Dw2. *Hum Immunol* 1986;17:94–101.
119. Rosenberg SA, Lotze MT, Muul LM, et al. Observations on the systemic administration of autologous lymphokine-activated killer cells and recombinant interleukin-2 to patients with metastatic cancer. *N Engl J Med* 1986;313:1485–1490.
120. Trowsdale J, Duncan Campbell R. Physical map of the human HLA region. *Immunol Today* 1988;9:34–35.

3
Association of Insulin-Dependent Diabetes Mellitus and Other Autoimmune Diseases

FREDDA GINSBERG-FELLNER

Introduction

The clinical association of type I insulin-dependent diabetes mellitus with many other autoimmune diseases and diseases of the other endocrine organs has been known for many years[1-3] even though, to date, a total understanding of these related phenomena is lacking. The endocrine diseases include goitrous and non-goitrous forms of Hashimoto's thyroiditis, pernicious anemia, Grave's disease, Addison's disease, and less common diseases of other endocrine glands.

Although today we group these diseases under the label autoimmune endocrinopathies, and several hypotheses concerning their etiologies have been advanced, it remains likely that more than one cause will finally be determined not only for each of the clinical disease states but also for their interrelationships. Organ-specific autoimmune diseases can be divided into various groups[4] including (1) those in which specific antireceptor antibodies have been demonstrated, (2) those in which antibodies against the entire endocrine gland have been found, and (3) those in which nonspecific antibodies have been demonstrated. In the latter group, it is highly likely that more specific autoimmune phenomena will be defined in the near future, whereas in the second group, more than one organ-specific antibody is already known, although the specific antigen is still largely undefined. Anti-receptor diseases include Grave's disease, certain types of atrophic thyroiditis, certain forms of insulin-resistant diabetes mellitus (usually associated with acanthosis nigricans), and myasthenia gravis. Type I, insulin-dependent diabetes mellitus (IDDM), now classified as an autoimmune disease because it manifests both humoral and cell-mediated immune features, has many aspects in common with Addison's disease, certain types of primary hypogonadism, diabetes insipidus, hypophysitis, Hashimoto's thyroiditis, celiac disease (in which antigliadin antibodies are often found), hypoglucagonemia, pemphigus, and vitiligo. In other organ- or cell-specific autoimmune diseases, including dermatitis herpetiformis, thrombocytopenic purpura, Crohn's disease, ulcerative colitis, and alopecia (totalis and areata), it is now becoming possible to identify autoantibodies with reactivities similar to ones already described or directed against specific organs. Other more diffuse autoimmune diseases affecting many parts of the body include rheumatoid arthritis, probably rheumatic fever, systemic lupus erythematosis, and Reiter's syndrome.

The mechanism (or mechanisms) responsible for the various so-called autoimmune diseases remains almost entirely unknown. Thus until the sequence of events that lead to the destruction of the organ or organs involved is elucidated, therapy aimed at ameliorating or curing these diseases remains, of necessity, either empiric or experimental. Fundamental to the understanding of the pathogenesis of autoimmune diseases is an explanation for the question of why one's own organs would be perceived as foreign by the body's immune system. Possible answers include antigenic modification, or the genetic activation of certain autoreactive T-lymphocytes or perhaps other lymphocytes to attack certain organs. Investigations into both of these hypotheses are currently ongoing.

Although it has long been known that certain of these diseases, such as type I diabetes mellitus and Hashimoto's thyroiditis, cluster in families, we still do not know if the same antigenic stimulant is required to elicit both diseases and/or if the same subset of lymphocytes is involved in the final destructive processes in both the thyroid and the pancreatic islets of Langerhans.

There is some evidence that, at least in certain cases, selected antibodies will react with tissues from more than one gland.[5,6] Thus certain monoclonal antibodies derived from normal donors as well as from patients with autoimmune diseases of various types may react with tissue from the thyroid, pancreas, pituitary, etc. Although this suggests that the immune system is stimulated by a common antigen, no such antigen has been definitively identified and characterized, at least in humans. Nevertheless, isolation of such an antigen would be important, since innovative therapeutic avenues employing blocking antibody might thereby be opened. Similarly identification of the T-cell responsible for destruction of target tissue might allow production of a monoclonal antibody, perhaps coupled to a toxin, that would be specific for this clone of T-cells. Work in this area is active, although extremely difficult since it is almost impossible to obtain tissue from which such cells can be extracted and the concentration of these cells in the peripheral blood is extremely low.

How does autoimmunity against so many apparently different types of tissues develop? Although there is still no conclusive answer to this question, evidence has accumulated over the past decade that suggests that genetic factors, predominantly controlled by the major histocompatability (HLA) loci, as well as by certain environmental factors, including viruses, stress, and, perhaps, toxins, all play important roles. This chapter focuses on possible etiologic factors contributing to the development of other endocrinopathies in patients with insulin-dependent diabetes mellitus and in their first-degree relatives. It also discusses the differential susceptibility of different endocrine glands to autoimmune attack, as well as possible differences in the mode of inherited susceptibility. Finally the capacity of each of these glands to partially regenerate after damage will be assessed.

IDDM and Thyroid Autoimmunity

Thyroid autoantibodies occur frequently in patients with type I diabetes and in their family members.[7-9] It has been estimated that more than 70% of type I diabetics have evidence of thyroid autoimmunity, with females being affected to a greater extent than males.[10] Almost 20 years ago, Irvine et al. reported the incidence of antibodies to cytoplasm of thyroid and gastric parietal cells in the sera of 1054 diabetics without clinical thyroid disease or pernicious anemia and in 871 control subjects.[11] The diabetic subjects ranged in age from 5 to 75 years of age; 521 were females and 533 males. At the time of study, 184 of 219 females and 204 of 225 males under the age of 39 years were being treated with insulin and were probably classic IDDM patients, although this study was performed before the earliest reports that defined type I diabetic patients according to the prevalence of islet-cell autoantibodies and HLA types. Antibodies against gastric parietal cells, thyroid cytoplasm, and thyroglobulin were increased in all the diabetics compared to the age-matched controls, with women being antibody-positive significantly more often than men. Furthermore antibody-positivity was much more common in those subjects who were insulin-dependent, the only way known at that time to classify the disease we now refer to as IDDM. By the relatively crude methods then available, 85 of 171 female and 32 of 195 male insulin-dependent subjects were antibody-positive.

Our own studies in families of type I diabetics have allowed us to expand these data and to suggest and/or confirm possible modes of inheritance, and linkage, of certain of these thyroid disorders to the HLA system. Our studies differ from many other studies in that entire families were evaluated, and antimicrosomal and antithyroglobulin antibodies were measured by a sensitive ELISA assay developed by Roman et al.[12]; this enabled us to use the presence of the antibodies rather than reported clinical thyroid disease as a measure of autoimmunity to the thyroid gland. In our first study, the prevalence of antimicrosomal and antithyroglobulin antibodies was measured in 84 HLA-typed families having a type I diabetic child.[13] The control groups consisted of 43 children, ranging in age from 1 month to 20 years and a mean age of 10.8 ± 6.5 (1 SD) years and a female:male ratio of 1.1:1, and 45 adults with an age range of 24 to 57 years and mean age of 37.1 ± 8.1 years and a female:male ratio of 1.1::1. None of the control population had any personal or family history of autoimmune thyroid disease or type I diabetes. We found that 160 of the 407 (39%) individuals (the entire study population) had antimicrosomal antibodies; 143 of 407 (35%) individuals had antithyroglobulin antibodies; 201 of 407 (49%) individuals were positive for one of the thyroid autoantibodies; and 102 of 407 (25%) individuals were positive for both autoantibodies. To control for the effects of age and the influence of type I diabetes, prevalence figures were also calculated separately for parents, diabetic children, and nondiabetic siblings, and these prevalence figures were categorized by sex, HLA-DR antigen typing, and HLA sharing in diabetic and nondiabetic siblings. Study females had a higher prevalence of thyroid autoantibodies than did controls ($p = .04$). Of the study males, 65 of 197 (33%) also were antibody-positive, versus only 7% of control males (p<.005). Thus in this selected population, the expected ratio of affected females to males of 4:1 was significantly reduced to 1.4:1. Similar results were also obtained if the data were divided according to the prevalence figures for antimicrosomal and antithyroglobulin antibodies alone. The diabetic children had a prevalence of antimicrosomal antibodies of 46% and a prevalence of antithyroglobulin antibodies of 40% (p<.001 versus control children). Their nondiabetic sibs also had increased prevalence rates: 28% had antimicrosomal antibodies (p<.001 versus control children) and 23% had antithyroglobulin antibodies (p<.01 versus control children). In our study, HLA-DR5 was the only DR antigen positively associated with the presence of antimicrosomal antibodies. Further analysis revealed that this association was limited to the males in the study population. Additionally, children from families in which both parents had autoantibodies had a significantly higher prevalence of the autoantibodies than did the children from all the other families. Finally, the prevalence of thyroid antibody positivity was significantly increased in sibs who were HLA-identical, or haploidentical with the index sibling, compared to control children. In summary, these data demonstrated both an increased prevalence and an earlier onset of thyroid autoantibody production in the study population compared to the control population, and particularly in the diabetic children in the study population.

It is of great interest that HLA-DR3 and/or -DR4, the DR antigens most closely associated with type I diabetes,[14-16] were not significantly associated with thyroid autoantibody production. The limited HLA-DR5 association between HLA-DR5 and thyroid antibody-positivity in males in these families therefore is most likely related to thyroiditis and not to IDDM. The relationship of thyroiditis to HLA-DR5 has been noted by other groups of investigators.[17,18]

Clustering of thyroid antibody production, which we believe to be a sensitive indicator of autoimmune destruction of the thyroid gland, within families is also demonstrated when the families of type I diabetic probands are simultaneously studied for the presence of both thyroid autoantibodies and islet cell cytotoxic (ICSA) and cytoplasmic (ICA), antibodies.[19,20] One of the earliest demonstrations of the presence of ICA in diabetic subjects was in families known to have polyendocrine diseases.[21] Retrospective examination of our own earlier studies of families of IDDM probands revealed that ICA-positivity in nondiabetic siblings and parents was associated in more than 80% of the cases with other autoimmune diseases, particularly thyroid disorders, but also pemphigus, myasthenia gravis, rheumatoid arthritis, multiple sclerosis, Crohn's disease, and ulcerative colitis.[22] In 31 families with an IDDM proband, we found a significant correlation between the presence of ICSA and antithyroid antibodies; 7.6% of all IDDM family members who were nondiabetic had both types of antibodies.[23] It is also interesting that this subgroup with dual antibody-positivity were all female. We had previously shown, by segregation analysis, that the tendency to produce ICSA probably segregates as a dominant trait in these families,[24] and this probably also is true for the production of thyroid autoantibodies. This is of great interest, for clearly the presence of thyroid autoantibodies in the general (control) population is much higher than that of either ICA or ICSA, and the presence of clinical thyroid disease (at least 10% to date in our nondiabetic population and in the general population) is higher than that of IDDM (most recent estimates 0.2%)[25,26] in the general population and in the relatives of IDDM probands (<5%).[27] Although it is tempting to speculate from these findings that the tendency to produce thyroid autoantibodies is inherited in a dominant fashion, and we believe that the tendency to develop IDDM probably is inherited in a recessive manner,[28,29] the data are not conclusive. In certain animal models for IDDM, most notably the BB rat, there also is a marked tendency to develop thyroiditis.[30] It it not known whether the same mechanisms operate in both these animal models and in human subjects. However it is noted that most investigators use the presence of clinical disease, rather than just the presence of autoantibodies as we have done, as a marker for autoimmune thyroid dysfunction. For example, in the study by Becker et al.[31] 85 children age 5–16 years with IDDM, hypothyroidism, and hyperthyroidism were each found in 2% of their study population. Another 8% had chemical hypothyroidism, defined as T4 <4.8 μg/dl and TSH >10 mIU/l. Thyroid autoantibodies, however, were found in 49% of the IDDM youngsters, a figure quite similar to ours. In this study, hemagglutination was used to measure the antibody titers, rather than the more sensitive ELISA method we employ.

Other Endocrinopathies Associated with Type 1 Diabetes Mellitus

In the above studies of Becker et al., testicular, adrenal, and parietal cell antibodies were found in only a few of the patients tested.[31] No antibodies to ovary or parathyroid glands were detected. No euthyroid patients had antiadrenal antibodies and adrenal function in all of these children was normal. Whether any of the adrenal antibody-positive subjects progressed to Addison's disease is not known but, it is probable, given similar situations with other autoantibodies, that not all will. Prospective studies are of great importance.

Complement-fixing adrenocortical antibodies may predict progression to overt Addison's disease and are often associated with the HLA haplotype B8/DR3, one of the haplotypes commonly also associated with IDDM.[32] Patients with pernicious anemia and thyroiditis have an increased prevalence of both parietal and intrinsic factor antibodies, as well as an increased prevalence of antithyroid autoantibodies. Pernicious anemia with parietal and intrinsic factor autoantibodies also develops in IDDM patients, albeit at a lower rate and at a later time than does thyroiditis. Certain cases of hypergonadotrophic hypogonadism also appear to be associated with antibodies against ovarian hormone-producing cells; a similar picture in boys who have antibodies against the Leydig cells is even rarer.[33] In our IDDM population of more than 1000 cases, we have yet to find a case of hypogonadism secondary to autoantibodies, although there is no doubt that this entity exists. Perhaps much older, type I diabetics will have a higher prevalence rate of such antibodies. Also IDDM associated with hypoparathyroidism of the autoimmune type is extremely rare, and may be associated with Addison's disease and/or mucocutaneous candidiasis. In our population, we have identified two young women who appear to have idiopathic hypoparathyroidism and IDDM. One also has had thyrotoxicosis, which eventually led to a thyroidectomy; the parathyroids, visualized during surgery, were small but were not biopsied. (After the patient developed a persistent, low-grade hypocalcemia, the slides from the thyroidectomy were reviewed, but no parathyroid tissue was found.) Most recently antiparathyroid antibodies have been identified in her serum, albeit in low titer. Hyperthyroidism had developed in this female at age 3, and she had been maintained on antithyroid drugs with a moderate degree of success until age 12, when she developed IDDM. The second patient had hypoparathyroidism and candidiasis before developing IDDM in her mid-teens.

Recently Bottazzo and his colleagues[34] have used indirect immunofluorescent techniques to identify, in IDDM patients and their relatives, autoantibodies that react with growth hormone and with prolactin secreting cells of the pituitary. To date however, few data concerning the clinical course of these individuals have been published, so it cannot be stated with any degree of certainty whether any clinical manifestations are associated with these autoantibodies. Also more recently, autoantibodies against the adrenal medulla, definitely differing from those directed at the adrenal cortex have been identified.[35] To date again, it is not clear if these autoantibodies are associated with clinical disease.

How Does Autoimmune Disease Develop?

Clearly this is a most important question and one that has received a great deal of attention, especially over the past few years. The finding that endocrine organs and T-lymphocytes can be activated by similar stimuli, including interferon and viral infections,[36-40] suggests that environmental factors, particularly repeated viral infections, may be extremely important, especially when they are superimposed on a genetically susceptible background, a major component of which appears to reside close to the DR and DQ loci on chromosome 6. In thyroiditis, alopecia areata, and IDDM, for example, HLA-DR antigens not normally expressed in the affected organs[41] can be identified by newer techniques. The potential role of viral infections in triggering autoimmune diseases is supported by many rodent studies[42-44] and by immunologic findings associated with AIDS.[44] The AIDS virus actively destroys one portion of the cellular immune system, the CD4+ lymphocytes, and viral particles can be recovered from these cells as well as from macrophages, an earlier participant in the activation of the immune system.

Pertinent to the association between thyroiditis and type 1 diabetes are the results of our ongoing studies concerning the influence of documented in utero infection with the RNA virus rubella. Infection occurring during the critical first trimester of pregnancy leads to the development of diabetes in more than 20% of the affected population, an amazing rate of penetrance.[45-48] Our data demonstrate that the diabetes developing in young adults and adolescents whose mothers were infected with rubella in the 1965–1967 epidemic in New York City have the same classic relationships to HLA-DR antigens as does classic IDDM. Thus HLA-DR3 and -DR4 are significantly increased in the congenital rubella patients with diabetes, and HLA-DR2 is significantly decreased. In addition both ICA and ICSA have been found in sera samples obtained from these patients long before they developed diabetes. At least one-third of these patients also have antibodies against thyroid tissue. Although to date we have not been able to demonstrate antibodies against the adrenal cortex or gonads, individuals with congenital rubella are often extremely short as adults and may have hypogonadism. We will continue to follow these patients and we would not be surprised if autoantibodies against pituitary or gonadal cells were found. We have also found anti-insulin autoantibodies[49] in at least 10% of the congenital rubella patients without diabetes, although some of them already have glucose intolerance and/or hypo- or hyperinsulinemia, the latter a most unexpected finding that is currently being evaluated. Other groups have isolated rubella virus genomes from joint fluid of patients after rubella infection[50] and viral particles detected by immunofluorescence have been detected in thyroid tissue.[51] In our preliminary studies, we detected rubella viral particles in lymphocyte samples obtained from many of these patients. We are also attempting to isolate viral particles from thyroid cells of our patients after needle aspiration of the gland, and from fresh specimens of pancreas from neonates or older individuals who succumb to the IDDM or to its major complications. Although the number of cases of congenital rubella occurring in newborns fortunately is now very low, studies of such infants should continue to be helpful in elucidating the role of the immune system in triggering diabetes and thyroiditis, and in providing new data on temporal patterns.

Since T-cells normally function as important defenses against viral illness, it is perhaps not surprising that they have been implicated in the development of various autoimmune diseases. It is conceivable that the mechanisms that are thought to lead or contribute to the development of autoimmune disease in a number of tissues are similar in nature and that the organ, or organs, targeted for disease is determined by other genetic or environmental influences. In addition pathologists have long noted that endocrine organs by can be infiltrated by specific cells suggestive of both new and latent inflammatory-like processs,[52] whereas more recent studies have identified different subsets of T-lymphocytes in tissue specimens.[53] If T-lymphocytes and target organ endocrine cells are both infected by specific viruses, it will be of great importance to determine which cell is first infected, or, perhaps, if simultaneous infection can occur during critical periods of physical development or illness. Clearly viral incorporation into any cell could change the character of that cell such that either it is perceived as nonself by the body or, in the case of T-lymphocyte or macrophage subpopulations, it is so altered that it becomes a virtual killer-cell.

Can Autoantibodies Affect More Than One Gland? Are They Mono- or Polyclonal?

Recent investigations by Notkins and coworkers[5,6,54] strongly suggest that multiple organ-reactived monoclonal autoantibodies can be produced in vitro. These studies were undertaken in light of the documented association of multiple autoimmune diseases in our type 1 diabetic probands and their family members. The low titers of these autoantibodies and the polyclonal nature of serum made it difficult to determine if a variety of autoantibodies were reacting with a different organ, or if the autoimmune response were restricted, with only a limited number of common antigenic agents in different organs. To make these determinations, human hybridomas that synthesize monoclonal antibodies were created by fusing peripheral blood lymphocytes from patients with insulin-requiring diabetes mellitus and other autoimmune abnormalities with either mouse or human myeloma cells. Nine hybridomas that made autoantibodies were obtained and seven out of nine reacted with antigens in several endocrine organs. Of the five human monoclonal antibodies, all of them reacted with anterior pituitary, thyroid follicle, gastric mucosa, pancreatic islets and ducts, only two reacted with smooth muscle, and four reacted with cultured human fibroblasts. Thus 5.7% of immunoglobulin producing human hybridomas synthesized autoantibodies and 78% of these were multiple organ-reactive. These results clearly demonstrated that it is possible for some autoanibodies to react with antigens of multiple organs. Although there are several possible explanations for this multiple organ-reactivity, we felt the most likely interpretation was that the antibodies either react with the same epitope residing in different tissues or with common antigenic agents on different molecules in multiple organs, or with both of them. The organ reactivity of these monoclonal antibodies was similar to that observed with sera from patients with polyendocrine disease. In analogous experiments, monoclonal autoantibodies that reacted with antigens in multiple organs also were prepared from Epstein-Barr, virus-transformed lymphocytes. Both IgG and IgM antibodies reactive with normal tissues were produced. Four of ten IgM monoclonal antibodies were more thoroughly examined and were found to react with antigens in several organs, including thyroid, pancreas, smooth muscle, and nerve. Thus virus-transformed lymphocytes appear capable of triggering the production of autoantibodies reactive with uninfected target cells. In subsequent studies, IgG murine, monoclonal, anti-idiotypic antibodies were generated against five of the human monoclonal autoantibodies. By using these anti-idiotypic antibodies as probes, sera from patients with several autoimmune diseases, including IDDM and Hashimoto's thyroiditis, were screened for the expression of idiotopes; although certain idiotopes were recognized, there was no significant difference between normal controls and patients with IDDM or thyroiditis. Nevertheless, these experiments suggested that certain autoantibody-associated idiotopes are expressed in the B-cell repertoire; they also provided support for the concept of multiple organ antigen-sharing in vivo.

Summary and Conclusions

Other autoimmune diseases, most notably Hashimoto's thyroiditis, occur in an astonishingly high percentage of individuals with type 1 diabetes mellitus. Why these diseases occur together more often in certain individuals is not known, but it is apparent that genetic, viral, and common antigenic determinants on target organs may all play significant roles. Further understanding of the reasons behind these associations is obviously needed. Such information not only will provide insight into the pathogenesis of autoimmune diseases but also may ultimately be crucial for the development of rational, specific immunotherapies for these controllable but largely incurable conditions.

Acknowledgments. Portions of the work described were generously supported, in part, by grants from the National Institutes of Health, DK19631 and DK39286, the American Diabetes Association, the Juvenile Diabetes Foundation, the March of Dimes Birth Defects Foundation, and the William T. Grant Foundation.

References

1. Nissley SP, Drash AL, Blizzard RM, et al. Comparison of juvenile diabetics with positive and negative organ specific antibody titers: Evidence for genetic heterogeneity. *Diabetes* 1973;22:63–65.
2. Bottazzo GF, Mann JI, Thorogood M, et al. Autoimmunity in juvenile diabetics and their families. *Br Med J* 1978;2:165–168.
3. Frey HMM, Vogt JH, Nerup J. Familial poly-endocrinopathy. *Acta Endocrinol* 1973; 72:401–416.
4. Winter WE, Maclaren NK. Autoimmune endocrinopathies, in *Pediatric Endocrinology 1985*. Lifshitz F (ed): New York: Marcel Dekker, 1985, 521–540.
5. Satoh J, Prabhakar BS, Haspel MV, et al. Human monoclonal autoantibodies that react with multiple endocrine organs. *N Engl J Med* 1983;309:217–220.
6. Garzelli C, Taub FE, Scharff JE, et al. Epstein-Barr virus-transformed lymphocytes produce monoclonal autoantibodies that react with antigens in multiple organs. *J Virol* 1984;52:722–725.
7. Fialkow PJ, Zavala C, Nielsen R. Thyroid autoimmunity: Increased frequency in relatives of insulin-dependent patients. *Ann Int Med* 1975;83:170–176.
8. Riley WJ, Maclaren NK, Lezotte, DC, et al. Thyroid autoimmunity in insulin-dependent diabetes mellitus: The case for routine screening. *J Pediatr* 1981;98:350–354.
9. Nerup J, Binder C. Thyroid, gastric and adrenal autoimmunity in diabetes mellitus. *Acta Endocrinol* 1973;72:279–286.
10. Goldstein DE, Drash, AL, Gibbs J, et al. Diabetes mellitus: The incidence of circulatory autoantibodies against thyroid, gastric, and adrenal tissue. *J Pediatr* 1970;77: 304–306.
11. Irvine WJ, Scarth L, Clarke BF, et al. Thyroid and gastric autoimmunity in patients with diabetes mellitus. *Lancet* 1970;ii:163–168.

12. Roman SH, Korn E, Davies TF. Enzyme-linked immunosorbent microassay and hemagglutination compared for detection of thyroglobulin and thyroid microsomal autoantibodies. *Clin Chem* 1984;30:240–251.
13. Roman SH, Davies TF, Witt ME, et al. Thyroid autoantibodies in HLA-genotyped type 1 diabetic families: Sex-limited DR5 association with thyroid microsomal antibody. *Clin Endocrinol* 1986;25:23–33.
14. Rubinstein P, Suciu-Foca N, Nicholson JF. Genetics of juvenile diabetes mellitus: A recessive gene closely linked to HLA-D and with 50 percent penetrance. *N Engl J Med* 1977;297:1036–1040.
15. Christy M, Green A, Christau B, et al. Studies of the HLA system and insulin-dependent diabetes mellitus. *Diabetes Care* 1979;2:209–214.
16. Barbosa J, Chern M, Reinsmoen N, et al. HLA-Dw antigens in unrelated juvenile, insulin-dependent diabetes. *Tissue Antigens* 1979;14:426–436.
17. Farid NR, Sampson L, Moens H, et al. The association of goitrous autoimmune thyroiditis with HLA-DR5. *Tissue Antigens* 1981;17:265–268.
18. Farid NR, Hawe BS, Walfish PG. Increased frequency of HLA DR3 and 5 in the syndrome of painless thyroiditis with transient thyrotoxicosis: Evidence for an autoimmune etiology. *Clin Endocrinol* 1983;19:699–704.
19. Frasier SD, Penny R, Snyder R, et al. Antithyroid antibodies in Hispanic patients with type 1 diabetes mellitus. *Am J Dis Child* 1986;140:1278–1280.
20. Riley WJ, Winer A, Goldstein D. Coincident presence of thyrogastric autoimmunity at onset of type 1 (insulin-dependent) diabetes. *Diabetologia* 1983;24:418–421.
21. Irvine WJ, McCallum CJ, Gray RS, et al. Pancreatic islet cell antibodies in diabetes mellitus correlated with the duration and type of diabetes, coexistent autoimmune disease and HLA type. *Diabetes* 1977;26:138–147.
22. Ginsberg-Fellner F, Dobersen MJ, Witt ME, et al. HLA antigens, cytoplasmic islet cell antibodies, and carbohydrate tolerance of families of children with insulin-dependent diabetes mellitus. *Diabetes* 1982;31:292–298.
23. Ginsberg-Fellner F, Witt ME, Toguchi Y, et al. Thyroid and islet cell autoimmunity co-exist in families of type 1 diabetics. *Pediatr Adolesc Endocrinol* 1986;15:105–110.
24. Ginsberg-Fellner F, Toguchi Y, Witt ME, et al. Pancreatic islet cell antibodies (ICSA) in families of children with type 1 diabetes mellitus; dominant inheritance. *Pediatr Res* 1984;18:221.
25. Kyllo CJ, Nuttall FQ. Prevalence of diabetes mellitus in school-age children in Minnesota. *Diabetes* 1978;27:57–60.
26. Growitz K, Howen CG, Thompson J. Prevalence of diabetes in Michigan school-age children. *Diabetes* 1976;25:122–127.
27. Srikanta S, Ganda OP, Rabizadeh A, et al. First degree relatives of patients with type 1 diabetes mellitus: Islet cell antibodies and abnormal insulin secretion. *N Engl J Med* 1985;313:461–464.
28. Rubinstein P, Ginsberg-Fellner F, Falk C. Genetics of type 1 diabetes mellitus: A single recessive predisposition gene mapping between HLA-B and GLO. *Am J Hum Genet* 1981;33:865–882.
29. Rubinstein P, Walker M, Ginsberg-Fellner F. The excess of DR3,4 in type 1 diabetes: What does it portend? *Diabetes* 1986;35:985–989.
30. Colle E, Guttmann RD, Seemayer TA. Association of spontaneous thyroiditis with the major histocompatability complex of the rat. *Endocrinology* 1985;116:1243–1247.

31. Becker D, Rabin B, Villapando S, et al. Juvenile diabetes mellitus (JDM) as part of an autoimmune endocrinopathy. *Diabetes* (Suppl. 1)1977;26:323.
32. Neufeld N, Maclaren NK, Blizzard RM. Two types of autoimmune Addison's disease associated with different polyglandular autoimmune (PGA) syndromes. *Medicine* 1981;60:355–362.
33. Elders MJ, Maclaren N, Riley W. Gonadal autoantibodies in patients with hypogonadism and/or Addison's disease. *J Clin Endocrinol Metab* 1081;52:1137–1142.
34. Mirakian R, Bottazzo GF, Cudworth AG, et al. Autoimmunity to anterior pituitary cells and the pathogenesis of insulin-dependent diabetes mellitus. *Lancet* 1982;i: 755–759.
35. Rabinowe SL, Brown FM, Whatts M, et al. Antisympathetic ganglia antibodies and postural blood pressure in IDDM subjects of varying duration and patients at high risk of developing IDDM. *Diabetes Care* 1989:12:1–6.
36. Karlsson FA, Totterman TH, Jansson R. Subacute thyroiditis and interferon gamma expressing T cytotoxic/suppressor cells in thyroid tissue and peripheral blood. *Clin Endocrinol* 1986;25:487–493.
37. Weetman AR, Volkman DJ, Burman KD, et al. The in vitro regulation of human thyrocyte HLA-DR antigen expression. *J Clin Endocrinol Metab* 1985;61:817–824.
38. Bottazzo GF, Pujol-Borrell R, Hanafusa T, et al. Role of aberrant HLA-DR expression and antigen presentation in induction of endocrine autoimmunity. *Lancet* 1983;ii:115–118.
39. Pujol-Borrell R, Hanafusa T, Chiovato L, et al. Lectin induced expression of DR antigen on human cultured follicular thyroid cells. *Nature* 1983;303:71–76.
40. Hanafusa T, Pujol-Borrell R, Chiovato L, et al. Aberrant expression of HLA-DR antigen on thyrocytes in Grave's disease; relevance for autoimmunity. *Lancet* 1983;ii: 1111–1114.
41. Chan JYC, Walfish PG. Activated (IA+) T-lymphocytes and their subsets in autoimmune thyroid diseases: Analysis by dual laser flow microfluorocytometry. *J Clin Endocrinol Metab* 1986;62:403–409.
42. Notkins AL, Yoon JW, Onodera T, et al. Virus-induced diabetes mellitus. *Persp Virol* 1981;11:141–162.
43. Onodera T, Ray UR, Melez KA, et al. Virus-induced diabetes mellitus. Autoimmunity and polyendocrine disease prevented by immunosuppression. *Nature* 1982;297:66–69.
44. Stevens CE, Taylor PE, Zana EA, et al. Human T-cell lymphotropic virus type III infection in a cohort of homosexual men in New York City. *JAMA* 1986;255:2167–2172.
45. Rubinstein P, Witt ME, Fedun B, et al. The HLA system in congenital rubella patients with and without diabetes. *Diabetes* 1982;31:1088–1091.
46. Ginsberg-Fellner F, Witt ME, Yagihashi S, et al. The congenital rubella syndrome as a model for type 1 diabetes mellitus (IDDM): Increased prevalence of islet cell surface antibodies. *Diabetologia* 1984;27:87–89.
47. Ginsberg-Fellner F, Witt ME, Fedun B, et al. Diabetes mellitus and autoimmunity in patients with the congenital rubella syndrome. *Rev Infect Dis* (Suppl. 1) 1985;70: S170–S176.
48. Ginsberg-Fellner F, Fedun B, Cooper LZ, et al. Interrelationships of congenital rubella and type 1 insulin-dependent diabetes mellitus, in *The Immunology of Diabetes Mellitus* Jaworski MA et al. (ed): Amsterdam, Holland: Elsevier Science Publishers, B.V., 1986, 279–286.

49. McEvoy RC, Cooper LZ, Fedun B, et al. Type 1 diabetes mellitus (IDDM) and autoimmunity in patients with congenital rubella syndrome (CRS): Increased incidence of insulin autoantibodies. *Pediatr Res* 1986;20:332A.
50. Chantler JK, Tingle AJ, Petty RE. Persistent rubella infection associated with chronic arthritis in children. *N Engl J Med* 1985;313:1117–1123.
51. Ziring PR, Gallo G, Finegold M, et al. Identification of rubella virus antigen in the thyroid of a child with congenital rubella. *J Pediatr* 1976;88:1065–1066.
52. Gepts W. Pathological anatomy of the pancreas in juvenile diabetes mellitus. *Diabetes* 1965;14:619–623.
53. Bottazzo GF, et al. In situ characterization of autoimmune phenomena and expression of HLA molecules in the pancreas in diabetic insulitis. *N Engl J Med* 1985;313:353–359.
54. Uchigata Y, Prabhakar BS, Salata K, et al. Human monoclonal multiple-organ reactive autoantibodies distinguished by mouse monoclonal anti-idiotypic antibodies: Expression of idiotopes in humans with and without autoimmune diseases. *J Immunol* 1987;138:4218–4221.

4
The Humoral Anti-Islet Immune Response: Cytoplasmic Islet-Cell Antibodies: Technical Aspects and Clinical Applications

HELGA GLEICHMANN AND GIAN FRANCO BOTTAZZO

Introduction

For many decades, insulin dependent diabetes mellitus (IDDM, type I diabetes) was regarded as a metabolic disorder with a typical acute clinical onset. In the last decade, however, a number of studies have led to substantial changes in previous concepts of the etiopathogenesis of the disease. Immunologic, genetic, and epidemiologic studies (for review, see Refs. 1–3) indicate that the majority of patients with IDDM have a long, clinically asymptomatic, prediabetic period. This phase is characterized by autoimmune phenomena and very subtle metabolic changes,[4] which may antedate the clinical manifestation of insulin deficiency for many months and sometimes years. Organ-specific autoimmune reactions most likely play an essential role in the progressive destruction of the insulin-producing beta-cells of the islets of Langerhans; unequivocal proof for a direct autoimmune attack against beta-cells has yet to be demonstrated, however, despite the finding of a constellation of immune abnormalities around and inside injured islets at the time of diagnosis.[5]

Recently Eisenbarth[6] has defined type I diabetes as a chronic autoimmune disease that probably evolves through six major stages. In genetically susceptible individuals (stage I), an as yet unknown initiating factor (stage II) triggers a variety of immunologic abnormalities (stage III). These abnormalities precede the progressive loss of insulin secretion (stage IV), until clinical diabetes becomes manifest (stage V). At this point some residual beta-cell function may persist, but almost complete destruction of the beta-cells follows (stage VI), especially in selected lobules of the pancreas (for review, see Ref. 7).

These pathogenetic concepts have come from numerous investigations that were initially inspired by detailed histologic examinations of the insulitis picture in the pancreas of IDDM patients,[8] they were stimulated by a description of cell-mediated immune phenomena directed toward pancreatic extracts,[9] and they were further strengthened by the first description of cytoplasmic islet-cell antibodies (ICA) in IDDM patients with coexistent autoimmune endocrinopathies.[10] These findings were soon corroborated[11] and extended by several investigators, who described ICA in the serum of 50 to 85% of newly diagnosed IDDM patients.[12] Interestingly most of the studies found that there was a gradual decline in the prevalence of these antibodies following diagnosis,[13-16] regardless of the ethnic background of the patients studied.[17]

Since the first report,[10] ICA assays have been used in more clinical investigations of the diabetic syndrome. The ICA assay has helped to define heterogeneity within type I diabetes[18] and to identify a subgroup of adult, type II diabetic patients with secondary failure to oral hypoglycaemic agents, who are considered to have a latent form of type I diabetes.[19-21] In addition the assay has been employed to monitor residual beta-cell function and persistence of ICA after diagnosis,[22,23] to assess the therapeutic effect of islet[24] or segmental pancreatic grafts,[25] and to evaluate the effect of immunomodulating drugs in patients treated around the time of diagnosis.[26]

The most important information, however, has emerged from prospective family[27-31] and twin[32,33] studies, which convincingly have shown that the ICA test can identify individuals at risk of developing insulin deficiency. Most significantly, ICA may precede the clinical manifestation of diabetes by several years. The frequency of ICA in unaffected, first-degree relatives of diabetic families has been variously estimated as 4.2%,[27] 1.8%,[29] and 0.9.%[31] The presence of complement-fixing (CF-)ICA, detectable in about 2% of first-degree relatives has been reported to be more closely related to impending insulin deficiency than conventional ICA detected with anti-IgG serum.[27,34] Variations in the prevalence of ICA reported in these family studies can be ascribed only partly to differing followup periods of observation, two,[31] five,[27] and six years,[29] respectively. More likely methodologic variables relating to the definition of "presence" or "absence" of ICA in individual sera account for these differences. This interpretation is supported by the analysis of the results presented at the First International Workshop on the Standardization of Cytoplasmic ICA.[35,36]

The purpose of this review is to discuss the influence of several methodologic variables introduced in recent years for the determination of ICA and to address the question of why standardization of the ICA test is important. We also will discuss the clinical value of ICA, as it has emerged from studies in several centers worldwide.

Assays for ICA Determination

Originally ICAs were detected by the indirect immunofluourescence (IFL) technique.[10] Briefly the patient's serum is layered on an unfixed cryostat section of human pancreas from a blood group O donor, usually obtained from a cadaveric kidney donor. Binding of ICA to islet cells is then revealed with an antihuman IgG reagent, conjugated to fluorescein isothiocyanate (FITC), and detected with an ultraviolet (UV) microscope. If ICAs are present in the serum, only the cytoplasm of the Langerhans islet cells are stained; the nuclei and the surrounding exocrine tissue are not stained (Fig. 4.1).

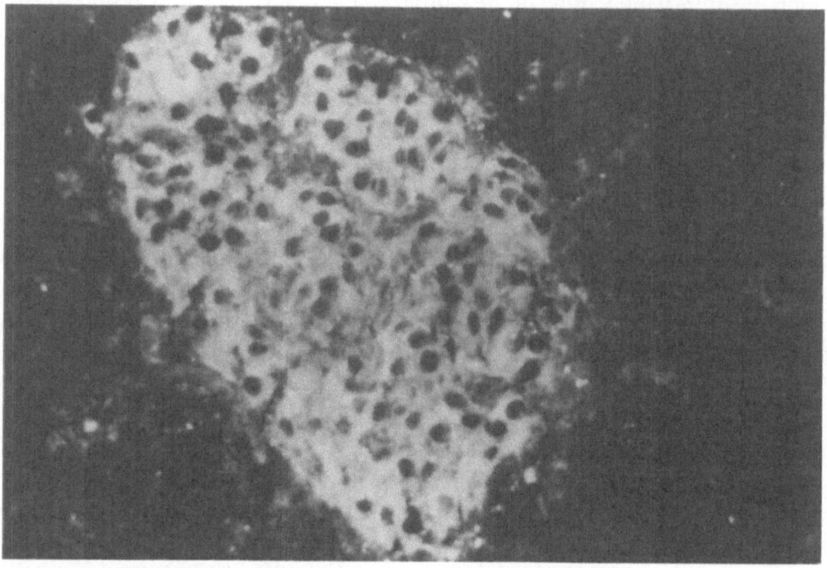

FIGURE 4.1. Cytoplasmic islet cell autoantibodies demonstrated by indirect immunofluorescence technique. An unfixed cryostat section of human pancreas from a blood group O cadaveric kidney donor is treated successively with patient serum and a fluorescein-conjugated antihuman IgG. The cytoplasm, but not the nuclei of the Langerhans islet cells, is stained (× 650).

Over a period of time, various modifications have been introduced in an attempt to improve the classic indirect IFL assay. Bouin's fixed or paraffin-embedded human pancreas from blood group O donors have been employed instead of unfixed sections,[28,37,38] and monkey pancreas has been substituted for human glands.[39] Dehydration of the sections by fixation in acetone was introduced to enhance adherance of the tissue to the microscope slides.[40]

Prolonging incubation of the sera to 24 hours, in the presence of aprotinin—improved the test's sensitivity.[41] Inhibition of residual serum proteinase activity prevented proteolytic degradation of the tissue and increased titers of ICA without a major loss of specificity. Application of protein A (prepared from *Staphylococcus aureus* and then labeled with FITC as a nonantibody, second-step reagent), was expected to reduce many of the problems encountered with tne routinely available FITC conjugated animal antisera.[42] In addition photometric quantitation of islet-cell staining, using a microfluorometric apparatus, was introduced to achieve a more objective final scoring of positive or negative ICA reactions than scoring by visual assessment.[42]

Immunohistochemical staining procedures have also been developed, utilizing either glucose oxidase[43] or peroxidase-labeled protein A.[17] The advantage of these modifications is to allow reading under the light as opposed to the UV microscope, thus eliminating problems caused by fluorescence fading. In order to facilitate the recognition of islets, counterstaining of the sections with monoclonal antibodies has been explored. One assay used monoclonal antibodies specific for distinct islet cell structures and directly conjugated with tetra-methyl rhodamine B isothiocyanate[42]; another assay used monoclonal antibodies, which are reactive with human proinsulin and can be detected by the simultaneous application of Texas-red-labeled antimouse immunoglobulins.[44] Identification of pancreatic islets independent of ICA-positivity reduces the risk of false negative ICA readings.

In a different assay, CF-ICAs are measured by first loading sections with the patient's serum, followed by incubation with fresh normal human serum to supply complement before applying a FITC-conjugated antiserum to human C3.[45] More recently ICA have been detected by a three-layer immunofluorescence method making use of biotin–avidin system.[46] In this case biotinylated antihuman IgG is used as a second-step reagent followed by FITC-conjugated avidin D.

Approaching Standardization of the Cytoplasmic ICA Test

Although some laboratories compared ICA results obtained by various methods[39-41] with those obtained by the originally described, indirect IFL assay,[10] no laboratory was able to indicate the degree of intra- or interassay precision, sensitivity, or specificity.

Despite the widespread application and the evident diagnostic value of the ICA assay, reliable interlaboratory comparisons of results from worldwide studies would be extremely difficult to achieve without standardization of the conventional or modified assays. It was for this reason that International Workshops on the Standardization of Cytoplasmic ICA were organized.[35,47]

From the data presented during the first two workshops, it became evident that, by introducing reference sera, each laboratory could calibrate its own standard curve. By using these standard curves, it might eventually be feasible to express ICA results in arbitrary units, thus permitting reliable interlaboratory comparison of the various results that surely will continue to emerge from future investigations.

In the First International Workshop, each of the 26 participating laboratories received 13 coded serum samples.[35,36] Participants were asked to determine end-point titers of ICA according to a standard protocol based on the conventional IFL assay carried out on cryostat sections of human pancreatic specimens.[10] The results were submitted along with those obtained by any of eight specified methodologic modifications each laboratory was invited to devise. These included:

1. Acetone fixation of sections
2. Heat inactivation of samples at 56 °C prior to testing
3. Prolonged incubation (24 hours) of samples with the tissue substrate in the presence of aprotinin

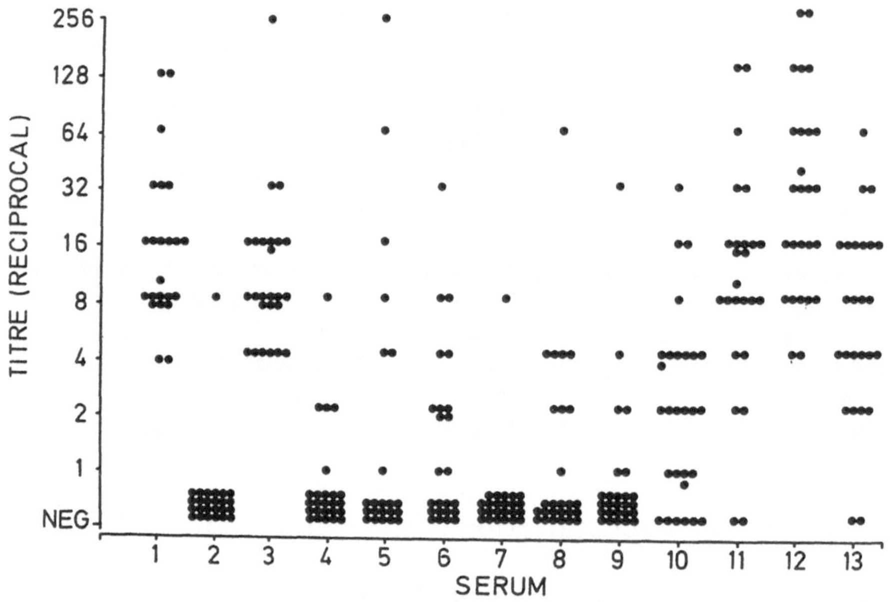

FIGURE 4.2. End-point titers of ICA obtained by 24 laboratories according to a standard protocol with 13 coded sera. (Reproduced from Ref. 35 with permission.)

4. Use of FITC staphylococcal protein A as a second-step reagent, instead of conventional FITC-labeled, antihuman IgG serum
5. Counterstaining of sections with monoclonal antibodies specific either for human proinsulin or for distinct cytoplasmic determinants of human islet cells
6. Immunohistochemical staining with glucose oxidase
7. Substitution of monkey pancreas as test substrate
8. Determination of CF-ICA

The initial qualitative analysis of the data submitted for the standard protocol was based on scoring for the presence or absence of ICA in individual test samples. An interlaboratory concordance of 90% was found for seven samples with high-titer ICA or no ICA. As expected agreement for sera with relatively low-titer ICA was less, and varied between 52 and 79%.

The quantitative analysis also revealed a broad range of end-point titers (Fig. 4.2). These differences could be ascribed to several factors, but they probably related mainly to variability in the expression of the antigenic determinants in the pancreatic islets. The results were apparently not influenced by the choice of the antiserum in each laboratory, the application of various fluorescein–isothiocyanate–protein ratios, or the final concentration of antisera used.

Three samples contained other autoantibodies, alone or mixed with ICA, that is, antinuclear and antimitochondrial antibodies, the presence of which can cause difficulties in interpretating ICA staining. Fortunately the majority of the participants correctly identified these interfering autoantibodies. Antinuclear antibodies, which only stain the large nuclei of islet cells, were recognized more easily than the reactive mitochondria, which exhibit a fine granular pattern in the cytoplasm of both islet and exocrine cells.

End-point titers scored for ICA by individual laboratories using the standard protocol were compared with those obtained for the modified assays. The following conclusions were drawn:

1. In the majority of determinations, acetone fixation of the sections prior to incubation with the test sample did not affect the end-point titer. However it was recommended that exposure be limited to three minutes to avoid disrupting the cellular structure of the pancreas.
2. Counterstaining of the sections with monoclonal antibodies to facilitate islet recognition also did not substantively influence the end-point titers of ICA in the majority of determinations. However this modification is particularly useful for teaching purposes in better identification of unstained islets
3. In support of the original report,[41] prolonged incubation for 24 hours of the test sample with the pancreatic section in the presence of aprotinin improved the sensitivity of the test, as reflected by a two- to sixteen-fold increase in end-point titers of ICA, compared with the results obtained with the standard protocol. Increased titers were reported in 81% of 70 determinations previously scored positive for ICA with the standard protocol. Of the 36 determinations that did not show ICA by the standard test, 15 (42%) were defined as ICA-positive, and the reciprocal titers ranged from 2 to 16. This change from ICA-negativity to ICA-positivity does not necessarily indicate loss of specificity because 11 of the 13 circulated samples were derived from newly diagnosed diabetic patients[49]; increased sensitivity of the assay is believed to be responsible.

   Similar results were obtained in an independent study of the aprotinin modification (H. Gleichmann, unpublished). Of 157 sera collected from diabetic patients, 112 (71%) were initially scored ICA-positive by the conventional method, but they increased in titer when tested in the presence of aprotinin (Fig. 4.3). Only in five (3%) sera was ICA no longer detectable with the more sensitive assay. Further analysis confirmed that no major loss in substrate specificity occurs when the aprotinin modification is used. Sera from 147 nondiabetic blood donors and 170 diabetic patients, all of which were originally scored ICA-negative by the conventional method, were incubated for prolonged periods of time in the presence of aprotinin. Although 19 (11%) of the sera from the diabetic patients became ICA-positive, all of the sera from the blood donors remained ICA-negative.

   During the workshop discussion, several participants commented that they had improved sensitivity by prolonged incubation of the serum without

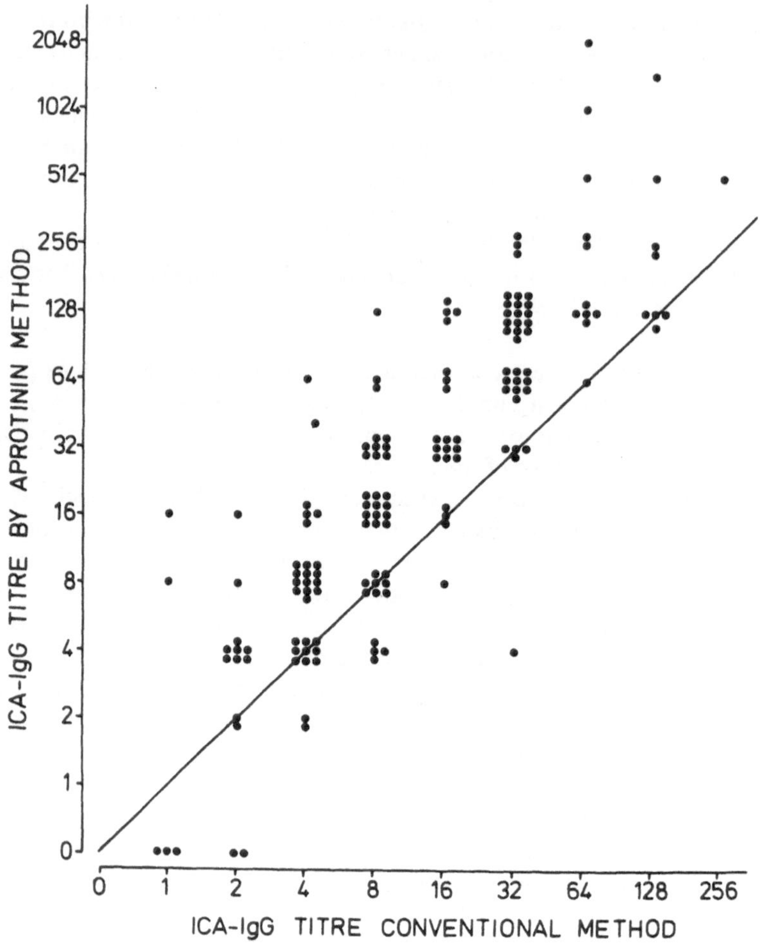

FIGURE 4.3. Comparison of end-point titers of ICA obtained by conventional indirect immunofluorescence and after incubation for 24 hours in the presence of aprotinin.

diluting it in aprotinin. However because the quality of tissue morphology also was reduced, it is not clear whether the addition of the proteinase inhibitor is necessary to stabilize the pancreatic substrate.

4. Heat inactivation of samples prior to testing for the presence of ICA lead to a two- to fourfold decrease in end-point titers.
5. The use of FITC staphylococcal protein A as the second-step reagent decreased the titer in approximately 50% of the positive determinations by the conventional method.
6. Since immunohistochemical staining procedures are well established for a number of diagnostic histologic examinations, a comparable method would be useful for ICA determination. However despite the limitations of data pre-

sented at the workshop, the results obtained with the glucose oxidase modification indicated the need for further methodologic improvements, especially in relation to the problem of eliminating an often observed, disturbing high background staining.

7. In sections of monkey pancreas, there was a possible increase in sensitivity, resulting in higher end-point ICA titers, compared with those obtained when human glands were used. Although few data were available, it appeared that the content of cytoplasmic antigenic determinants in the islets may vary between pancreases from different monkeys relative to that in human tissue. Despite these reservations, it is highly desirable to continue the search for alternative sources of pancreas, to obtain the same quality characteristics offered by human pancreas. It is important to find a substitute for the human substrate, especially now that transplantation programs are reducing the supply of human tissue.
8. CF-ICA were detected in approximately 50% of samples that scored positive by the conventional method. The presence of CF-ICA tended to correlate with higher-titer ICA, thus confirming a previously reported trend.[20,21,49,50]

Although not assayed in the workshop, it was generally agreed that sections cut from Bouin's-fixed pancreatic blocks could no longer be recommended. Abandoning the use of this substrate will help avoid the interference caused by the stain binding to the insulin-specific antibodies that are often present in the sera of diabetic and prediabetic individuals (see review Ref. 51).

It is well known that several factors may affect routine IFL assays to identify a variety of circulating autoantibodies, including ICA, for diagnostic purposes. These factors may be responsible for the problems in determining whether a given serum is ICA-positive or ICA-negative and resolving the wide differences in end-point titers reported. To overcome the variables, inherent in the new assays, it was suggested that standard curves be introduced in an attempt to reduce the scatter within each individual assay and to minimize the drift that may occur with time,[35] with the ultimate aim of improving interlaboratory agreement and intralaboratory precision and accuracy.

The value of constructing standard curves to help standardize autoantibody determination has been convincingly demonstrated for the detection of antinuclear antibodies that, like ICA, are determined by an IFL assay.[52] Figure 4.4 shows that incremental, individual standard curves could be generated for the majority of the laboratories but that quantitative interlaboratory differences in results are still observed. A subsequent analysis of these preliminary results indicated that expression of end-point titers for ICA in common units could improve interlaboratory agreement further.[36]

The data presented at the First International Workshop prompted the recommendation that a Second International Workshop be convened to determine precision within laboratories and to assess further the value of reference sera, standard curves, and the definition of ICA in common units. This would allow each laboratory to assess the validity of its own method regardless of the modi-

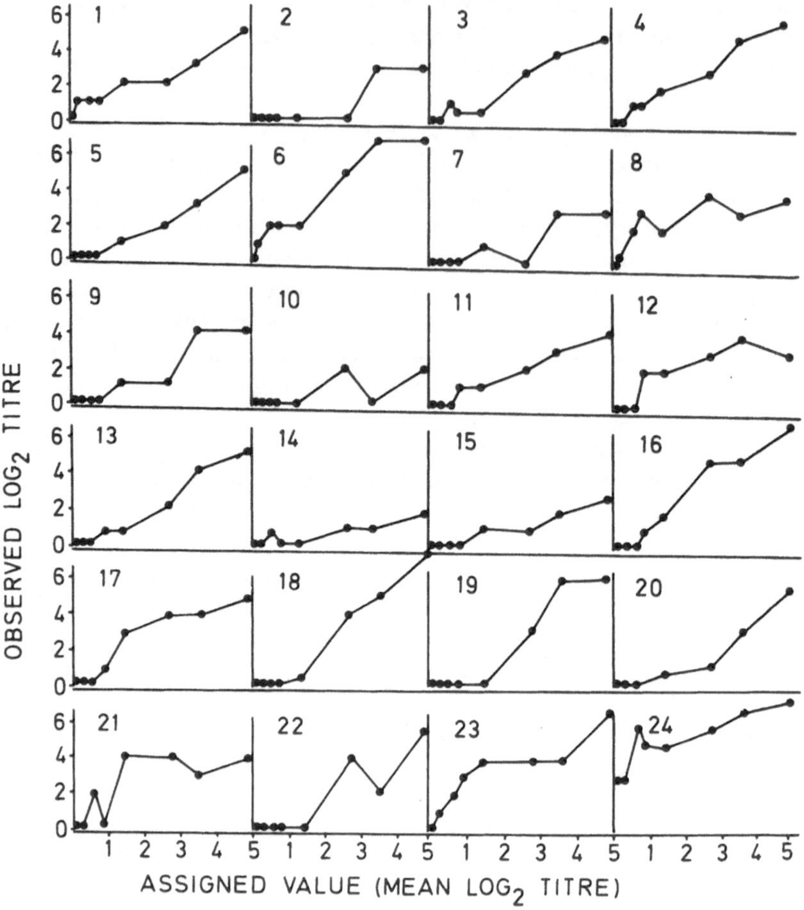

FIGURE 4.4. Standard curves based on titers reported from 24 laboratories according to a standard protocol with coded sera. These curves were computed from sera providing interim standards covering linearly increasing values from low to high titers of ICA. (Reproduced from Ref. 35 with permission.)

ications introduced. Forty laboratories participated; each received two batches of 22 coded samples. Each batch contained four dilutions of the standard serum in duplicate to test for precision and, in addition, a number of sera from healthy blood donors and diabetic patients. The second batch was identical to the first and was mailed three to six months after the results of the first set were submitted.

In general analysis of the data confirmed that the availability of reference sera enabled the establishment of standard curves that could be used to convert ICA results to common arbitrary units. It was agreed to call the ICA units "Juvenile Diabetes Foundation" (JDF-U).[47] Once again this approach was found to improve interlaboratory agreement and intralaboratory precision.[53] Thus the establish-

TABLE 4.1. ICA as marker of impending insulin deficiency in unaffected, first-degree relatives of diabetic patients (prospective family studies).

Members Studied (*N*)	Duration of study (years)	ICA-positive members				% (*N*) of members		References
				With IDDM		With IDDM		
		N	%	*N*	%	ICA +	ICA −	
719	9	30	4.2	14	46.7	1.9 (14)[a]	0.3 (2)	55 (updated 1987)
2200	6	39	1.8	8	20.5	0.4 (8)	0.0	29
1723	2	16	0.9	2	12.5	0.1 (2)	0.06 (1)	31

[a] All but one patient had CF-ICA.

ment of an international standard ICA-positive serum should ultimately make ICA determinations reliable and allow interlaboratory comparisons to be made with confidence in future studies. To reach this goal, it was concluded that it was necessary to prepare a serum pool to be distributed in the Third International Workshop, which would be available as the International Standard for the final standardization of ICA. Furthermore it was recommended that, by means of the International Standard, each laboratory would prepare its own supply of a comparable reference serum that would be used routinely to plot a standard curve before and during each ICA assay in order to express ICA concentration in JDF-U.

Until the cytoplasmic antigens recognized by ICA are purified and characterized biochemically, it must be emphasized that the different immunofluorescence and immunohistochemical assays, reviewed here, will remain the methods of choice in diabetes research, especially in the effort to recognize ongoing beta-cell destruction in the prediabetic period.

Clinical Application of the ICA Test

Despite the several methodologic factors that seem to influence results with current assays for the determination of ICA, convincing data have emerged from several clinical investigations evaluating the practical value determining of ICA. In particular these serologic markers have been used in prospective family studies and in studies of patients with type II diabetes, women with gestational diabetes, and normal populations. Although individual laboratories have reported different amounts of ICA for each of these groups, ICA nonetheless has proved to be a reliable marker that, when present in susceptible individuals, indicates a relatively high, insulin dependency risk. These studies have also demonstrated that the absence of ICA does not necessarily exclude the subsequent progression to overt disease.

FAMILY, TWIN, AND GENERAL POPULATION STUDIES

As previously mentioned, and summarized in Table 4.1, the percentage of identified, ICA-positive, unaffected first-degree relatives of insulin-dependent diabetic

TABLE 4.2. Prevalence of ICA in prospective studies of general populations (healthy persons without a family history of diabetes).

Members studied (*N*)	Duration of study (years)	ICA-positive individuals				Individuals with IDDM (*N*)		References
				With IDDM				
		N	%	*N*	%	ICA +	ICA −	
2512	6	10	0.4	0	0	0	0	29
1112	4	6	0.5	0	0	0	0	54

patients in three independently performed, prospective family studies[29,31,55] were 0.9%, 1.8%, and 4.2%; the percentage of ICA-positive family members who developed diabetes later were 12.5%, 20.5%, and 46.7%. The overall incidence of first-degree relatives who progressed to diabetes was 0.2%, 0.4%, and 2.2% in these family groups. It is noteworthy that, in all these studies, the overall incidence of the disease was higher in ICA-positive than in ICA-negative family members (0.1%, 0.4%, and 1.9% versus 0.06%, 0.0%, and 0.3%). Thus even though ICA has value as a predictor of insulin deficiency, its presence alone is not an absolute indicator. The differences in the reported percentages of ICA-positive individuals in these family studies may be attributable to at least three variables: a different period of followup (2, 6, and 9 years), populations from distinct geographic areas, and sensitivity of the ICA test.

In two prospective studies of a general population for four and six years,[29,54] no ICA-positive individuals developed overt diabetes (Table 4.2). Here the percentage of ICA-positive reactions was comparable, being 0.4% in one group and 0.5% in the other. It has not been established whether all persons with ICA will finally develop overt diabetes, since it has been shown that this immunologic abnormality can, in some instances, precede functional impairment of the pancreatic beta-cells for at least a decade.[32]

Conflicting results have been reported on the issue of "persistent" versus "fluctuating" of circulating ICA levels measured in a prospective manner. An initial report indicated that, out of the 10 unaffected first-degree relatives of diabetic families who were CF-ICA–positive upon entry to the study, one remained persistently positive after 5 years, 3 lost positivity, and 6 tended to fluctuate.[55] All of these individuals remained positive for conventional ICA that, in the absence of CF-ICA, also fluctuated. Interestingly three relatives who were ICA-negative at the time of initial testing became CF-ICA–positive during the followup period. Remarkably one of them became glucose intolerant one year after the ICA first appeared. Overall 10 of 13 (77%) CF-ICA–positive individuals had either persisting or fluctuating CF-ICA in their serum. Although fluctuation of ICA in prospective family studies has been disputed by some workers,[56] other workers have described this phenomenon in about 25% of ICA-positive individuals during a six-year followup investigation.[57]

In the four-year prospective study of a general population in Japan, ICA appeared transiently for one year,[54] and similar results have been reported in

TABLE 4.3. Prevalence of ICA in diabetic patients (apparently type II); incidence of secondary failure to therapy with antidiabetic drugs.

Patients	Patients who were [N (%)]		Patients requiring insulin [N (%)]		Type and duration	References
(N)	ICA +	ICA −	ICA +	ICA −	(years of study)	
160	23 (14.4)	137 (85.6)	14 (60.8)	11 (8.0)	Prospective (2)	19
106	17 (16.0)	89 (84.0)	10 (58.8)	2 (2.2)	Prospective (3)	58
67	18 (26.9)	49 (73.1)	16 (89.0)	29 (59.2)	Cross-sectional	20
154	22 (14.3)	132 (85.7)	9 (40.9)	32 (24.2)	Prospective (2)	21
96	32 (33.3)	64 (66.6)	7 (21.8)	0 (0)	Prospective (3)	59

unaffected identical twins discordant for the disease over a period of five years.[33] Further evaluation of the latter data shows that this is especially true with reference to conventional ICA, that is, ICA without complement-fixing ability (Johnston et al., submitted). It is worth emphasizing at this point that persistence of CF-ICA or a high titer of IgG ICA seems to predict a higher risk for future beta-cell decompensation. In fact most of the prediabetic individuals identified in the various groups studied (vide infra) as possessing these immunologic characteristics eventually progressed, though slowly, to overt symptomatology. It remains to be seen if the observed fluctuations in ICA pattern represent a real change in autoantibody concentration in the serum or whether they represent intraassay variations. Again the ongoing ICA standardization program should help in elucidating these critical and still controversial problems.

TYPE II DIABETES

Among adults who have noninsulin-dependent (type II) diabetes, by clinical criteria, the presence of ICA may predict a secondary failure of treatment with oral hypoglycemic agents. The rate of beta-cell decompensation is much faster in patients with ICA than in type II diabetic patients who do not have these antibodies. As shown in Table 4.3, the percentage of ICA-positive patients in three studies was approximately 15%[19,58,21]; it was 27% and 33% in two studies.[20,59] These differences are most likely attributable to the clinical criteria determining patient selection, especially in the latter two studies.

Moreover type II diabetic patients who were ICA-positive required insulin therapy significantly sooner than did those patients who were ICA-negative. The mean ± SEM time interval from diagnosis of diabetes to insulin requirement was 3.7 ± 0.9 years in the former group and 8.4 ± 1.1 years in the latter group (Table 4.4).[20] A similar accelerated course of beta-cell dysfunction was also observed in other studies, in which a faster decline of basal and stimulated C-peptide level was observed for insulin-requiring patients.[60] In a recent study reported from Japan, however, 10 (30%) of type II diabetic patients who were originally positive for ICA had lost these antibodies after 15 months.[59] Tests were carried out by both the complement-fixation modification[45] and the biotin–avidin assay[46];

TABLE 4.4. ICA-positive, type II diabetic patients benefited for significantly shorter times from treatment with antidiabetic drugs than ICA-negative patients.

Clinical classification	ICA status	Patients (N)	Time since diagnosis (years, $\bar{x} \pm$ SEM)	
Insulin dependent	Positive	16	3.7	(0.9)[a]
	Negative	29	8.4	(1.1)[a]
Noninsulin dependent	Positive	2	1.3	(0.5 – 2)
	Negative	20	4.8	(1.1)

[a] $p < 0.01$ (from Ref. 20, with permission.)

these individuals did not show any apparent deterioration of beta-cell function during the followup period. However as in the other studies, persistently ICA-positive probands showed earlier signs of beta-cell decompensation.

When HLA-DR phenotypes were analyzed,[20] it was found that HLA-DR3 and the heterozygous DR3/DR4 and DR3/DRw6 frequencies were significantly increased in the ICA-positive type II diabetic patients compared to the ICA-negative subgroup and to a normal control population (Table 4.5), whereas the HLA-DR5 frequency was greatly reduced. This latter negative association has also been reported in type I diabetic patients.[61] Similar genetic data were reported in Finnish ICA-positive, type II diabetic patients.[21,62] In the latter study, however, HLA-DR5 increased in the ICA-positive cases, although the increase was more pronounced in type II diabetic patients without ICA and with no signs of secondary failure. Differences between the two studies may be explained by the ethnic background of the populations, but the results nevertheless indicate a greater heterogeneity within type II diabetes than was previously suspected. Overall these frequencies are comparable to those originally reported in juvenile type I diabetic patients[63]; this reinforces the idea that the ICA-positive type II diabetic subgroup represents a pathogenetically related but apparently slow form of type I diabetes. It remains to be established whether this older diabetic population with an increased predisposition to insulin dependency also tends to develop subclinical or clinically overt autoimmune polyendocrinopathy,[64] as suggested in some studies wherein females with a higher prevalence of thyrogastric autoantibodies predominated in the ICA-positive type II diabetic subgroup.[21]

Using figures from acturial tables, Irvine and his colleagues[19] calculated that 86% of ICA-positive patients initially controlled with diet or oral hypoglycemic agents will become insulin deficient within five years. Therefore repeated clinical workups of this ICA-positive subgroup, with regular monitoring for the persistence of ICA, are highly desirable so that insulin treatment can be started without delay, when ICA levels rise. When prevalence and the prognostic value of ICA in type I diabetic family members and in patients with type II diabetes are compared, the ICA assay seems to be the test of choice in the latter group, particularly if one considers that they represent approximately 90% of all patients with diabetes.

TABLE 4.5. HLA-DR3 and heterozygous DR3/4 and DR3/w6 phenotypes were in excess in ICA-positive, adult, type II diabetic patients who required insulin later.

DR antigens	HLA-DR phenotype frequencies (%) in ICA-positive subjects			
	ICA +	ICA −	Controls	*P*
1	12.5	7.7	21	
2	6.3 II	30.8	31	II 0.04
3	75.0 II[a]	38.5[a]	30 II	II < 0.0005 a 0.05
4	50.0	38.5	31	
5	0.0 II[a]	23.1[a]	27 II	II 0.02 a 0.04
W6	25.0	23.1	19	
7	18.8	30.8	22	
W8	0.0	0.0	4	
W10	0.0	0.0	1	
3/4	31.3 II	15.4	6	II < 0.001
3/W6	25.0 II	0.0	3	II < 0.006
N	16	13	100	

[a] From Ref. 20, with permission.

GESTATIONAL DIABETES

ICA also has been found to predict a future need for insulin therapy in gestational diabetes. Of 50 pregnant women with impaired glucose tolerance, 5 (10%) had ICA and 3 (60%) required insulin treatment in the first postpartum year in one study.[65] An even higher incidence of ICA-positive, insulin-requiring pregnant women (15 out of 47 and 8 out of 11, respectively) was reported in another series.[66]

Synopsis, Recommendations, and Conclusions

It is generally accepted that type I (insulin-dependent) diabetes mellitus has a clinically asymptomatic, prediabetic phase of several years during which circulating ICA can be detected. This concept is substantiated by histopathologic,[8] clinical,[67] and metabolic[4,68] observations. Detection of ICA in adult type II (noninsulin-dependent) diabetic patients identifies a subgroup that may clinically and pathogenetically represent a slow form of type I diabetes. In gestational diabetes, the presence of ICA signals early insulin deficiency, thus indicating that ICA can indicate ongoing beta-cell damage.

For the detection of ICA, several modifications of the originally described indirect IFL test have been developed. However these assays have different degrees of sensitivity, which has made direct interlaboratory comparisons difficult. Major progress has been made in the attempt to standardize ICA assays. Standard curves to be established with reference sera can convert end-point titers for ICA into arbitrary units. The general use of standard sera should finally permit reliable interlaboratory comparisons of ICA results and improve precision. By this approach, three major aims can be pursued in future investigations:

1. The absolute prevalence of ICA in diabetic patients in various ethnic groups and in control populations could be assessed.
2. Serum ICA could be determined precisely to predict an impending insulin deficiency in predisposed individuals, so that the probable time clinically overt diabetes might develop could be calculated. A cut-off with respect to the diagnostic value of ICA could be evaluated only in prospective studies of large populations that are at risk of developing the disease.
3. The physiologic prevalence of ICA can be more accurately assessed by using highly sensitive test systems. This approach would contribute to the identification and understanding of the important antigenic determinants present in beta-cells that are recognized by disease-specific autoantibodies.

These outlined goals are expected to elucidate further the pathogenetic process resulting in beta-cell destruction. Identification of the immunocyte subpopulations that are primarily involved in the destructive process, which is currently considered to be autoimmune in nature, is also needed. Such information will permit elaboration of therapeutic protocols that specifically abrogate or even prevent chronic destruction of pancreatic beta-cells. Because ICA can reflect florid insulitis, they may be used to trace known or unknown environmental factors that might initiate the disease process. Furthermore it should be possible to characterize, on a molecular basis, relevant antigenic determinants recognized by ICA so that precise quantitative assays can be developed and made available as new tools for more sophisticated clinical and research purposes. It can be predicted that certain autoantibodies with specificity for particular islet antigens may be more closely associated with the disease process.[69]

At the Third International Workshop on ICA Standardization[70] it was concluded to make available recommendations based on the use of standard sera for future scientific works. Furthermore, available standards will be used to calibrate proficiency sera being available on a regular basis to laboratories using ICA assays.

Acknowledgment. We are grateful to the members of the International Diabetes Workshops Committee for their efforts in launching the standardization program. The Juvenile Diabetes Foundation (United States), Novo Research Institute (Denmark), Nordisk (Denmark), and SciMedx (United States), generously supported these International Workshops. We thank Anni Wachowiak and Marian Pine for their excellent secretarial assistance.

References

1. Bottazzo GF, Pujol-Borrell R, Gale EAM. Autoimmunity and Type I diabetes: Bringing the story up to date, in Alberti KGMM, Krall LP (eds.): *The Diabetes Annual/3*. Amsterdam: Elsevier Science Publishers B.V., 1987, 15–38.

2. Kaldany A, Busick EJ, Eisenbarth GS. Diabetes mellitus and the immune system, in Marble A, Krall LP, Bradley RF, et al. (eds): *Joslin's Diabetes Mellitus*. Philadelphia, PA: Lea & Febiger, 1985, 51–64.
3. Baekkeskov S, Christie M, Lernmark Å. Islet cell antibodies, in McGregor A, (ed.): *Immunology of Endocrine Diseases*. Lancaster, PA: MTP Press, 1986, 73–88.
4. Tarn AC, Smith CP, Spencer KM, Bottazzo GF, Gale EAM. Type I (insulin-dependent) diabetes: A disease of slow clinical onset? *Br Med J* 1987;294:342–345.
5. Bottazzo GF, Dean BM, McNally JM, MacKay EH, Swift PGF, Gamble DR. In situ characterization of autoimmune phenomena and expression of HLA molecules in the pancreas in diabetic insulitis. *N Engl J Med* 1985;313:353–360.
6. Eisenbarth GS. Type I diabetes mellitus. A chronic autoimmune disease. *N Engl J Med* 1986;314:136–168.
7. Foulis AK, Bottazzo GF. Insulitis—1986, in Lefèbre P, Pipelleers D (eds): *The Pathology of the Endocrine Pancreas in Diabetes*. Berlin: Springer-Verlag, (in press).
8. Gepts W. Pathologic anatomy of the pancreas in juvenile diabetes mellitus. *Diabetes* 1965;14:19–33.
9. Nerup J, Andersen OO, Bendixen G, Egeverg J, Poulsen JE. Anti-pancreatic cellular hypersensitivity in diabetes mellitus. *Diabetes* 1971;20:424–427.
10. Bottazzo GF, Florin-Christensen A, Doniach D. Islet cell antibodies in diabetes mellitus with autoimmune polyendocrine deficiences. *Lancet* 1974;ii:1279–1282.
11. MacCuish A, Barnes EW, Irvine WJ, Duncan LJP. Antibodies to pancreatic islet cells in insulin-dependent diabetics with coexistent autoimmune disease. *Lancet* 1974;ii: 1529–1531.
12. Lendrum R, Walker G, Gamble DR. Islet cell antibodies in juvenile diabetes mellitus of recent onset. *Lancet* 1975;i:880–883.
13. Lendrum R, Walker G, Cudworth AG, Theophanides C, Pyke DA, Bloom A, Gamble DR. Islet cell antibodies in diabetes mellitus. *Lancet* 1976;2:1273–1276.
14. Irvine WJ, McCallum CJ, Gray RS, Campbell CJ, Duncan LJP, Farquhar JW, Vaughan H, Morris P. Pancreatic islet-cell antibodies in diabetes mellitus correlated with the duration and type of diabetes, coexistent autoimmune disease, and HLA type. *Diabetes* 1977;26:138–147.
15. Del Prete GF, Betterle C, Padovan D, Erle G, Toffolo A, Bersani G. Incidence and significance of islet cell autoantibodies in different types of diabetes mellitus. *Diabetes* 1977;26:909–915.
16. Richens ER, Quilley J, Hartog M. Immunological features of juvenile onset diabetic patients correlated to HLA type. *Clin Exp Immunol* 1979;36:198–204.
17. Takahashi A, Tsujihata M, Yokota A, Yamaguchi Y, Ueda Y, Akazawa S, Miyake S, Nagataki S. A new method of detection of islet cell antibodies (ICA) using peroxidase-labeled protein A, and incidence of ICA in Type I (insulin-dependent) diabetes. *Diabetologia* 1986;29:378–382.
18. Bottazzo GF, Mirakian R, Dean BM, McNally JM, Doniach D. How immunology helps to define heterogeneity in diabetes mellitus, in Köbberling J, Tattersal R (eds.): *The Genetics of Diabetes Mellitus*. New York: Academic Press, 1982;79–80.
19. Irvine WJ, Javers JSA, Feek CM, Prescott RJ, Duncan LJP. The value of islet cell antibody in predicting secondary failure of oral hypoglycaemic agent therapy in diabetes mellitus. *J Clin Lab Immunol* 1979;2:23–26.
20. Gleichmann H, Zörcher B, Greulich B, Gries FA, Henrichs HR, Bertrams J, Kolb H. Correlation of islet cell antibodies and HLA-DR phenotypes with diabetes mellitus in adults. *Diabetologia* 1984;27:90–92.

21. Groop LC, Bottazzo GF, Doniach D. Islet cell antibodies identify latent Type I diabetes in patients aged 35–75 years at diagnosis. *Diabetes* 1986;35:237–241.
22. Madsbad S, Bottazzo GF, Cudworth AG, Dean B, Faber O, Binder C. Islet-cell antibodies and beta-cell function in insulin-dependent diabetics. *Diabetologia* 1980;18: 45–47.
23. Marner B, Agner T, Binder C, Lernmark Å, Nerup J, Mandrup-Poulsen T, Walldorff S. Increased reduction in fasting C-peptide in association with islet cell antibodies in Type I (insulin-dependent) diabetic patients. *Diabetologia* 1985;28:875–880.
24. Gunnarsson R, Bottazzo GF, Freedman ZR, Lernmark Å, Zuhlke B, Groth OG. Does specific immunity to islet cell tissue develop in diabetes receiving pancreatic grafts. *Horm Metab Res* (Suppl) 1982;108–111.
25. Sibley RK, Sutherland DER, Goetz FC, Michael AF. Recurrent diabetes mellitus in the pancreas iso and allograft: A light and electron microscopic and immuno-histochemical analysis. *Lab Invest* 1985;53:132–145.
26. Mandrup-Poulsen T, Nerup J, Stiller CR, Marner B, Bille G, Heinrichs D, Martell R, Dupré J, Keown PA, Jenner MR, Rodger NW, Wolfe B, Graffenried BV, Binder C. Disappearance and reappearance of islet cell cytoplasmic antibodies in cyclosporin-treated insulin-dependent diabetics. *Lancet* 1985;i:599–602.
27. Gorsuch AN, Spencer KM, Lister J, McNally JM, Dean BM, Bottazzo GF, Cudworth AG. Evidence for a long prediabetic period in Type I (insulin-dependent) diabetes mellitus. *Lancet* 1981;2:1363–1365.
28. Ginsberg-Fellner F, Dobersen MJ, Witt ME, Rayfield EJ, Rubinstein P, Notkins A. HLA-antigens, cytoplasmic islet cell antibodies, and carbohydrate tolerance in families of children with insulin-dependent diabetes mellitus. *Diabetes* 1982;31:292–298.
29. Riley WJ, Spillar R, Waltz JB, Brody B. Predictive value of islet cell autoantibodies. *Diabetes* 1984;33 (Suppl. 1):175.
30. Betterle C, Zanette F, Pedini B, Presotto F, Rapp LB, Monciotti CM, Rigon F. Clinical and subclinical organ-specific autoimmune manifestations in Type I (insulin-dependent) diabetic patients, and their first degree relatives. *Diabetologia* 1984;26: 431–436.
31. Srikanta S, Ganda OP, Rabizadeh A, Soeldner JS, Eisenbarth GS. First-degree relatives of patients with Type I diabetes mellitus. Islet-cell antibodies and abnormal insulin secretion. *N Engl J Med* 1985;313:461–464.
32. Srikanta S, Ganda OP, Eisenbarth GS, Soeldner JS. Islet cell antibodies and beta-cell function in monozygotic triplets and twins initially discordant for Type I diabetes mellitus. *N Engl J Med* 1983;308:322–325.
33. Millward BA, Alviggi L, Hoskins PJ, Johnston C, Heaton D, Bottazzo GF, Vergani D, Leslie RDG, Pyke DA. Immune changes associated with insulin dependent diabetes may remit without causing the disease: A study in identical twins. *Br Med J* 1986; 292:793–796.
34. Betterle C, Caretto A, Tiengo A, Trevisan A. Complement-fixing islet-cell antibodies in Type I diabetes and in susceptible patients with autoimmune diseases. *Lancet* 1980;i:1418–1419.
35. Bottazzo GF, Gleichmann H. Immunology and Diabetes Workshop: Report of the first international workshop on the standardization of cytoplasmic islet cell antibodies. *Diabetologia* 1986;29:125–126.
36. Gleichmann H, Bottazzo GF. Progress toward standardization of the cytoplasmic islet-cell antibody assay. *Diabetes* 1987;36:578–584.

37. Dobersen MJ, Bell AM, Jenson AB, Notkins AL, Ginsberg-Fellner F. Detection of antibodies to islet cells and insulin with paraffin-embedded pancreas as antigen. *Lancet* 1979;2:1078.
38. Huang SN, Minassian H, More JD. Application of immunofluorescent staining on paraffin sections improved by trypsin digestion. *Lab Invest* 1976;35:383–390.
39. Crossley JR, Lendrum BD, Elliott RB. Rhesus-monkey pancreas for measuring islet-cell antibodies. *Lancet* 1977;i:1267.
40. Barbosa J, Chavers B, Dunsworth T, Michael A. Islet cell antibodies and histocompatibility antigens (HLA) in insulin-dependent diabetics and their first-degree relatives. *Diabetes* 1982;31:585–588.
41. Pilcher C, Elliott RB. Improved sensitivity of islet cell cytoplasmic antibody assay in diabetics. *Lancet* 1984;i:352.
42. Srikanta S, Rabizadeh A, Omar MAK, Eisenbarth GS. Assay for islet cell antibodies. Protein A-monoconal antibody method. *Diabetes* 1985;34:300–305.
43. Krell J, Rabin BS. Comparison of an immunohistochemical and immunofluorescence procedure to detect antibody to pancreatic islet cells. *Diabetes* 1984;33:709–711.
44. Madsen OD, Landin Olsson M, Bille G, Sundkvist G, Lernmark Å, Dahlqvist G, Ludvigsson J. A two-colour immunofluorescence test with a monoclonal human pro-insulin antibody improves the assay for islet cell antibody. *Diabetologia* 1986;29: 115–118.
45. Bottazzo GF, Dean BM, Gorsuch AN, Cudworth AG, Doniach D. Complement-fixing islet cell antibodies in Type I diabetes: Possible monitors of active beta-cell damage. *Lancet* 1980;i:668–672.
46. Kobayashi T, Sugimoto T, Itoh T, Kosaka K, Tanaka T, Suwa S, Sato K, Tsuji K. The prevalence of islet cell antibodies in Japanese insulin-dependent and non-insulin dependent diabetic patients studied by indirect immunofluorescence and by a new method. *Diabetes* 1986;35:335–340.
47. Bonifacio E, Dawkins RL, Lernmark Å. Immunology and diabetes workshops: Report of the second international workshop on the standardization of cytoplasmic islet cell antibodies. *Diabetologia* 1987;30:273.
48. Ludvigsson J, Heding L, Lieden G, Marner B, Lernmark Å. Plasmaphoresis in the initial treatment of insulin-dependent diabetes mellitus in children. *Br Med J* 1983;286:176–179.
49. Riley W, Neufeld M, MacLaren N. Complement fixing islet cell antibodies: A separate species? *Lancet* 1980;i:1133.
50. Bruining GJ, Molenaar J, Tuk CW, Lindemann J, Bruining HA, Marner B. Clinical time-course and characteristics of islet cell cytoplasmic antibodies in childhood diabetes. *Diabetologia* 1984;26:24–29.
51. Wilkin T, Armitage M. Markers for insulin dependent diabetes: Towards early detection. *Brit Med J* 293;1323–1326.
52. Bonifacio E, Hollingsworth PN, Dawkins RL. Anti-nuclear antibody: Precise and accurate quantitation without dilution. *J Immunol Meth* 1986;91:249–255.
53. Bonifacio E, Lernmark Å, Dawkins RL, et al. Serum exchange and use of reference standards have improved precision of measurement of islet cell antibodies. *Immunol Meth* 1988;106:83–88.
54. Notsu K, Oka N, Note S, Nabeya N, Kuno S, Sakurami T. Islet cell antibodies in Japanese population and subsets with Type I (insulin-dependent) diabetes. *Diabetologia* 1987;28:660–662.

55. Spencer KM, Tarn A, Dean BM, Lister J, Bottazzo GF. Fluctuating islet-cell autoimmunity in unaffected relatives of patients with insulin-dependent diabetes. *Lancet* 1984;i:764–766.
56. Srikanta S, Eisenbarth GS. Disappearing anti-islet antibodies? *Lancet* 1984;i:1176–1177.
57. Riley W, MacLaren N. Islet-cell antibodies are seldom transient. *Lancet* 1984;i:1351–1352.
58. Di Mario U, Irvine WJ, Borsey DQ, Kyner JL, Weston J, Galfo C. Immune abnormalities in diabetic patients not requiring insulin at diagnosis. *Diabetologia* 1983;25:392–395.
59. Kobayashi T, Itoh T, Kosaka K, Sato K, Tsuji K. Time course of islet cell antibodies and beta-cell function in non-insulin-dependent stage of Type I diabetes. *Diabetes* 1987;36:510–517.
60. Groop LC, Pelkonen R, Koskimies S, Bottazzo GF, Doniach D. Secondary failure to treatment with oral anti-diabetic agents in non-insulin-dependent diabetes. *Diabetes Care* 1986;9:129–133.
61. Henson V, MacLaren N, Winter W, Riley W, Rotter J, Wakeland EK. Molecular genetics of insulin-dependent diabetes mellitus. *Mol Biol Med* 1986;3:129–136.
62. Groop L, Miettinen A, Groop PH, Meri S, Koskimes S, Bottazzo GF. Organ-specific autoimmunity and HLA-DR antigens as markers for beta-cell destruction in patients with maturity onset diabetes. *Diabetes* 1988;37:99–103.
63. Cudworth AG, Wolf E. The genetic susceptibility to type I (insulin-dependent) diabetes mellitus. *Clin Endocrinol Metab* 1982;11:389–408.
64. Bottazzo GF, Doniach D. Polyendocrine autoimmunity, in Volpe R (ed): *Autoimmunity and Endocrine Disease*. New York: Marcel Dekker, 1985;375–403.
65. Steel JM, Irvine WJ, Clarke BF. The significance of pancreatic islet cell antibody and abnormal glucose tolerance during pregnancy. *J Clin Lab Immunol* 1980;4:83–85.
66. Ginsberg-Fellner F, Dobersen M, Mark EM, Nechemias C, Hausknecht RU, Rubinstein P, Notkins AL. Islet cell antibodies in gestational diabetes mellitus: Prognostic value. *Diabetes* 1980 (Suppl) 2:53A.
67. Steel JM, Gray RS, Clarke BF. Obstetric history of diabetics: Its relevance to the aetiology of diabetes. *Br Med J* 1979;1:1303–1305.
68. Rosenbloom A, Hunt SS, Rosenbloom EK, MacLaren NK. Ten year prognosis of impaired glucose tolerance in siblings of patients with insulin-dependent diabetes. *Diabetes* 1982;31:385–387.
69. Gleichmann H, Bottazzo GF. Islet cell and insulin autoantibodies in diabetes. *Immunol Today* 1987;8:167–168.
70. Boitard C, Bonifacio E, Bottazzo GF, Gleichmann H, Molenear J. Immunology and diabetes workshops: Report on the third international (stage 3) workshop on the standardization of cytoplasmic islet cell antibodies. *Diabetologia* 1988;31:451–452.

5
The Humoral Anti-Islet Response: Biochemical Characterization

HEIKE BÄRMEIER, MICHAEL CHRISTIE, BETSY HEROLD, KEVAN HEROLD, AND ÅKE LERNMARK

Introduction

The humoral immune response to islet cells in insulin-dependent diabetes includes a number of autoantibodies that may be detected before as well as at the clinical onset of the disease. Numerous assays that may be used to detect antibody reactivity to different preparations of islet-cell material have been described.[1-6] In many of these analyses, demonstration of the presence of an autoantibody depends on a complex assay system rather than on a well-characterized interaction between specific antibodies and antigens. Typically the assay systems are qualitative or, at best, semiquantitative, which hampers our understanding of the possible pathogenetic importance of islet-cell antibodies. It also is not clear how autoantibody formation is initiated. Do autoantibodies appear after presentation of islet antigens as the primary event? Do they appear following an undefined injury to beta-cells? The latter may include attacks by virus, bacteria, environmental antigens, or possible cytotoxic lymphocytes that have been activated to detect beta-cell antigens.

In an attempt to understand the specificity of the autoimmune reaction toward islet beta-cells and the series of events that trigger and maintain this response, it is important to define the molecules implicated in these processes. Biochemical studies are therefore being conducted in a number of laboratories to characterize both the autoantibodies themselves and the molecules they recognize. In the following discussion, we review the progress made in defining the humoral anti-islet response biochemically.

Autoantigens Detected in Human Islets

Autoantibodies against islet cells were first detected by indirect immunofluorescence using frozen sections of human pancreas[7,8] and later on the surface of dispersed, normal[9,10] or tumorous[11,12] islet cells in vitro. These tests detected a large number of autoantibodies that recognized both specific and common islet-cell antigens. In the first analysis, islet-cell cytoplasmic antibodies (ICA) were found not to be beta-cell–specific, since an indirect immunofluorescence reaction could be discerned in the cytoplasm of all islet cells.[7,8] The antibodies that define the immunofluorescence reaction on the surface of islet cells (ICSA) may similarly bind to different antigens. Some of these antigens may be beta-cell–specific, whereas other antigens may be present on the surface of all islet cells and perhaps of other cells as well. It is also possible that some antigens participate in both the ICA and the ICSA reactions.

Little progress has been made toward isolating the autoantigens detected by the ICA reaction. Isolation might be effected through subcellular fractionation of human pancreas or human pancreatic islets; the reactive fraction would then be dried onto glass slides and used in the indirect immunofluorescence test. Such an approach to a solid-phase immunofluorescence assay has not yet been taken, but although the detection system would mimic that of the indirect immunofluorescence assay on frozen sections of human pancreas, it would probably require large amounts of islet cells or pancreas material.

Another approach to isolating islet-cell antigens would be to add various antigen preparations to sera known to be ICA-positive. After absorption to allow the antigen to neutralize the ICA, the ICA sera would be titrated in the indirect immunofluorescence ICA assay on frozen sections of human pancreas. The extent to which the ICA reaction is blocked would demonstrate the presence of an antigen involved in the ICA reaction.

Colman and coworkers[13,14] have used this approach in an attempt to identify possible autoantigens in lipid fractions of human pancreas extracted in chloroform-methanol. The fractions tested by the ICA blocking assay have indicated, so far, that a GM-2 glycolipid-containing fraction inhibits the ICA immunofluorescence reaction. It was also reported that treating the frozen sections with proteolytic enzymes did not affect the binding of ICA-positive patient sera[13]; treating the sections with chloroform-methanol, however, did.

Because it is difficult to obtain human pancreatic islets in sufficient quantities to prepare islet-specific lipid fractions to be used in the blocking assay progress toward identifying a relevant glycolipid antigen has been slow. How the ICA immunofluorescence reaction is blocked is also important, since it is not known whether the ICA reaction as such reflects interaction with multiple antigens. If it does, one would not expect the ICA immunofluorescence reaction in positive sera to be totally abolished by just one antigen preparation. It should be possible, however, to measure a graded response either by photomultiplier-assisted measurements of fluorescence intensity[14] or by a careful checkerboard titration of the antigen preparation and serum. This is important, since a major question that needs to be answered is why the ICA reaction covers all endocrine islet cells but the pathognomonic sign is the disappearance of pancreatic beta-cells. Further advances may come from a cross-species approach because it is known that good correlation between ICA reactions are obtained on frozen sections of rat and human pancreas.[15]

The specificity of an autoimmune reaction directed against one particular cell type, such as the pancreatic beta-cell, would seem to be by the ability of autoantibodies to react with a cell-specific autoantigen. In view of evidence indicating that T-lymphocytes are important to the autoimmune reaction against the beta cells, it has been proposed that T-cells might be used for autoantigen isolation. T-lymphocytes are important at two levels. One is at the level of the T-helper cell, which will proliferate and signal the differentiation of both cytotoxic T-lymphocytes and antibody-producing B-lymphocytes. The T-cell receptor on the T-helper lymphocytes proliferates in response to a fragment of an antigen presented by a major histocompatibility class II antigen expressed on the surface of an antigen-presenting cell. The other T-lymphocyte is at the level of cytotoxic/suppressor T-cells, which are able to kill their target cells, provided the T-cell receptor recognizes an antigenic epitope presented by a major histocompatibility class I molecule. It would probably be difficult to use the T-cell receptor to isolate specific islet cell autoantigens.

Nevertheless, the concept that beta-cell–specific, islet-cell autoantibodies are present in insulin-dependent diabetes and are responsible for the specific destruction of beta-cells through one or more innate immune functions is vital. The autoimmune reactions that have been detected in several other disorders by using the autoantibody as a biochemical tool to define antigen suggest that autoantigens might determine the cell- or tissue-specificity of an autoimmune reaction (Table 5.1).

Autoantigens May Define Disease Specificity in Human Autoimmunity

In a number of autoimmune diseases affecting tissues rather than isolated organs or cells, and therefore called systemic autoimmune diseases, various autoantigens have been detected in human tissues by immunoblotting techniques (Table 5.1). Several autoantigens in systemic lupus erythematosus involving specific single- and double-stranded DNA, histones, ribonuclear particles, etc., have been identified.[16-25] Molecular cloning of two of these autoantigens using antibodies to detect the gene product has recently been reported.[25,26] In Sjoegren's syndrome, another autoimmune disorder with systemic characteristics, many lupus-type autoantibodies may be present, but a new set of autoantibodies directed against disease-related autoantigens seems to be responsible for the clinical expression of the disease.[24,30,31] The La antibody found in 94% of patients with Sjoegren's syndrome is directed against a nucleoprotein that has been cloned and used to detect a subgroup of ANA (the legend to Table 5.1 lists these abbreviations) more specifically associated with primary Sjoegren's syndrome.[28,31] Similar observations have been made in progressive systemic sclerosis (PSS)[32-36]; CREST,[37] which includes rheumatoid arthritis[38,39]; and ankylosing spondylitis.[40] One autoantigen detected by anti-centromeric antibodies has been characterized by molecular cloning.[41]

In a number of diseases that are associated with organ specificity, including insulin-dependent diabetes, there is a more distinct pattern of autoantibodies

TABLE 5.1. Autoantibodies and autoantigens detected by immunoblotting.

Disease	Antibody	Antigen detected	References
		Systemic Disease	
SLE	Anti-DNA	ds + 34-,33-,17-,16-,14-K LAMP	16
	Anti-histones	H1	17,18
		2A (in Pr-induced LE)	18
		2B	17,18
		3 (in Hy-induced LE)	18
		4 (in Hy-induced LE)	18
	Anti-Sm	Complex snRNPart	19
		Free Sm 14-K	20
		U-proteins: B′ 29-K,B 28-K, D 22-K	21,22,23
		A′,B,B′,D,E,F,G	20
		U2: U2RNA + A′,B″ (only in U2)	21
		+ D,E,F,G	23
		26-, 13-K protein	24
		E	25
		B″	26
	Anti-Me	100-,65-,21-,16-K protein	27
	Anti-U1 RNP	Complex nRNP + U1 RNA	19
		RNP 65- -70-K	24
		U1-protein: 68-K, A 32-K, C 22-K	21,22
		and B′/B	23
	SS-A/Ro	60-,94-K proteins	28,29
SS	SS-B/La	43-,45-,50-K proteins	28,30,31
PSS	AMA	72-K mitochondrial protein	32
	Anti-Scl-70	100-K nuclear topoisomerase I	33
		100-,95-K nuclear protein	34
	ACA	72-,19.5-K centromeric protein	35
	ANA	13 polypeptides (14- -210-K)	36
CREST	ACA-A	17-K protein	37
	ACA-B	80-K (true mass 65-K) protein	37,41
	ACA-C	140-K protein	37
RA		48-K protein	38
		26-,39-,40-,43-K proteins	39
Ankylosing spondylitis		70-,19-,14-K proteins	40
		Tissue- or Cell-Specific Disease	
Myasthenia gravis	Anti-ACh-receptor	ACh-receptor (alpha-chain)	42,71
Myositis	Anti-Jo-1	Histidyl-tRNA-synthetase Dimer 160-K	43,44
Grave's disease	LATS-P	TSH-receptor 197-K	45
	Antimicrosomal	Microsomal protein 105-K, 107-K	46,53
	Anti-TPO	100-,107-K TPO	47,48,50
	Anti-TG	260-,230-,180-,142-K proteins	46
	Anti-ATRA1	ATRA1 protein	49

TABLE 5.1. *Continued.*

Disease	Antibody	Antigen detected	References
Hashimoto	Anti-microsomal	Microsomal protein 105--107-K	46,51,53
	Anti-TPO	100-,107-K TPO	47,48,50
	Anti-TG	Thyroglobulin 320-,200-K	52
		260-,230-,180-,142-K proteins	46
	Anti-ATRAl	ATRAl protein	49
ITP	Anti-Pl	gP III	54
		gP IIa (100-K)	55
PBC	AMA	70- -72-K, 45- -47-K mitochondrial proteins	56,32
		70-K mitochondrial protein	57
		62-, 48-K proteins	58
CAH	LKMA	50-K membrane protein	59
Goodpasture syndrome	Anti-GBM	6 GBM proteins (26--58-K)	60
		GBM proteins	61
		Collagen IV alpha-chain	62
		M2 (subunit)	63
		26-K protein	64
Nephritis	Anti-TBM	58-K protein	65
IgA-Nephropathy		69-K nuclear protein	66
Ulcerative colitis		40-K colonic protein	67
Pemphigus vulgaris		130-,80-K epidermal BM protein	68
Pemphigus foliaceus		160-K desmosome core protein	68
EBA		800-K glycoprotein	69
Uveoretinitis		Retinal S-protein 48-K	70

Key to abbreviations:

ACA	anti-centromeric antibody
Ag	antigen
AMA	anti-mitochondrial antibody
ANA	anti-nuclear antibody
ATRA	autoimmune thyroiditis-related antigen
BM	basal membrane
BSA	bovine serum albumin
CAH	chronic active hepatitis
CREST	calcinosis, Raynaud's phenomenon, esophageal dysmotility, sclerodactylia, teleangiektasia
ds	double-stranded
EBA	epidermyolysis bullosa acquista
GBM	glomerular basal membrane
gP	glycoprotein
H	histone
Hy	Hydralazin
ITP	idiopathic thrombocytopenia
Jo-l	name of antigen (refers to first patient in whom described)
La	name of antigen (refers to first patient in whom described)

TABLE 5.1. *Continued.*

LAMP	lupus-associated membrane protein
LATS-P	long-acting, thyroid-stimulating (human)
LKMA	liver-kidney–associated antibody
Me	name of antigen (refers to first patient in whom described)
MCTD	mixed connective tissue disease
n	nuclear
PBC	primary biliary cirrhosis
Pl	platelet
Pr	Procainamid
PSS	progressive systemic sclerosis
RA	rheumatoid arthritis
RF	rheumatic factor
RNPart	ribonuclear particles
Ro	name of antigen (refers to first patient in whom described)
sc	small cytoplasmic
Scl	name of antigen
SLE	systemic lupus erythematosus
Sm	name of antigen (refers to first patient in whom described)
sn	small nuclear
SS	Sjoegren's syndrome
TBM	tubular basal membrane
TG	thyroglobulin
TPO	thyroid peroxidase
TSH	thyroid-stimulating hormone
U	uridine-rich

reacting with presumptive cell- or organ-specific autoantigens (Table 5.1). Specific antigens identified by using patient sera in immunoblotting analyses have been reported in myasthenia gravis,[42] myositis,[43,44] Graves' and Hashimoto's diseases,[45-53] idiopathic thrombocytopenia,[54,55] primary biliary cirrhosis,[32,56-58] chronic active hepatitis,[59] Goodpasture's syndrome,[60-64] interstitial tubular nephritis,[65] IgA nephropathy,[66] ulcerative colitis,[67] pemphigus vulgaris,[68] pemphigus foliaceus,[65] epidermolysis bullosa aquista,[69] and uveoretinitis.[70] A more complete characterization by molecular cloning of human autoantigens has been reported for the alpha-chain of the nicotinic acetylcholine receptor,[71] thyroid peroxidase, and candidate autoantigens in thyroiditis.[49,50]

Thus studies to identify autoantigens associated with different autoimmune diseases suggest that each disorder may be linked to a particular autoantigen. In addition to antibodies, these autoantigens also may direct immune system cells to the autoantigen. In some of these diseases there is close correlation between the presence of the specific autoantibodies and disease activity. The lack of standardized tests for islet cell antibodies precluded such analyses in diabetes.[72] In insulin-dependent diabetes, a disease-specific autoantigen, the M_r 64000 islet protein has been described,[73,74,75] disease activity appears to be closely associated with antibodies against this protein.

It is of interest that several of the diseases listed above, such as Graves' and Hashimoto's, are associated with certain HLA types and that insulin-dependent diabetes occurs more often in persons with these HLA types than in the general population (cf. Refs. 5 and 72). It is therefore possible that patients with insulin-dependent diabetes or some of the organ-specific autoimmune disorders listed in Table 5.1 suffer from a syndrome of hyperautoreactivity. We have suggested that such hyperautoreactivity may be familial.[76] The disease specificity in individuals who have more than one autoimmune disorder would then be reflected by autoantibodies against specific autoantigens. It is of interest to note that some of the autoantigens listed in Table 5.1 seem to represent proteins that are required for the function of individual cell types. One such example is thyroid peroxidase.[46-51] In other instances, the autoantigen is a specific product of a cell; these include thyroglobulin[46,52] or insulin.[77,78] The reader is referred to Chapter 6 in this volume for further information on insulin autoantibodies.

It would be a simple task to explain the loss of the pancreatic beta-cells in insulin-dependent diabetes if such antigen-specific effectors as cytotoxic T-cells or islet-cell antibodies or both allowed the immune system to detect only pancreatic beta-cells. Sera from diabetic patients provide an opportunity to use islet-cell antibodies as biochemical reagents to detect and perhaps to isolate antigen(s). The first approach to biochemical characterization of islet-cell autoantigens was to label cultured human islet cells with radioactive amino acids metabolically. The labeled islets were solubilized in detergent, and the extract incubated with serum from diabetic and control individuals.[73] The antigen–antibody complexes formed were absorbed to protein A-Sepharose, and the immunosorbent was carefully rinsed to remove nonspecifically bound radioactive islet-cell proteins. The immunosorbent was then boiled in SDS to denature and solubilize bound antibodies and labeled proteins. Following centrifugation, the supernatant was analyzed by gel electrophoresis and subsequent autoradiography to identify the labeled islet proteins which had formed complexes with the antibodies in sera from diabetic patients and normal controls. It was found that the autoantibodies in diabetic sera specifically recognized a human islet protein of about M_r 64000.[73] These studies were extended to include an analysis of 14 subjects followed prospectively until the clinical onset of insulin-dependent diabetes.[74] Antibodies against the M_r 64000 protein were consistently detected in 11 of these subjects. In one patient, the M_r 64000 antibodies were detected as early as 8 years before clinical onset. Although the study group was small, it was noted that M_r 64000 antibodies may occur independently of ICA, since M_r 64000 antibodies were detected before ICA in two of the patients.[74] These findings suggested that antibodies for the M_r 64000 protein may be an early marker for an autoimmune reaction toward beta-cells and that the presence of these autoantibodies may predict a later onset of insulin-dependent diabetes.

The M_r 64000 molecule appears to be hydrophobic[74]; it is also associated with subcellular fractions of islet cells enriched in plasma membrane enzyme markers. It will be necessary to characterize completely the M_r 64000 antigen and to clone its gene to determine its function in the beta-cell. Since recent approaches to molecular cloning of a large number of proteins have taken advantage of bacterial expression systems and immunoblotting to detect expressed gene products, attempts are being made to identify islet-cell antigens by immunoblotting after transfer of proteins from polyacrylamide gels to nitrocellulose paper (Western blotting), rather than by the immunoprecipitation technique described above. Radiolabeling and immunoprecipitation of islet-cell antigens have been described in detail.[75]

Western Blotting Analysis of Islet Autoantigens

The above studies suggest that autoantibodies found in the sera of patients with autoimmune diabetes may recognize a specific islet autoantigen. The immunoprecipitation techniques[75] used in these studies are important tools in characterizing these antigens, but they also present a number of technical problems that limit their use in large-scale population studies. Furthermore only antigens recognized by antibodies that bind well to the adsorbant (e.g., protein A) can be analyzed. We have recently used the technique of Western or immunoblotting to study islet-reactive antibodies found in the sera of patients with insulin-dependent diabetes. This technique involves the separation of islet membrane proteins on a polyacrylamide gel followed by transfer to a nitrocellulose membrane. This method, described in detail by Towbin et al.,[79] is now widely used, as exemplified in several of the investigations listed in Table 5.1. Islet proteins recognized by antibodies may be identified by first incubating sera with a strip of the nitrocellulose overnight at 4 °C, washing, and then identifying bound immunoglobulins with an alkaline phosphatase–conjugated second antibody.

We initially applied this technique by studying autoantibodies reactive with a detergent phase preparation of rat islets in 12 patients with newly diagnosed, insulin-dependent diabetes; our controls were 10 healthy blood donors. A number of common bands could be detected using serum from diabetic and control subjects (Fig. 5.1). For example a band corresponding to albumin-reactive antibodies was seen frequently, but these immunoglobulins were not unique to diabetic or control populations. A band corresponding to a protein of M_r 64000 could be identified in 9 of 12 diabetic subjects, but in none of the 10 controls (unpublished observations). The presence of immunoglobulins with this reactivity in the subjects studied with the Western blotting technique corresponded closely to results obtained with immune precipitation, as discussed above. When samples were run on more than one occasion, identically patterned bands were obtained. In addition several of the patients with newly diagnosed diabetes had undergone plasmapheresis,[80] and their sera had been found positive for antibodies against a human islet M_r 64000 protein detected by immunoprecipitation.[74] This approach for obtaining an antigen preparation enriched in the M_r 64000 islet-cell protein is being employed as the starting material for further Western blotting analyses.

Since the approach for detecting islet-cell autoantigens by Western blotting involves reagents from two different species, rat and man, it will be necessary to subject all test sera to appropriate absorptions before applying them to nitrocellu-

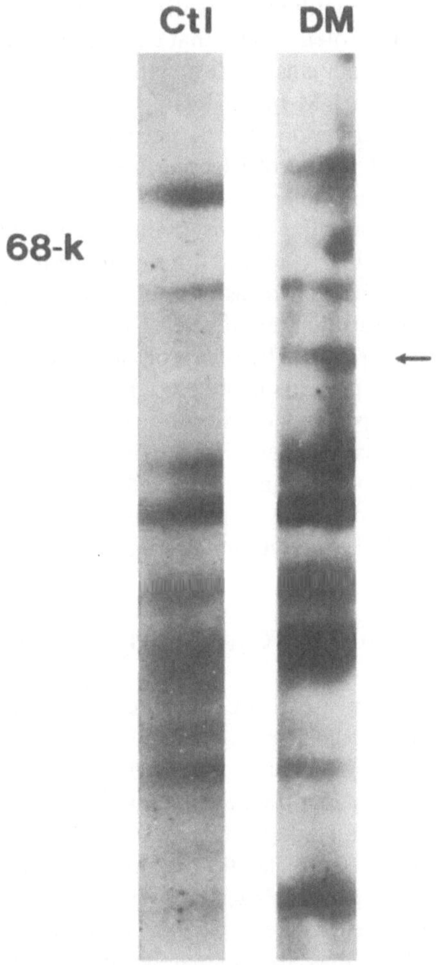

FIGURE 5.1. Immunoblot analyses of rat islet cell proteins with sera from either a healthy control (ct1) or a patient with insulin-dependent diabetes mellitus (DM). A particulate fraction was prepared from isolated rat islets, electroblotted onto nitrocellulose, blocked for non-specific binding and cut into strips. The strips were incubated first with human serum and after washing with an alkaline phosphatase-labelled second antibody to detect bound human IgG. An M_r 64000 component detected by the diabetic serum is indicated (arrow) along with a size marker (68K).

lose strips containing precious rat islet-cell material. It will also be important to control antibody reactivity to other determinants, such as albumin[82,83] or immunoglobulin.[84] Once an immunoblotting assay is established, however, there will be several advantages to using this technique to study islet autoantigens. Western blotting is a method ideally suited to screening large populations for the presence of antibodies of interest. We would be able to screen, for example, 50 serum samples from one gel that includes membranes from 18,000 rat islets. In addition the technique is rapid; results are obtained within 24 hours. It is also possible to standardize this technique to minimize the interlaboratory variability that has plagued islet autoantibody determinations by other techniques. Strips of nitrocellulose may be stored frozen and used later without detectable loss of protein antigenicity.

Autoantibodies identified by this technique would not be limited to those that bind protein A and that can immunoprecipitate membrane proteins. Antibodies from humans or from animal models of insulin-dependent diabetes such as human IgM or rat IgG_{2b} may also be identified by this method. Such ease of identification may be particularly important if patients in very early stages of development of islet autoimmunity, before isotype switching has occurred, can be identified.

Finally the rapidity with which results are obtained also makes the technique particularly useful for monoclonal antibody production. Antibodies reactive with islet antigens of interest may be produced by fusing cells from humans or animals that manifest islet-reactive autoantibodies. Similarly because antigenicity is retained by islet proteins after electrophoresis and transfer to nitrocellulose, these proteins may be used to immunize animals to obtain antibodies that react with proteins of interest.

There is preliminary evidence that Western blotting may be used to identify and to screen serum samples for islet autoantibodies. The ease with which this procedure may be performed, and other features suggest that it will be a valuable tool in islet autoimmunity studies.

Analyzing Circulating Immune Complexes in Diabetes

An increase in circulating immune complexes is yet another immune abnormality reported in insulin-dependent diabetes. Immune complexes have been reported in 25 to 40% of patients with newly diagnosed, insulin-dependent diabetes. The frequency of immune complexes, however, tends to decrease later.[85-87] But at later stages and, depending on the degree of insulin antibody formation, immune complexes subsequently may increase. The role of immune complexes in insulin-dependent diabetes pathogenesis still remains to be determined. In other diseases, immune complexes have been shown to be pathogenetic. Immune complexes may also regulate antibody synthesis[88] to affect the function of cells that are expressing Fc or complement receptors.[89-92] It has also been shown that immune complexes that form during antigen excess may be immunogenic,[92,93] whereas complexes formed when excess antibody is present may induce an antiidiotypic response, with immune suppression.[94,95] It would be of interest to determine the composition of immune complexes found in the serum or plasma of newly diagnosed, insulin-dependent diabetic patients; the objective would be the identification of antigen(s). Similar analyses have been reported in a number of autoimmune disorders (Table 5.2). A variety of methods have been used to isolate immune complexes and to study their antigens, and specific components have been described in rheumatoid arthritis,[38,39,96,97] ITP,[98] PBC,[99,100] CAH,[101] MCTD,[102] IgA nephropathy,[103] and ankylosing spondylitis.[40] Often these antigens are the same as or similar to those detected by immunoblotting (Table 5.1).

Ghalambor et al.[104] characterized immune complexes in insulin-dependent diabetes by a series of precipitations following separation by column chroma-

TABLE 5.2. Analyses of antigen in immune complexes associated with autoimmune diseases.

Disease[a]	Antigen in immune complexes	References
RA	Haganutzi-Deicher Ag	96
	19S IgG RF, 7S IgM RF	97
	48-K protein	38
	26-,39-,40-,43-K proteins	39
ITP	Platelet Ag	98
PBC	Mitochondrial Ag	99
	Ro Ag (scRNP)	100
CAH	Histones	101
	UIRNP	
MCTD	nRNP	102
IgA-Nephropathy	BSA	103
Ankylosing spondylitis	70-,19-,14-K proteins	40

[a] For a key to the abbreviations, see the legend for Table 5.1.

tography. The antibodies in the purified immune complexes were found to react primarily with antigens in islet cells and islet-cell subcellular fractions. Iodination of the immune complexes, followed by dissociation of the antigen, and a subsequent test of the reactivity of the iodinated components showed binding to islet-specific antibodies. This binding could not be disrupted by insulin, which suggests that immune complexes isolated from the plasma of diabetic patients contain distinct islet-specific antigens. Further studies, including the analysis of isolated immune complexes by Western blotting, will be necessary to determine whether circulating immune complexes contain an antigen of pancreatic, beta-cell origin. In preliminary studies (Herold et al., submitted), we have determined that a M_r 64000 protein is detected in diabetic immune complexes immunoblotted with diabetic but not with control sera.

Summary

The specific loss of insulin-producing cells in insulin-dependent diabetes might be explained by the presence of cytotoxic T-lymphocytes or specific beta-cell autoantibodies or both. An islet-cell– and, possibly, a beta-cell–specific M_r 64000 molecule has been identified by immune precipitation with diabetic sera of detergent-solubilized islet-cell proteins. Thus M_r 64000 autoantibodies are seen at the time of clinical diagnosis, although they may appear several years earlier. The M_r 64000 molecule may, therefore, be a marker for beta-cell autoimmunity and may be used to predict a later onset of insulin-dependent diabetes. The M_r 64000 antigen may be a membrane molecule, but whether it is an important receptor molecule for the specific properties of the beta-cell, such as glucose-stimulated insulin biosynthesis, storage, or release, remains to be seen. A preliminary analysis of immune complexes reported to be present in the serum of diabetic patients suggests that a M_r 64000 component may be present in such immune complexes. If this M_r 64000 protein is shown to be identical to the islet component, it might indicate that the M_r 64000 protein appears in the circulation, perhaps after beta-cell destructive events have taken place. Other workers have reported that the islet-cell antibody reaction detected by indirect immunofluorescence on sections of frozen human pancreas may be decreased or abolished by preabsorbing the sera with glycolipid-enriched fractions of human pancreas. Although many patients with newly diagnosed, insulin-dependent diabetes would seem to have titers of ICA high enough to detect an isolated, specific, islet-cell antigen, a lack of islet-cell antigen preparations has hampered progress toward the isolation and characterization of disease-specific, islet-cell, autoantigen(s).

Acknowledgment. We thank our colleagues and students for their contributions to the studies on islet-cell antigens in our laboratory and Augusta Litwer for help in preparing this manuscript. The studies from the authors' laboratories were supported in part by the National Institutes of Health and the Juvenile Diabetes Foundation International.

References

1. Bottazzo GF, Todd I, Mirakian R, et al. Organ-specific autoimmunity: A 1986 overview. *Immunological Reviews 1986, 94*. Munksgaard, Copenhagen: 1986, 137–169.
2. Eisenbarth GS. Type I diabetes mellitus. A chronic autoimmune disease. *N Engl J Med* 1986;314:1360–1368.
3. Lernmark Å, Li S, Baekkeskov S, et al. Islet-specific immune mechanisms. *Diabetes/Metab Reviews* 1987;3:959–980.
4. Baekkeskov S, Christie M, Lernmark Å. Islet cell antibodies, in McGregor A (ed): *Immunology of Endocrine Diseases*, Lancaster: MTP Press 1986, 78–88.
5. Lernmark Å. Islet cell antibodies. *Diab Med* 1987;4:285–292.
6. Drell DW, Notkins AL. Multiple immunological abnormalities in patients with type 1 (insulin-dependent) diabetes mellitus. *Diabetologia* 1987;30:132–143.
7. Bottazzo GF, Florin-Christensen A, Doniach D. Islet cell antibodies in diabetes mellitus with autoimmune polyendocrine deficiencies. *Lancet* 1974;ii:1279–1283.
8. MacCuish AC, Jordan J, Campbell CJ, et al. Antibodies to islet-cell in insulin-dependent diabetics with coexistent autoimmune disease. *Lancet* 1974;ii:1529–1533.
9. Lernmark Å, Freedman ZR, Hofmann C, et al. Islet cell surface antibodies in juvenile diabetes mellitus. *N Engl J Med* 1978;299:375–380.
10. Dobersen MJ, Scharff JE, Ginsberg-Fellner F, et al. Cytotoxic autoantibodies to beta-cells in the serum of patients with insulin-dependent diabetes mellitus. *N Engl J Med* 1980;303:1493–1498.

11. MacLaren NK, Huang SW, Fogh J. Antibody to cultured human insulinoma cells in insulin-dependent diabetes. *Lancet* 1975;i:997–1000.

12. Eisenbarth GS, Morris MA, Scearce RM. Cytotoxic antibodies to cloned rat islet cells in serum of patients with diabetes mellitus. *J Clin Invest* 1981;67:403–408.

13. Nayak RC, Mak O, Rabizadeh A, et al. Cytoplasmic islet cell antibodies: Evidence that the target antigen is a sialoglycoconjugate. *Diabetes* 1985;34:617–619.

14. Colman PG, Nayak RC, Connelly J, et al. Islet cell antibodies: Clinical utility and target antigen(s), in Jaworski MA, Molnar GD, Rajoste RV, et al. (eds): *The Immunology of Diabetes Mellitus*. Amsterdam: Elsevier, 1986, 345–350.

15. Dib SA, Colman PG, Dotta F, et al. Expression of cytoplasmic islet cell antigens by rat pancreas. *Diabetes*. 1987;36:982–985.

16. Jacob L, Lety MA, Bach JF, et al. Human systemic lupus erythematosus sera contain antibodies against cell-surface protein(s) that share(s) epitope(s) with DNA. *Proc Natl Acad Sci USA* 1986;83:6970–6974.

17. Hardin JA, Thomas JO. Antibodies to histones in systemic lupus erythematosus: Localization of prominent autoantigens on histones H1 and H2B. *Proc Natl Acad Sci USA* 1983;80:7410–7414.

18. Portanova JP, Arndt RE, Tan EM, et al. Antihistone antibodies in idiopatic and drug-induced lupus recognized distinct intrahistone regions. *J Immunol* 1987;138:446–451.

19. Lerner MR, Steitz JA. Antibodies to small nuclear RNAs complexed with proteins are produced by patients with systemic lupus erythematosus. *Proc Natl Acad Sci USA* 1979;76:5495–5499.

20. Tsay GJ, Chan EK, Peebles CL, et al. An immunoassay differentiating sera with antibodies to Sm alone, antibodies to Sm/RNP complex, and antibodies to RNP alone. *Arthritis Rheum* 1987;30:389–396.

21. Pattersson I, Hinterberger M, Mimori T, et al. The structure of mammalian small nuclear ribonucleoproteins, identification of multiple protein components reactive with anti-(Ul) ribonucleoprotein and anti-Sm autoantibodies. *J Biol Chem* 1984; 259:5907–5914.

22. Williams DG, Stocks MR, Charles PJ, et al. Antibodies to La, Jo-1, nRNP and Sm detected by multi-track immunoblotting using a novel filter holder: A comparative study with counterimmunoelectrophoresis and immunodiffusion using sera from patients with systemic lupus erythematosus and Sjoegren's syndrome. *J Immunol Meth* 1986;91:65–73.

23. Homma M, Mimori T, Takeda Y, et al. Autoantibodies to the Sm antigen: Immunological approach to clinical aspects of systemic lupus erythematosus. *J Rheumatol* 1987;14:188–193.

24. Billings PB, Hoch SO. Isolation of intact Sm/RNP antigens from rabbit thymus. *J Immunol* 1983;131:347–351.

25. Stanford DR, Rohleder A, Neiswanger K, et al. DNA sequence of a human Sm autoimmune antigen, the multigene family contains a processed pseudogene. *J Biol Chem*1987;262:9931–9941.

26. Habets WJ, Sillekens PT, Hoet MH, et al. Analysis of a cDNA clone expressing a human autoimmune antigen: full-length sequence of the U2 small nuclear RNA-associated B″ antigen. *Proc Natl Acad Sci USA* 1987;84:2421–2425.

27. Treadwell EL, Boak AM, Kovacs SA, et al. The autoimmune antigen Me is distinct and related to undifferentiated connective tissue disease. *Arthritis Rheum* 1987; 30:1239–1246.

28. Bachmann M, Schroeder HC, Wagner HG, et al. Purification and characterization of the Ro and La antigens. *Hoppe Seylers Physiol Chem* 1986;367:671–680.
29. Mamula MJ, Fox OF, Yamagata H, et al. The Ro/SS-A autoantigen as an immunogen. *J Exp Med* 1986;86:1889–1901.
30. McNeilage LJ, Whittingham S, Jack I, et al. Molecular analysis of the RNA and protein components recognized by anti-La (SS-B) autoantibodies. *Clin Exp Immunol* 1985;62:685–695.
31. Whittingham S, Naselli G, McNeilage LJ, et al. Serological diagnosis of primary Sjoegren's syndrome by means of human recombinant La (SS-B) as nuclear antigen. *Lancet* 1987;i:1–3.
32. Alderuccio F, Toh BH, Barnett AJ, et al. Identification and characterization of mitochondrial autoantigens in progressive systemic sclerosis: Identity with the 72,000 dalton autoantigen in primary biliary cirrhosis. *J Immunol* 1986;137:1855–1859.
33. Shero JH, Bordwell B, Rothfield NF, et al. High titers of autoantibodies to topoisomerase I (Scl-70) in sera from scleroderma patients. *Science* 1986;231:737–740.
34. Alderuccio F, Barnett AJ, Campbell JH, et al. Scl-95/100: Doublet of endothelial marker autoantigens in progressive systemic sclerosis. *Clin Exp Immunol* 1986;64: 94–100.
35. McNeillage LJ, Whittingham S, McHugh N, et al. A highly conserved 72,000 dalton centromeric antigen reactive with autoantibodies from patients with progressive systemic sclerosis. *J Immunol* 1986;137:2541–2547.
36. Reimer G, Rose KM, Scheer U, et al. Autoantibody to RNA polymerase I in scleroderma sera. *J Clin Invest* 1987;79:65–72.
37. Earnshaw WC, Rothfield N. Identification of a family of human centromere proteins using autoimmune sera from patients with scleroderma. *Chromosoma* 1985;91: 313–321.
38. Melsom RD, Smith PR, Maini RN. Demonstration of an unidentified 48 kD polypeptide in circulating immune complexes in rheumatoid arthritis. *Ann Rheum Dis* 1987;46:104–109.
39. Inman RD, Hamilton NC, Redecha PB, et al. Electrophoretic transfer blotting analysis of immune complexes in rheumatoid arthritis. *Clin Exp Immunol* 1986;63:32–40.
40. Rødahl E, Iversen OJ. Analysis of circulating immune complexes from patients with ankylosing spondylitis by gel electrophoresis and immunoblotting using antiserum against a psoriasis associated retrovirus-like particle. *Ann Rheum Dis* 1986;45: 892–898.
41. Earnshaw WC, Sullivan KF, Machlin PS, et al. Molecular cloning of cDNA for CENP-B, the major human centromere autoantigen. *J Cell Biol* 1987;104:817–829.
42. Williams CL, Lennon VA, Momoi MY, et al. Serum antibodies and monoclonal antibodies secreted by thymic B-cell clones from patients with myasthenia gravis define striational antigens. *Ann NY Acad Sci* 1987;505:168–179.
43. Mathews MB, Bernstein RM. Myositis autoantibody inhibits histidyl-tRNA synthetase: A model of autoimmunity. *Nature* 1983;304:177–179.
44. Walker EJ, Jeffrey PD. Purification of bovine liver histidyl-tRNA synthetase, the Jo-1 antigen of polymyositis: Size of the whole enzyme and its characteristic proteolytic fragments. *Hoppe Seylers Physiol Chem* 1987;368:531–537.
45. Islam MN, Briones-Urbina R, Bako G, et al. Both TSH and thyroid-stimulating antibody of Graves' disease bind to an Mr 197,000 holoreceptor. *Endocrinology* 1983;113:436–438.

46. Banga JP, Pryce G, Hammond L, et al. Structural features of the autoantigens involved in thyroid autoimmune disease: The thyroid microsomal/microvillar antigen. *Mol Immunol* 1985;22:629–642.
47. Kotani T, Umeki K, Matsunaga S, et al. Detection of autoantibodies to thyroid peroxidase in autoimmune thyroid diseases by micro-ELISA and immunoblotting. *J Clin Endocrinol Metab* 1986;62:928–933.
48. Ohtaki S, Kotani T, Nakamura Y. Characterization of human thyroid peroxidase purified by monoclonal antibody-assisted chromatography. *J Clin Endocrinol Metab* 1986;63:570–576.
49. Hirayu H, Seto P, Magnusson RP, et al. Molecular cloning and partial characterization of a new autoimmune thyroid disease-related antigen. *J Clin Endocrinol Metab* 1987;64:578–584.
50. Seto P, Hirayu H, Magnusson RP, et al. Isolation of a complementary DNA clone for thyroid microsomal antigen, homology with the gene for thyroid peroxidase. *J Clin Invest* 1987;80:1205–1208.
51. Banga JP, Mirakian R, Hammond L, et al. Characterization of monoclonal antibodies directed toward the microsomal/microvillar thyroid autoantigen recognized by Hashimoto autoantibodies. *Clin Exp Immunol* 1986;64:544–554.
52. Weetman AP, Nutman TB, Burman KD, et al. Heterogeneity of thyroid autoantigens identified by immunoblotting. *Clin Immunol Immunopathol* 1987;43: 333–342.
53. Hamada N, Grimm C, Mori H, et al. Identification of a thyroid microsomal antigen by Western blot and immunoprecipitation. *J Clin Endocrinol Metab* 1985;61:120–128.
54. Devine DV, Rosse WF. Identification of platelet proteins that bind alloantibodies and autoantibodies. *Blood* 1984;64:1240–1245.
55. Beardsley DS, Spiegel JE, Jacobs MM, et al. Platelet membrane glycoprotein IIIa contains target antigens that bind anti-platelet antibodies in immune thrombocytopenias. *J Clin Invest* 1984;74:1702–1707.
56. Frazer IH, MacKay IR, Jordan TW, et al. Reactivity of anti-mitochondrial autoantibodies in primary biliary cirrhosis: Definition of two novel mitochondrial polypeptide autoantigens. *J Immunol* 1985;135:1739–1745.
57. Gershwin ME, MacKay IR, Sturgess A, et al. Identification and specificity of a cDNA encoding the 70 kD mitochondrial antigen recognized in primary biliary cirrhosis. *J Immunol* 1987;138:3525–3531.
58. Manns M, Gerken G, Kyriatsoulis A, et al. Two different subtypes of antimitochondrial antibodies are associated with primary biliary cirrhosis: Identification and characterization by radioimmunoassay and immunoblotting. *Hepatology* 1987;7: 893–899.
59. Alvarez F, Bernard O, Homberg JC, et al. Anti-liver-kidney microsome antibody recognizes a 50,000 molecular weight protein of the endoplasmic reticulum. *J Exp Med* 1985;161:1231–1236.
60. Pusey CD, Dash A, Kershaw MJ, et al. A single autoantigen in Goodpasture's syndrome identified by a monoclonal antibody to human glomerular basement membrane. *Lab Invest* 1987;56:23–31.
61. Pressey A, Pusey CD, Dash A, et al. Production of a monoclonal antibody to autoantigenic components of human glomerular basement membrane. *Clin Exp Immunol* 1983;54:178–184.

62. Cederholm B, Wieslander J, Bygren P, et al. Patients with IgA nephropathy have circulating anti-basement membrane antibodies reacting with structures common to collagen I, II, and IV. *Proc Natl Acad Sci USA* 1986;83:6151–6155.
63. Butkowski RJ, Langeveld JP, Wieslander J, et al. Localization of the Goodpasture epitope to a novel chain of basement membrane collagen. *J Biol Chem* 1987;262: 7874–7877.
64. Wieslander J, Barr JF, Butkowski RJ, et al. Goodpasture antigen of the glomerular basement membrane: Localization to noncollagenous regions of type IV collagen. *Proc Natl Acad Sci USA* 1984;81:3838–3842.
65. Fliger FD, Wieslander J, Brentjens JR, et al. Identification of a target antigen in human antitubular basement membrane nephritis. *Kidney Int* 1987;31:800–807.
66. Nomoto Y, Suga T, Miura M, et al. Characterization of an acidic nuclear protein recognized by autoantibodies in sera from patients with IgA nephropathy. *Clin Exp Immunol* 1986;65:513–519.
67. Das KM, Sakamaki S, Vecchi M, et al. The production and characterization of monoclonal antibodies to a human colonic antigen associated with ulcerative colitis: Cellular localization of the antigen by using the monoclonal antibody. *J Immunol* 1987;139:77–84.
68. Koulu L, Kusumi A, Steinberg MS, et al. Human autoantibodies against a desmosomal core protein in pemphigus foliaceus. *J Exp Med* 1984;160:1509–1518.
69. Woodley DT, O'Keefe EJ, Reese MJ, et al. Epidermolysis bullosa acquisita antigen, a new major component of cutaneous basement membrane, is a glycoprotein with collagenous domains. *J Invest Dermatol* 1986;86:668–672.
70. Pfister C, Chabre M, Plouet J, et al. Retinal S antigen identified as the 48k protein regulating light-dependent phosphodiesterase in rods. *Science* 1985;228:891–893.
71. Barkas T, Mauron A, Roth B, et al. Mapping the main immunogenic region and toxin-binding site of the nicotinic acetylcholine receptor. *Science* 1987;235:77–80.
72. Nerup J, Lernmark Å. Autoimmunity in diabetes mellitus. *Am J Med* 1981;70: 135–141.
73. Baekkeskov S, Nielsen JH, Marner B, et al. Autoantibodies in newly diagnosed diabetic children immunoprecipitate specific human pancreatic islet cell protein. *Nature* 1982;198:167–169.
74. Baekkeskov S, Landin M, Kristensen JK, et al. Antibodies to a 64000 M_r human islet cell antigen precede the clinical onset of insulin-dependent diabetes. *J Clin Invest* 1987;79:926–934.
75. Baekkeskov S. Radiolabeling and immune precipitation of islet cell antigens, in Larner J, Pohl SL, (eds): *Methods in Diabetes Research, Vol I.* New York: Wiley, 1985, 245–258.
76. Hagglof B, Rabinovitch A, MacKay P, et al. Islet cell and other organ-specific autoantibodies in healthy first-degree relatives to insulin-dependent. *Acta Faediatr Scand* 1986;75:611–618.
77. Palmer JP, Asplin CM, Clemons P, et al. Insulin antibodies in insulin-dependent diabetics before insulin treatment. *Science* 1983;222:1337–1339.
78. Srikanta S, Ricker AT, McCulloch, et al. Autoimmunity to insulin, beta cell dysfunction, and development of insulin-dependent diabetes mellitus. *Diabetes* 1986;35: 139–142.
79. Towbin H, Staehelin T, Gordon J. Electrophoretic transfer of proteins from polyacrylamide gels to nitrocellulose sheets: Procedure and some applications. *Proc Natl Acad Sci USA* 1979;7:4350–4354.

80. Marner B, Lernmark Å, Ludvigsson J, et al. Islet cell antibodies in insulin-dependent (type 1) diabetic children treated with plasmapheresis. *Diabetes Res* 1985;2: 231–236.
81. Christie MR, Landin-Olsson M, Sundkvist G, et al. Antibodies to a M_r 64000 islet cell protein in Swedish children with newly diagnosed type 1 (insulin-dependent) diabetes. *Diabetologia* 1988;31:597–602.
82. Gregor I, Iberg N, Berger W, et al. Albumin-directed antibodies in diabetes: Demonstration of human serum albumin-directed IgM autoantibodies. *Diabetologia* 1986; 29:481–484.
83. Colman PG, Campbell IL, Kay TW, et al. 64000-M_r autoantigen in type 1 diabetes. *Diabetes* 1987;36:1432–1440.
84. DiMario U, Dotta F, Crisa L, et al. Circulating antiimmunoglobulin antibodies in recent-onset type 1 diabetic patients. *Diabetes* 1988;37:462–466.
85. Borsey DQ, DiMario U, Irvine WJ, et al. Humoral immunity in type 1 diabetes mellitus: A prospective study. *J Clin Lab Immunol* 1983;11:9–15.
86. Irvine WJ, DiMario U, Guy K, et al. Immune complexes in newly diagnosed insulin-dependent (type 1) diabetics. *J Clin Lab Immunol* 1978;1:183–186.
87. Contreas G, Lernmark Å, Mathiesen ER, et al. Immune complexes in insulin-dependent diabetes. *Biomed Biochim Acta* 1985;44:129–132.
88. Weigle WO. Cyclical production of antibody as a regulatory mechanism in the immune response. *Adv Immunol* 1975;21:87–111.
89. Bloch-Shtacher N, Hischosn K, Uhr JW. The response of lymphocytes from non-immune humans to antigen-antibody complexes. *Clin Exp Immunol* 1968;3:889–899.
90. Soderberg LSF, Coons AH. Complement-dependent stimulation of normal lymphocytes by immune complexes. *J Immunol* 1978;120:806–811.
91. Kaattari S, Scibienski RJ, Benhamini E. The immunoregulatory role of antigen-antibody complexes. *Immunology* 1980;40:9–16.
92. Terres G, Morrison SL, Habich GS, et al. Appearance of an early primed state in mice following the concomitant injections of antigen and specific antiserum. *J Immunol* 1971;108:1473–1477.
93. Caulfield MJ, Luce KJ, Shaffer D, et al. The antibody response to specific immune complexes is under genetic control and correlates with the expression of a recurrent idiotype. *J Exp Med* 1986;163:75–86.
94. Klaus GGB. Antigen-antibody complexes elicit anti-idiotypic antibodies to self-idiotopes. *Nature* 1978;272:265–266.
95. Caulfield MJ, Shaffer D. Immunoregulation by antigen/antibody complexes. *J Immunol* 1987;138:3680–3683.
96. Nishimaki T, Kano K, Milgrom F. Studies on immune complexes in rheumatoid arthritis. *Arthritis Rheum* 1978;21:639–644.
97. Moore TL, Dorner RW, Zuckner J. 19S IgM rheumatoid factor -7S lgG rheumatoid factor immune complexes isolated in patients with rheumatoid arthritis. *J Lab Clin Med* 1986;107:465–470.
98. Kurata Y, Hayashi S, Aochi H, et al. Analysis of antigen involved in circulating immune complexes in patients with idiopathic thrombocytopenic purpura. *Clin Exp Immunol* 1987;67:293–299.
99. Penner E, Goldenberg H, Albini B, et al. Immune complexes in primary biliary cirrhosis contain mitochondrial antigens. *Clin Immunol Immunopathol* 1982;22: 394–399.

100. Penner E. Demonstration of immune complexes containing the ribonucleoprotein antigen Ro in primary biliary cirrhosis. *Gastroenterology* 1986;90:724–727.
101. Penner E. Nature of immune complexes in autoimmune chronic active hepatitis. *Gastroenterology* 1987;92:304–308.
102. Negoro N, Kanayama Y, Takeda T, et al. A solid-phase radioimmunoassay for the detection of nRNP immune complexes. *J Immunol Meth* 1986;91:83–89.
103. Yap HK, Sakai RS, Woo KT, et al. Detection of bovine serum albumin in the circulating IgA immune complexes of patients with IgA nephropathy. *Clin Immunol Immunopathol* 1987;43:395–402.
104. Ghalambar AM, Pek SB, Schwartz SA, et al. Characterization of immune complexes from plasma of patients with insulin-dependent diabetes mellitus, in Larner J, Pohl SL (eds): *Methods in Diabetes Research Vol. I*, part C. New York, Wiley, 1985, 139–158.

6
The Humoral Anti-Islet Immune Response: Immunochemical Studies of Glycoconjugate Antigens

R.C. Nayak and G.S. Eisenbarth

Introduction

In 1974, two reports established the existence of islet-cell autoantibodies in patients having type I diabetes mellitus.[1,2] These circulating anti-islet–cell autoantibodies were detected by indirect immunofluorescence on a substrate of unfixed, fresh frozen, human pancreas sections. Since this initial demonstration, it has been shown that type I diabetes has a prodromal phase during which islet-cell autoantibodies can be demonstrated in the circulation[3-6] and that islet-cell antibodies are an important predictive hallmark of type I diabetes mellitus.

Islet-cell antibodies have been found to be only of the IgG class,[7-9] and preponderantly of the IgG1 subclass. Antibodies detected by binding to frozen sections have been commonly referred to as "cytoplasmic" islet-cell antibodies since it was assumed that the target antigens are located in the cytoplasm of the sectioned cells. This is an unsubstantiated assumption, since the resolution of fluorescence microscopy on frozen sections is not sufficient to distinguish cell surface reactivity from intracellular reactivity in the presence of antibodies that bind to both subcellular compartments independently. Consequently the definition of cytoplasmic ICA is ambiguous and reference to "cytoplasmic" ICA is probably best considered a convention rather than an indication of the cellular location of autoantigen(s).

Antigen(s) bound by islet-cell autoantibodies in the frozen section assay are largely uncharacterized, but we have begun to define biochemical and immunochemical features of autoantigens and of monoclonal antibody (mAb)-defined antigens of islets in frozen sections.[10-12] We will not discuss the immunochemistry of insulin as an autoantigen (see Chapter 8, this volume) but we will focus primarily on the role of glycolipid antigens (principally gangliosides) in type I diabetes mellitus.

What Are Gangliosides?

Gangliosides, a family of acidic glycosphingolipids (glycolipids), are characterized by the presence of sialic acid. They are unusual molecules insofar as they have both hydrophobic and hydrophilic domains (i.e., they are amphipathic) and they bear a strong negative charge (see Fig. 6.1). Gangliosides are primarily membrane components that are highly enriched in plasma membranes. The carbohydrate portion of the molecule consists of sialic acid, hexoses, and N-acetylated hexosamines. The hydrophobic moiety, known as ceramide, consists of long-chain fatty acids linked through an amide bond to the nitrogen atom on carbon-2 of the amino alcohol sphingosine. Oligosaccharides are linked through a glycosidic bond to carbon-1 of the sphingosine portion of the ceramide. Gangliosides were first identified in the brain more than half a century ago and have since been found to exist in all tissues although they vary in amount and pattern of expression (see Refs. 13–15 for a more extensive review of ganglioside structure and function).

The International Union of Pure and Applied Chemistry (IUPAC) has recently proposed a systematic nomenclature for glycolipids.[16] It is still common, however, to find the nomenclature of Svennerholm[17] used in the current literature. For example GM3 indicates G for the ganglio- series of glycolipids, M for monosialylated, and 3 for the order of migration on thin layer chromatography (TLC) plates under the conditions of Svennerholm's initial description.[17] Hence GM3, GM2, and GM1 are all monosialogangliosides. Similarly GD2, GD1a, and GD1b are all disialogangliosides, the GTs are all trisialogangliosides, the GQs are all tetrasialogangliosides, and so forth.

Immunochemical Characterization of Monoclonal Antibody-Defined, Islet-Cell Antigens

To determine the nature of islet-cell antigens, which are detected in situ by antibody binding to frozen sections, we considered conditions that might biochemically perturb antigen structure in tissue sections.[10,11] These treatments included periodate oxidation, Pronase® digestion, and immersion in acetone or 2:1 chloroform:methanol (C:M). Treated sections and control sections were then screened for reactivity of mAbs to islet cells.[10-12] The results of these procedures are summarized in Table 6.1, which shows that the antigens recognized by these mAbs can be divided into three categories according to their immunochemical properties. Thus mAbs HISL-1, 5, 9, 14, 4F2, and LC7/2 comprise one category insofar as their antigens are insensitive to periodate oxidation and C:M treatment; they are, however, sensitive to Pronase® digestion. These observations suggest that the epitope bound by these mAbs is borne on a protein (Pronase®-sensitive/C:M insensitive) and that the epitope is not on carbohydrate chains of glycoproteins (periodate resistant). The mAbs A2B5 and 3G5 recognize a second category of molecules that are periodate and C:M sensitive but insensitive to Pronase® digestion, which is consistent with these antigens being sialoglycolipids (i.e., gangliosides).

A third class of antigen, defined by mAb HISL-19, is sensitive to all three treatments applied to acetone-fixed, frozen sections. These properties are consistent with HISL 19 being a glycoprotein.

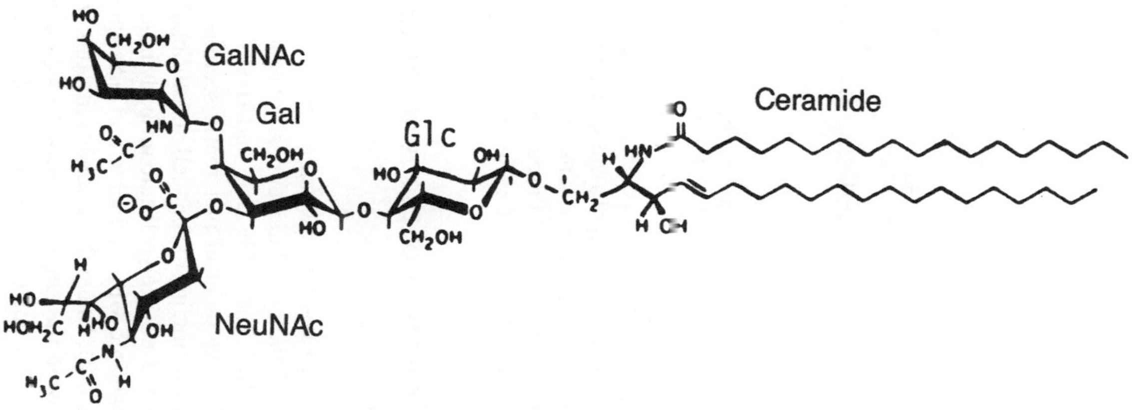

FIGURE 6.1. Structure of ganglioside GM2. Glc, glucose; Gal, galactose; GalNAc, N-acetyl-galactosamine; NeuNAc, N-acetylneuraminic acid.

TABLE 6.1. Preliminary biochemical characterization of anti-islet-cell monoclonal antibodies.[a,b]

mAb	PBS	$NaIO_4$	Acetone	(2.1) C:M	Pronase
HISL 1	+	+	+	+	↓ ↓
HISL 5	+	+	+	+	↓ ↓
HISL 8	+	−	+	−	ND
HISL 9	+	+	+	+	↓ ↓
HISL 14	+	+	+	+	↓ ↓
HISL 19	+	−	+	−	↓ ↓
4F2	+	+	+	+	↓
LC7/2	+	+	+	+	↓
3G5	+	−	+	−	+
A2B5	+	−	+	−	+

[a] From Nayak, Colman and Eisenbarth.[11]
[b] +, Positive binding; −, negative binding; ↓, reduced binding; ↓ ↓, greatly reduced binding; ND, not determined.

The treatment described above might be a rapid way of broadly identifying these antigens, which were previously defined only by their binding of mAbs. Furthermore these studies of anti-islet-cell mAbs have revealed a possible interesting correlation between periodate sensitivity and extractability in C:M (Table 6.1). All periodate-sensitive antigens were sensitive to C:M treatment, and periodate-insensitive antigens were insensitive to C:M treatment. These observations appear to be independent of Pronase® sensitivity which suggests that these antigens may be purified from C:M extracts.

Immunochemical Properties of the Autoantigen(s) of Pancreatic Islets

We used similar techniques to evaluate the binding of circulating autoantibodies from patients with type I diabetes to biochemically perturbed antigens in pancreatic sections (Table 6.2).[10] The mAbs 3G5 and HISL-14 were used as antibody-binding controls. Pretreatment of pancreatic sections with a solution of sodium periodate ablated the binding of islet-cell autoantibodies from the sera of three ICA-positive patients. Anti-islet–cell autoantibody binding was restored on sections that had been oxidized with periodate and subsequently reduced with sodium borohydride (Fig. 6.2). Under these conditions, this observation suggests that sialic acid residues were the targets of anti-islet–cell autoantibodies, since periodate oxidation cleaves C7 and C8 and/or C8 and C9 of the vicinal diol containing extracyclic chain of sialic acid residues (Fig. 6.3). Mild oxidation with periodate produces a sialic acid derivative (heptulosonic or octulosonic acid) with the C7/C8 alcohol functional group oxidized to an aldehyde. This extracyclic aldehyde clearly is involved in the ablation of antibody binding, since borohydride reduction, which converts the aldehyde back to a primary alcohol,

TABLE 6.2. Effects of various biochemical treatments of sections on antibody binding to human islet cells.[a,b]

	Patient			mAb 3G5 anti-ganglioside	mAb HISL-14 monoclonal anti-protein	Control serum	mAb P3X63 control
	1	2	3				
No treatment	+	+	+	+	+	−	−
Periodate	−	−	−	−	+	−	−
Periodate/borohydride	+	+	+	+	ND	−	−
Neuraminidase	−	−	−	−	+	−	−
Pronase®	+	+	+	+	+/−	−	−
Chloroform:methanol	−	−	−	−	+	−	−
Methanol	−	−	−	−	+	−	−

[a] From Nayak et al.[10] Reproduced with permission from the American Diabetes Association, Inc.
[b] +, positive binding; −, negative binding; +/−, marginal binding; ND, not determined.

restores antibody binding. These results further suggest that the C8/C9 carbon atoms are not essential to the structure of the "auto-epitope." An additional 17 ICA-positive sera were tested in this system, and their binding was found to be ablated by periodate oxidation of the pancreatic sections. Similarly binding of anti-islet cell mAb 3G5 could be ablated by periodate oxidation of pancreatic sections and restored by borohydride reduction (Table 6.2). The HISL-14 antigen, however, was unaffected by periodate oxidation, which indicates that sialic acid is not a crucial part of the HISL-14 antibody-binding site. Sialic acid involvement in the antibody-binding site(s) of islet-cell autoantibodies and mAb 3G5 was confirmed by loss of antibody binding to pancreatic sections after neuraminidase digestion (Table 6.2; Fig. 6.4).

The nature of the glycoconjugate bearing the immunoreactive sialic acid residue(s) on the autoantigen was investigated by immersing pancreatic sections in organic solvents in an attempt to remove the antigen(s). Both methanol and a 2:1 mixture of chloroform:methanol ablated the binding of ICA and mAb 3G5 to pancreatic sections but had no effect on HISL-14 binding, which suggests that the sialic acid residue is carried by a lipid rather than a protein core. Further evidence that the ICA and 3G5 determinants are attached to lipid is found in the observation that both 3G5 and the ICA antigen(s) are insensitive to the action of the protease Pronase® (Table 6.2). However HISL-14 binding is ablated by Pronase® digestion of pancreatic sections, which indicates that the antigen is a protein or linked to a protein.

The above evidence indicates that the autoantigen(s) detected in frozen sections of pancreas has the immunochemical properties of a sialoglycolipid.

A lectin that binds to sialic acid has recently been isolated from the hemolymph of the horseshoe crab *Limulus polyphemus*. This lectin binds to sialic acids of both glycoproteins and glycolipids and has been used to investigate the nature of the sialoglycoconjugate in frozen sections of rat pancreas.[18] It was shown that acinar and endocrine islet cells expressed *L. polyphemus* agglutinin (LPA) receptors. Intense linear staining of the apical surface of acinar cells, with fainter

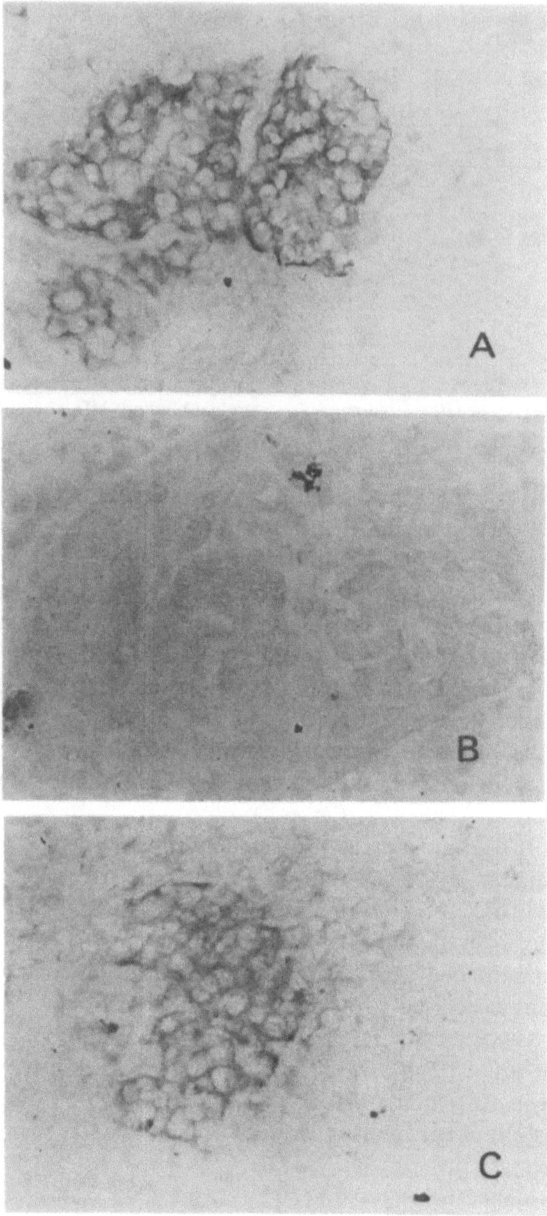

FIGURE 6.2. Indirect immunoperoxidase staining of human pancreas with islet-cell, antibody-positive serum. (**A**) Antibody binding to sections after incubation in control buffer; (**B**) absence of antibody binding to sections incubated in 0.05 M periodate solution–the putative islet is shown lightly stained with hematoxylin; (**C**) restoration of antibody binding to sections on subsequent reduction of periodate-treated sections by incubation in a 1 mg/ml solution of sodium borohydride. From Nayak et al.[10] Reproduced with permission from the American Diabetes Association, Inc.

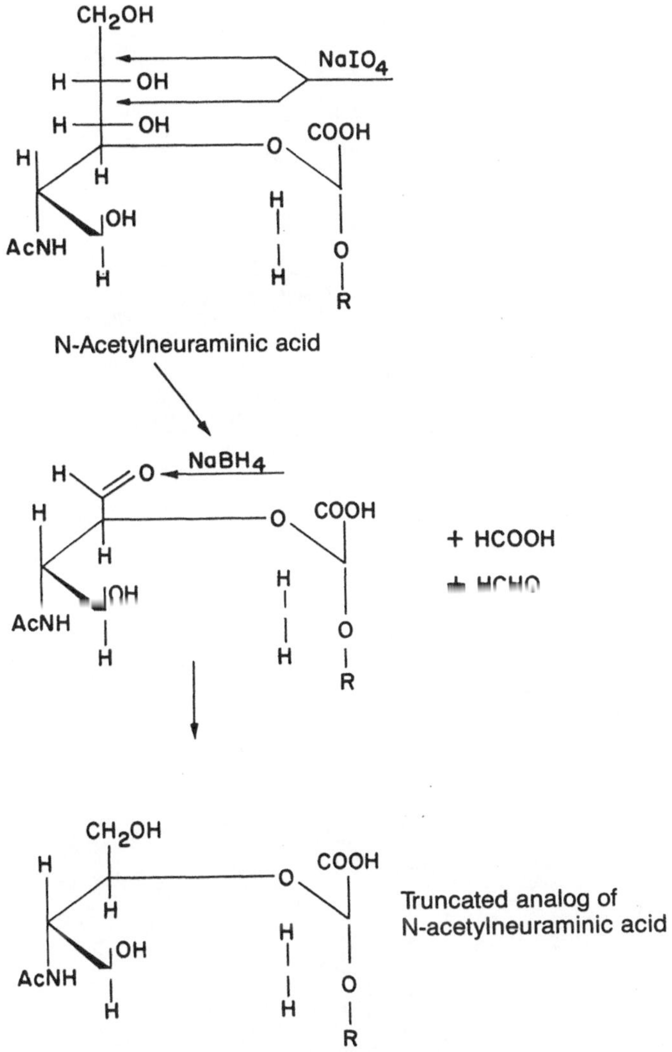

FIGURE 6.3. A reaction scheme for periodate oxidation and subsequent borohydride reduction of N-acetylneuraminic acid. Mild oxidation with sodium M-periodate has specificity for the vicinal diol-containing side chain of sialic acids. Such oxidation produces a heptulosonic acid with an aldehyde functional group at C7 and formic acid and formaldehyde. Reduction with sodium borohydride converts the C7 aldehyde to a primary alcohol, to yield a truncated analog of N-acetylneuraminic acid. From Nayak, Colman, and Eisenbarth.[11]

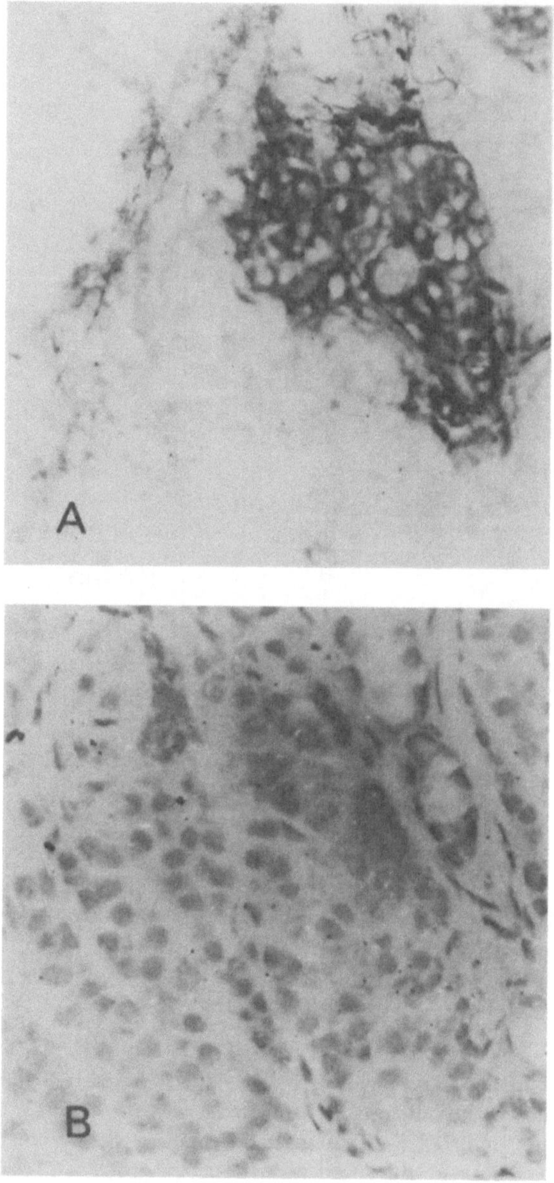

FIGURE 6.4. Indirect immunoperoxidase staining of human pancreas with islet-cell, antibody-positive serum. (**A**) Antibody binding to sections incubated in buffer lacking neuraminidase; (**B**) absence of antibody binding to sections incubated with neuraminidase: the putative islet is shown lightly stained with hematoxylin. From Nayak, Colman, and Eisenbarth.[11]

staining on the basal but not the lateral cell surfaces, was demonstrable. This LPA staining was specific for sialic acid in that (1) pretreatment of pancreas sections with neuraminidase or periodate, (2) competition with sialic acid, and (3) incubation with calcium-free buffers abolished LPA binding to the section.

Pretreatment of sections with proteases abolished LPA binding to the apical surface of acinar cells and enhanced LPA binding to the lateral cell surface. Extensive exposure to alcohols and chloroform/methanol mixtures following protease treatment markedly reduced LPA binding to the acinar cell periphery. These results suggest that LPA-binding sites on the acinar cell apical surface may be predominantly sialoglycoproteins, whereas the sites on the basolateral surface may be predominantly gangliosides. Acinar cell surface-LPA binding was not affected by digestion of tissue sections with hyaluronidase, heparinase, or collagenase, nor was it affected by exposure to 6 M guanidine hydrocholoride.

The islets of Langerhans in this study were always uniformly and heavily stained with LPA conjugates. The LPA binding to islets, however, was insensitive to proteases. This insensitivity was also observed with ICA binding to islets and suggests that the sialoglycoconjugates of islets may be predominantly gangliosides. The effect of lipid extraction or collagenase on islets was not determined in this study.

Extraction and Partial Purification of the ICA Glycolipid Antigen

The ability to ablate ICA binding to pancreatic sections by treatment with organic solvents suggests that the antigen(s) might be solubilized in these solvents, as would be expected of glycolipids. Consequently we attempted to solubilize ICA antigen(s) selectively by extracting frozen pancreas with organic solvents,[19] following the method of Svennerholm and Fredman.[20] This extraction procedure is summarized in Figure 6.5. Briefly frozen pancreas tissue was homogenized in three volumes of distilled water (w/v) using a Polytron Homogenizer. Eight volumes of methanol and four volumes of chloroform were added to the homogenate, which was extracted in this mixture for 30 minutes with stirring at ambient temperature (*ca* 20°C). The suspension was then centrifuged to separate particulate material from the supernatant. The pellet was re-extracted and the two supernatant extracts were combined. Water was added to the combined supernatant extracts to give a final chloroform:methanol:water ratio of 1:2:1.4; the mixture was mixed thoroughly and again centrifuged to effect a classic, two-phase "Folch" partition. The upper (methanol/water) phase contained the bulk of the polar glycolipids; the lower (chloroform) phase contained predominantly neutral lipids.

The fractions obtained by the above procedure were assessed for their ability to block the binding of ICA to pancreatic sections using the fluorescent protein A method; ICA binding was evaluated by quantitative photometry. This blocking assay demonstrated that the post-extraction residue (particulate pellet) and the lower-phase lipids could not block ICA binding to pancreatic sections, whereas

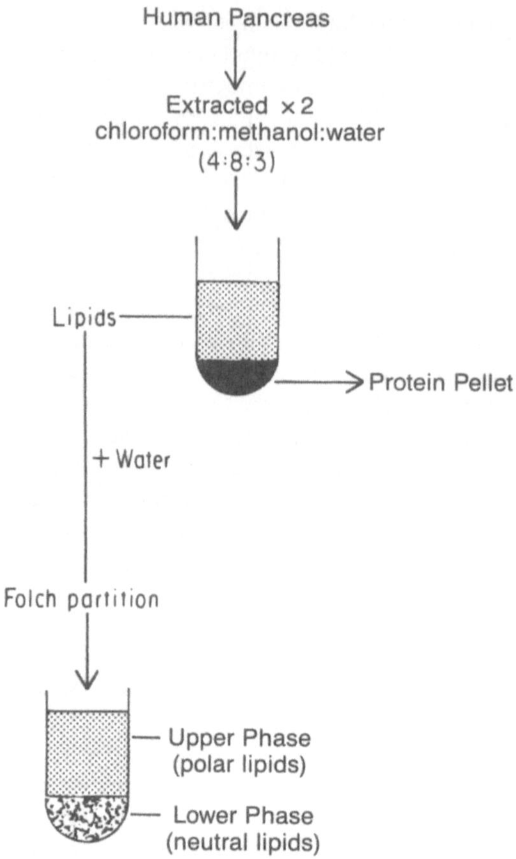

FIGURE 6.5. Preparation of the crude "upper-phase" glycolipid fraction of human pancreas.

the upper-phase glycolipids could block ICA binding in a dose-dependent manner,[11,19] as shown in Figure 6.6. A 50% blocking effect was observed with an amount of glycolipid derived from approximately 0.5 g of pancreas tissue; this represents approximately 7 to 8 μg glycoconjugate-bound sialic acid in the upper-phase glycolipid fraction.

The upper-phase extract was fractionated further on C18 Sep-Pak cartridges. Hydrophobic molecules bind to this 18-carbon chain matrix, and hydrophilic molecules can be washed out with aqueous solvents. The hydrophobic molecules are recovered by eluting with chloroform and methanol. The C18-bound material was found to contain the ICA blocking activity. The Sep-Pak–eluted glycolipids were then separated by preparative TLC. The chromatography plate was divided into fractions and the silica scraped off. Separated glycolipids were eluted with chloroform/methanol and tested for blocking ability. A fraction containing

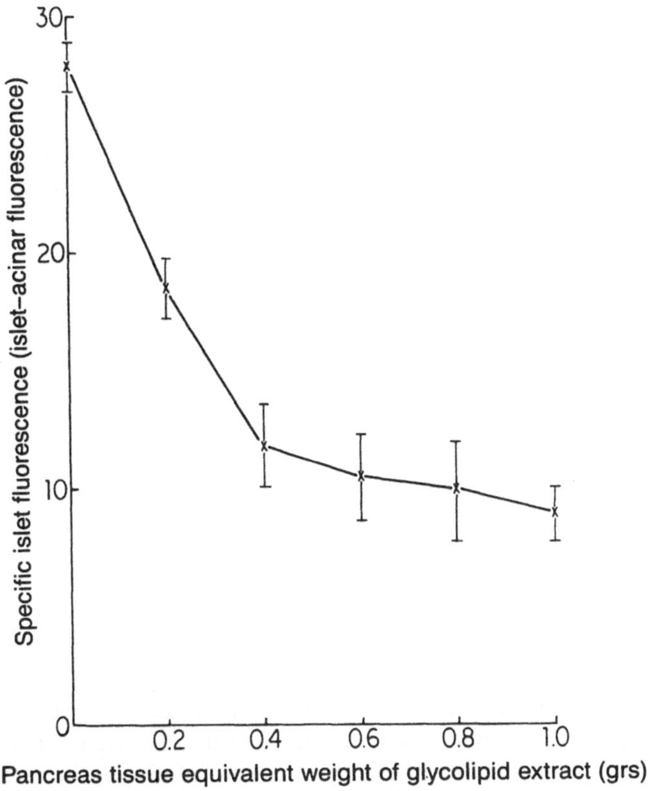

FIGURE 6.6. Absorption of islet-cell antibody binding to human pancreas sections with a crude human pancreatic glycolipid extract. Equal aliquots of serum from an islet-cell, antibody-positive patient were incubated with amounts of glycolipid extract equivalent to 0.2, 0.4, 0.6, 0.8, and 1.0 g of human pancreas dried down into polypropylene tubes. The incubation was performed for 24 hours at 4°C on a rocking table. Sera were then collected and assayed in a conventional, islet-cell, "cytoplasmic" antibody assay on sections of mouse pancreas. The fluorescence intensity was measured on a microfluorimeter; the specific islet fluorescence was calculated by subtraction of acinar fluorescence. From Nayak, Colman, and Eisenbarth.[11]

monosialogangliosides was found to contain the major ICA blocking activity. A similarly derived fraction from human liver could not block ICA binding to frozen sections. Consequently the glycolipid autoantigen appears to be a pancreatic monosialoganglioside.

Attempts were made to stain TLC-separated pancreatic glycolipids directly with ICA$^+$ sera, using immunoperoxidase to detect them indirectly. No qualitative differences could be found in the bands detected with normal or ICA$^+$ sera (Fig. 6.7). It is apparent from Figure 6.7 that antibodies to a number of glycolipids are present in both ICA$^+$ and ICA$^-$ sera. It is therefore possible that

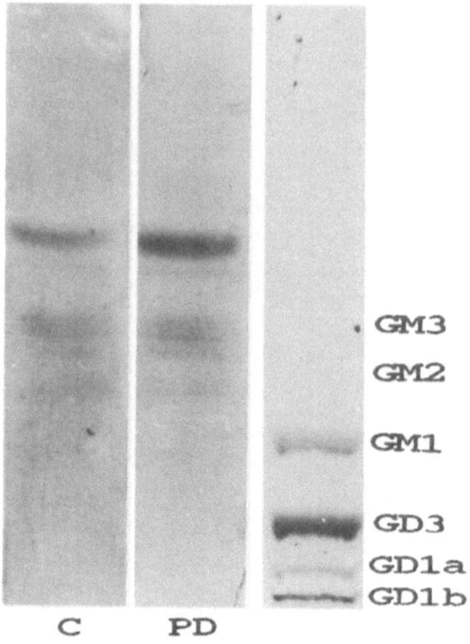

FIGURE 6.7. Immunoperoxidase staining with ICA⁺ serum on pancreatic glycolipids separated by thin layer chromatography. C, control ICA⁻ serum; PD, ICA⁺ prediabetic serum. GM3, GM2, GM1, GD3, GD1a, and GD1b were used as ganglioside markers and are shown to the right of PD.

"natural" antiglycolipid antibodies may obscure visualization of the islet ganglioside autoantigen. This potential interference with the direct detection of antiganglioside ICA may be obviated by purifying either the ICA or the autoantigen to homogeneity. Purification of the autoantibody is currently not feasible; although it is known that ICA are predominantly of the IgG1 isotype, so are 70% of the immunoglobulins in normal sera. No other known features of ICA (other than islet reactivity) can be used to enrich preparations selectively with this autoantibody. Purification of the autoantibody therefore depends on purification of the autoantigen, which may be possible with affinity purification techniques.

To facilitate purification of the ganglioside autoantigen, we used a high-performance liquid chromatography (HPLC) system to separate and analyze glycolipids in extracts of human pancreas.[21] The separation of mixtures of ganglioside markers was found to be reproducible and the peaks (detected by UV absorbance) corresponded to the appropriate ganglioside standard when analyzed by TLC (Fig. 6.8). Studies on extracts of human pancreas and human islets indicated that islet-specific UV peaks could be detected. The major gangliosides detected in glycolipid extracts of whole human pancreas were GM3 and GD3, and to a lesser extent, GD1a. In contrast isolated human islets were found to predominantly express GM3, an acidic glycolipid comigrating with GM2, and a ganglioside with mobility between GM1 and GM2 by both HPLC and TLC.

Attempts to recover ICA-blocking activity in HPLC fractions thus far have been unsuccessful, although the antigen for mAb 3G5 was recoverable in an HPLC fraction containing the GD3 peak (Fig. 6.9).

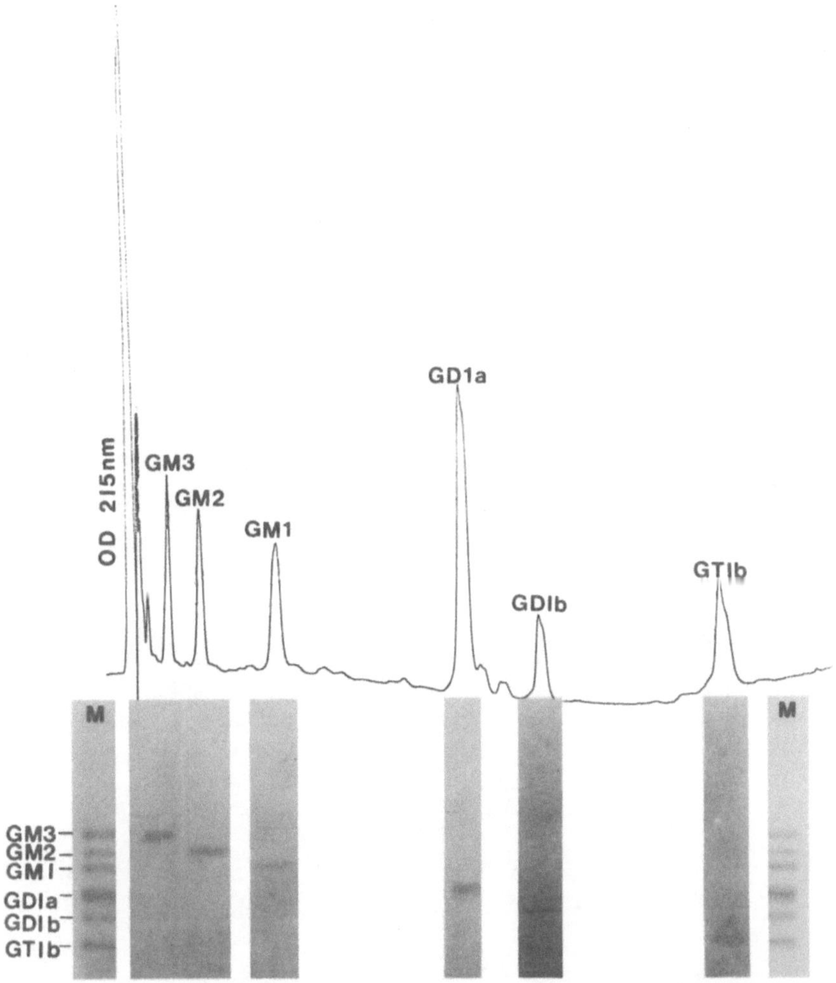

FIGURE 6.8. Separation of mixed ganglioside by high performance liquid chromatography. The peaks of the purified, mixed, bovine brain gangliosides were detected by UV absorbance (upper panel); each collected peak contained the appropriate ganglioside as established by subsequent thin layer chromatography (lower panel).

Glycolipid Antigens in Animal Models of Type I Diabetes

With the exception of the BB rat,[22] we are not aware of any studies in which glycolipids have been found to be the target antigen of circulating autoantibodies in animal models of type I diabetes. From studies performed with BB rats, it has been suggested that neutral glycolipids are targets of humoral autoimmunity. An autoantigen containing a fraction designated GL19/20 was derived by separating

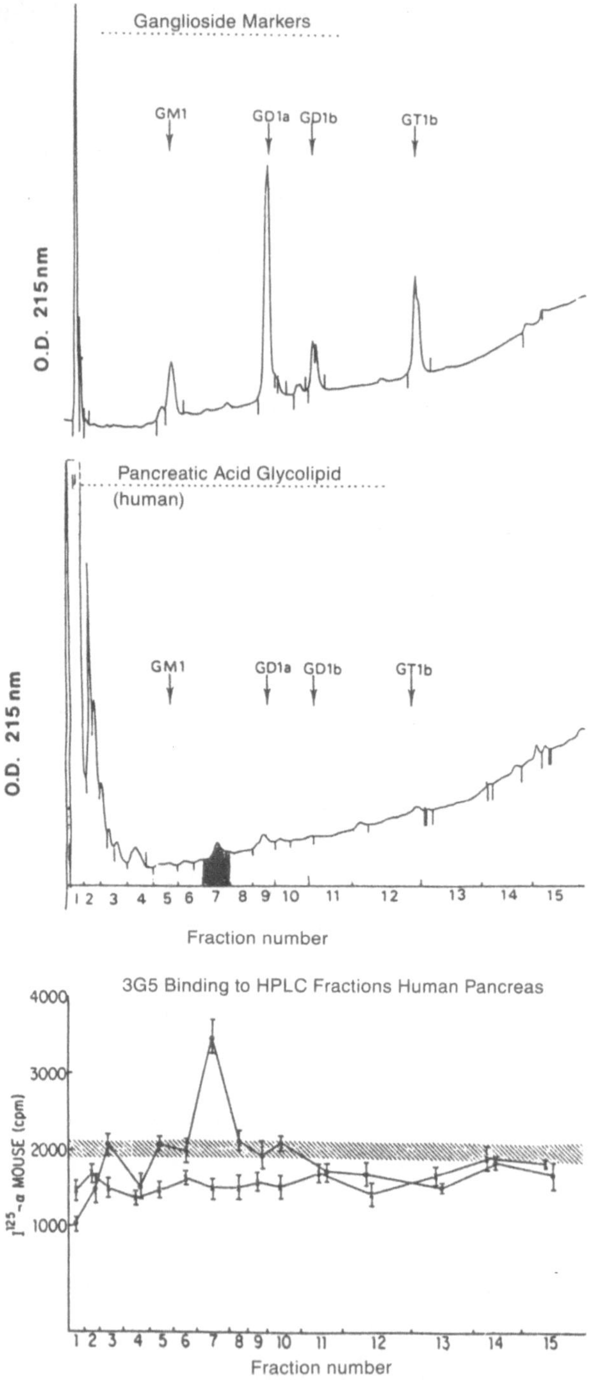

FIGURE 6.9. Recover of mAb 3G5 immunoreactive material in a HPLC peak. Chromatogram of ganglioside markers (top panel). Chromatogram of pancreatic acidic glycolipid preparation (middle panel). The shaded area (fraction 7) indicates the 3G5-binding fraction. Radiobinding assay results of mAb 3G5 binding to the HPLC fractions obtained in the middle panel (bottom panel).

a crude chloroform:methanol extract of rat islets by TLC and collecting and eluting lipids from strips of the silica-coated plate. The "autoantigenic" glycolipid ran at the solvent front (fractions 19 and 20; hence GL19/20). The GL19/20 material comigrated with cerebrosides in this TLC system and was resistant to mild base hydrolysis, which indicates that the antigen is not a saponifiable phospholipid. The GL19/20 material contained metabolically incorporated galactose, glucosamine, and N-acetylmannosamine. Serum antibodies from 17/17 diabetic BB/W rats bound to the GL19/20 glycolipids in a solid-phase radioimmunoassay, as compared to 1/17 diabetes-resistant BB/W rats. These antibodies were also detected in the sera of euglycemic diabetes-prone rats. However, the humoral autoimmunity to rat islet glycolipids in the BB rat appears to be heterogeneous, since glycolipids other than the GL19/20 glycolipids are bound, although less consistently than the GL19/20 antigens. The putative glycolipid autoantigen(s) recognized by BB rat autoantibodies is apparently distinct from the rat beta-cell–specific ganglioside defined by the monoclonal antibody R2D6 (see p. 128). Thus it appears that BB rats may mount an autoantibody response to neutral glycolipids of rat islets. Whether there is a consistent autoreactivity with islet-cell gangliosides in the BB rat is not known. Furthermore the class or subclass distribution of the antiglycolipid autoantibodies in the BB rat has not been reported. This may be of some importance, since islet-cell antibodies in human type I diabetes are exclusively of the IgG class,[7-9] and furthermore, it has been shown that IgM antibodies may bind both preferentially and nonspecifically to lipids immobilized on solid-phase matrices.[23]

Multiplicity of Antigens in Humoral Autoimmunity of Type I Diabetes

Insulin

Autoantibodies to insulin have been detected in prediabetic and recent-onset diabetic individuals prior to insulin therapy (see Chapter 8, this volume).

64-kD Protein

The immunochemistry of the 64-kD protein has recently been reviewed[11] (also see Chapter 3, this volume).

Diabetes-Associated Antigen of Islet-Cell Tumors: "Polar Antigen"

We recently reported the results of preliminary studies in which a novel diabetes-associated antigen[24] was identified. Frozen sections of the RINm38 tumor (solid tumor grown in NEDH rats) were used as a substrate to detect antibody binding by immunohistochemistry. Type I diabetic sera were seen to stain the outer pole of islet tumor cells specifically (see Fig. 6.10). This "polar" staining was seen on endocrine cells that were adjacent to blood vessels and fibrous tissue or that were

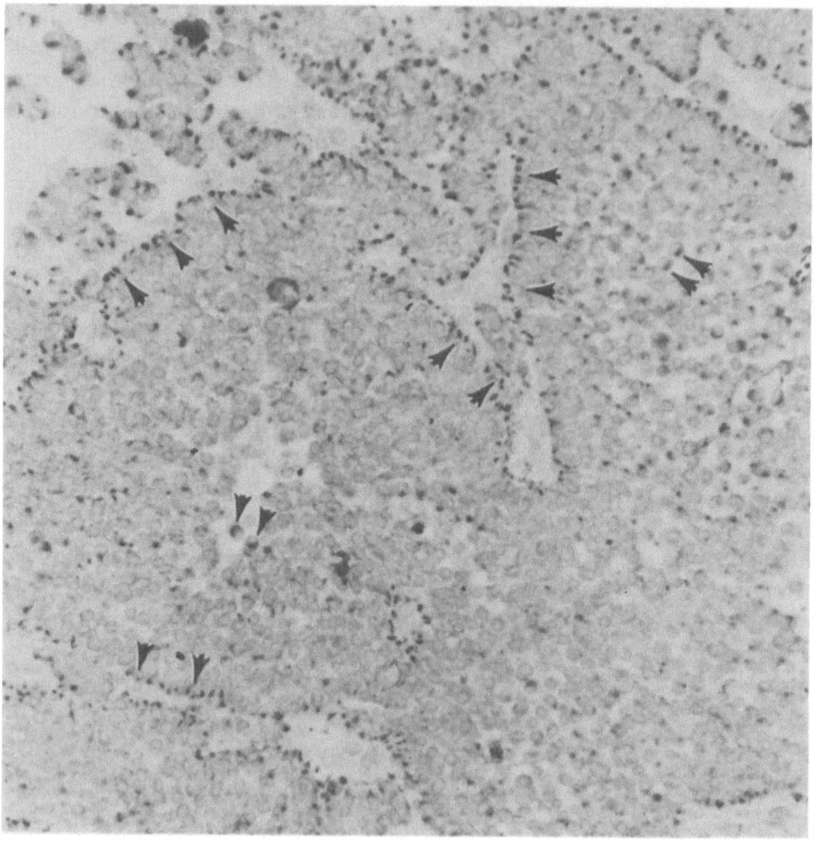

FIGURE 6.10. "Polar" antigen staining of RINm38 solid tumor with type I diabetic serum. Arrowheads, examples of "polar" staining pattern.

at the periphery of islet tumor clusters. The staining pattern was found in 38% (16/42) of new-onset diabetes (8 polar$^+$ of 21 ICA$^-$ sera and 8 polar$^+$ of 21 ICA$^+$ sera) and 25% of prospectively studied relatives (6/24). The polar staining pattern was not found in 60 sera from normal control subjects or in 15 sera from autoimmune thyroiditis patients.

We investigated the polar antigen(s) using the approach of biochemically perturbing antigens in frozen sections. The polar staining of the tumor is Pronase®-sensitive and neuraminidase-insensitive; it is unaffected by exposure to organic solvents. Indeed exposure to methanol improves the appearance of the stain by apparently reducing the background staining. The polar antigen would, therefore, appear to be a protein; it is otherwise uncharacterized. We are, however, in the process of identifying the antigen by the Western blot procedure.

Monoclonal Antibody-Defined, Islet-Cell Glycolipid Antigens

Monoclonal Antibody A2B5

The monoclonal antibody A2B5 was produced by immunizing mice with chick embryo retinal cells.[25] This antibody was initially reported to react with a plasma membrane antigen of neurons in retina, brain, spinal cord, and dorsal root ganglia. The antigen was shown by indirect immunofluorescence to be associated with most if not all the neuron cell bodies in chicken retina, but little or no antigen was detected on axons or dendrites, Müller cells, or retina pigment cells. The investigators[25] found that the A2B5 antigen was relatively stable at temperatures up to 100 °C, was insensitive to trypsin, exhibited the solubility properties of a ganglioside, and was destroyed by neuraminidase. Furthermore the cytotoxic activity of A2B5 against retina cells was found to be half maximally blocked by ganglioside GQ1c at a concentration of 2×10^{-7} M. The cytotoxic action of A2B5 could similarly be blocked by exogenous N-acetyl neuraminic acid with which the antibody reacts with lower avidity as reflected by the achievement of half maximal inhibition at a concentration of 5×10^{-3} M N-acetyl neuraminic acid. These observations suggest that the antigen for monoclonal antibody A2B5 is a complex ganglioside in plasma membranes of retinal neuron cell bodies.

Subsequently the monoclonal antibody A2B5 was shown to react with pancreatic islet cells,[26] as well as benign and malignant human insulinoma cells, human pheochromocytoma, human medullary carcinoma (thyroid), an islet-cell line (RINm5F), human melanoma, glioblastoma, and neuroblastoma (Table 6.3). The A2B5 antigen in these tissues has not been identified. The A2B5 antigen on RINm5F cells was found to be trypsin-insensitive and neuraminidase-sensitive, as was the antigen of retina cells. This, however, does not imply that the antigens of RINm5F cells and of retina cells are identical, since all gangliosides would be expected to show this pattern of biochemical sensitivity. The nature of the A2B5 antigen of RINm5F cells relative to the tetanus toxin receptor of these cells has been investigated.[26] Tetanus toxin binds with high affinity to gangliosides GD1b and GT1 has been shown to bind to pancreatic islet cells and the RINm5F tumor cell. Studies on RINm5F cells, employing both A2B5 and tetanus toxin, indicate that the tetanus toxin receptor and the A2B5 antigen of these cells are immunochemically distinct, since A2B5 cannot block tetanus toxin binding to these cells and tetanus toxin similarly cannot block A2B5 binding to these cells. Thus neither GD1b nor GT1 is the A2B5 antigen in RINm5F cells.

Additional studies on the nature of the A2B5 antigen were performed in neural tissues. Initially the ability of mAb A2B5 to bind a number of purified gangliosides was investigated: A2B5 was found to bind to GQ1c, but not to GQ1b, nor was it able to bind to such other gangliosides as GM1, GD1a, GD1b, GT1b, and Asialo GM1.[27] In another study, it was found that A2B5 bound to GQ1b, GD3, GD2, disialolactoneo-tetraosylceramide, and probably to GT1a when assayed by

TABLE 6.3. Monoclonal antibody tissue reaction patterns.[a]

Tissue	A2B5	3G5
Adult rat brain	+	+
Pituitary		
Anterior	+W	+
Posterior	+	−
Thymus epithelium	+	−
Adrenal		
Medulla	+	+
Cortex	−	−
Thyroid		
Follicular cell	−	−
C-cell	+	+
Melanoma cells	+	+
Glioblastoma (H79,H80)	+	NT
Gastric parietal cells	−	+W
T-lymphocytes	−	+
Microvascular pericytes	−	+
Kidney glomeruli	−	+
Kidney tubules	−	−
Pancreatic islets	+	+
Rat insulinoma		
RINm5F	+	+
RIN14B	−	+
Fibroblast	−	−
Skeletal muscle	−	−
Testicular Leydig cells	−	−

[a] +, positive staining; +W, weak positive staining; −, negative staining; NT, not tested.

immunostaining of TLC plates.[28] A third study suggested that A2B5 did not react with the gangliosides listed above but rather with contaminants in each preparation.[29] This study found multiple glycolipid antigens in the brain, none of which comigrated with any of the major brain gangliosides. They were instead found in the mono-, di-, tri-, and polysialoganglioside fractions. Thus the nature of the A2B5 antigen(s) in the brain is unknown. This may be due (at least in part) to the developmental stage of the brains examined, since it has been shown that the expression of A2B5-binding glycolipids is developmentally regulated.[29] A further possible source of error may be found in the propensity of IgM to bind nonspecifically to immobilized lipids.[23]

Similarly the identity of the A2B5 antigen in pancreatic islets is unknown. The ability of A2B5 to bind to gangliosides isolated from the rat insulinoma line RINm5F has been reported.[28] The investigators found that the major RINm5F cell gangliosides bound by A2B5 were GM3, GD3, and GT1b, with faint bands corresponding to GT1a and GD2 upon analysis by immunostaining of TLC plates. Degradation of RINm5F gangliosides with neuraminidase gave products that were consistent with the presence of the three major gangliosides (GM3, GD3, and GT1b) that previously had been tentatively identified by immunostaining of TLC plates. In our own preliminary studies (R.C. Nayak et al., unpublished observation), A2B5 was found to bind to a glycolipid that comigrated with GD3 in bovine pancreas extracts.

Monoclonal Antibody 3G5

The monoclonal antibody 3G5 was produced by immunizing mice with fetal rat brain.[30] This antibody was shown to react with all cells of pancreatic islets in human, rat, mouse, and bovine tissue; mAb 3G5 also reacted with the cell surface of isolated islet cells and with the rat insulinoma line RINm5F. The 3G5 antigen was found to have the properties of a ganglioside (neuraminidase-sensitive, extracted into chloroform-methanol, partitioned into a methanol–water phase, soluble in water, and nondialyzable) but was not further characterized in this study. The tissue distribution of 3G5 immunostaining suggested that mAb 3G5 may be a marker of neuroendocrine cell types[30]; however extended tissue distribution studies indicate that the tissue distribution is not restricted to neuroendocrine cell types and is, therefore, not easily explained (Table 6.3).

We are currently investigating the biochemical nature of the 3G5 antigen in the pancreas. The 3G5 antigen was shown to have similar properties to the islet cell glycolipid autoantigen (Table 6.2); however type I diabetic ICA-positive sera could not block 3G5 binding to pancreas sections (and vice versa), which suggests that these antigens are immunochemically distinct (R.C. Nayak et al., unpublished observation). Further biochemical studies have shown that the 3G5 antigen is most probably an O-acylated di-sialoganglioside, whereas the ICA antigen has properties that suggest it is a monosialoganglioside.[31]

The 3G5-positive subpopulation of T-cells (Table 6.3) appears to be of some significance in autoimmunity. It has been shown that T-cells bearing the 3G5 antigen increase linearly with age in normal subjects (Figs. 6.11, 6.12) to a greater extent than they are in age-matched controls in both amiodarone-induced thyroid autoimmunity and type I diabetes.[32,33] Since this T-lymphocyte subset clearly reflects the process of aging in man, it is tempting to speculate that an increase in this subset may be related to an increased prevalence of autoimmunity with age and that premature senescence of immunoregulatory or immune effector cells may play a role in the induction and/or maintenance of autoimmunity.

We have also recently found that the 3G5 antigen is expressed on microvascular pericytes in the retina (Fig. 6.13A,B) and the brain.[34] The 3G5 antigen was not expressed on retinal endothelial cells, retinal pigment epithelial cells, aortic smooth muscle cells, or aortic endothelial cells. The staining pattern of mAb 3G5 on retinal capillaries in trypsin-digested flat mounts of the retina is also consistent with pericyte staining (Fig. 6.13c). Thus within the microvasculature, mAb 3G5 is a pericyte marker. This antibody may prove to be useful, since pericyte loss is seen in diabetic retinopathy and is thought to be a diabetes-specific retinal lesion by some investigators. In nonprimate animal models of type I diabetes, the observation of pericyte loss is controversial; mAb 3G5 pericyte staining for quantitation, however, may be able to resolve this controversy. The mAb 3G5 has also

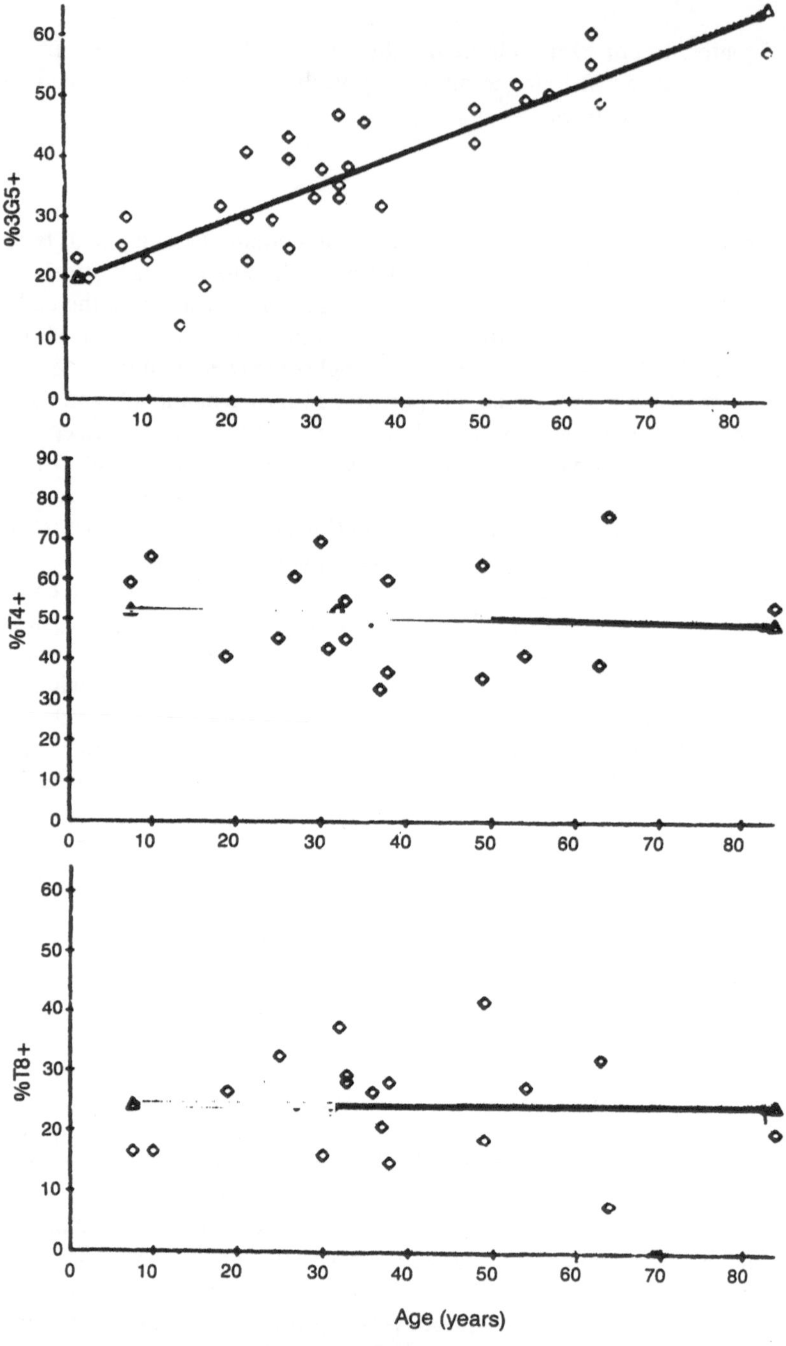

FIGURE 6.11. Correlation of human T-cell subsets with age. The proportion of 3G5+ peripheral blood T-cells increases linearly with age (top panel), but neither CD4+ T-cells (middle panel) nor CD8+ T-cells (bottom panel) increase significantly with age. Reproduced from *The Journal of Experimental Medicine*,[32] by copyright permission of The Rockefeller University Press.

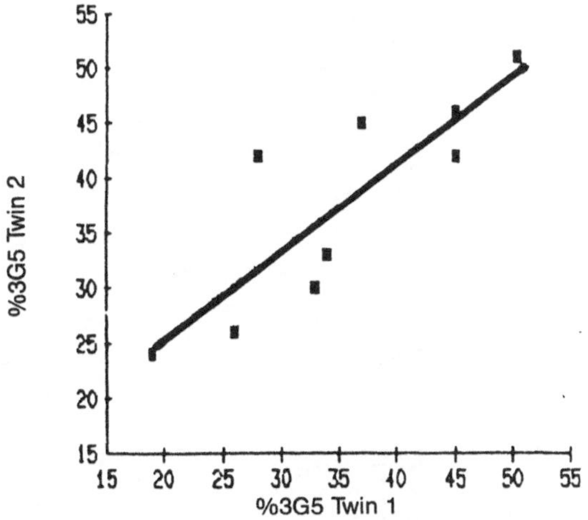

FIGURE 6.12. Levels of 3G5+ T-cells in monozygotic twins. A significant correlation exists in 3G5+ T-cell levels in pairs of monozygotic twins, which indicates a genetic regulation of the proportion of 3G5+ T-cells. Reproduced from *The Journal of Experimental Medicine*,[32] by copyright permission of The Rockefeller University Press

been found to stain renal glomeruli.[35] The type of cell that is stained within the glomeruli has not yet been identified, although staining is restricted to the glomerulus. The 3G5 antigen of pancreatic islets, renal glomeruli, and microvascular pericytes all migrate on TLC as a single band between GM1 and GM2, which indicates that these tissues express antigens in common. Since the pericyte and renal glomerulus are the primary sites of observed microvascular complications in diabetes, it is tantalizing to speculate that shared antigens also may be shared autoantigens; this implies that the immune system has a role in the generation and/or maintenance of diabetic microvascular disease.

Monoclonal Autoantibody E5C2

Monoclonal autoantibody E5C2[36] was generated by fusion of splenocytes from newly diagnozed diabetic BB rats. An IgM(k) antibody, E5C2, has been shown to be cytotoxic for neuraminidase-treated RINm5F rat insulinoma cells. This antibody does not bind to native primary rat islet cells, and it binds to only 5% of untreated rat insulinoma cells as shown by immunofluorescence. Pretreatment with neuraminidase increases antibody binding to primary islet-cell cultures from 0 to 45% and binding to rat insulinoma cells from 5 to 75%. These results suggest that the E5C2 antigen is normally a cryptic antigen of islets and that its crypticity is mediated by covalently bound sialic acid. The increase in antibody binding may be the result of binding to a determinant present in the neutral oligosaccharide chain of a surface glycoconjugate, since antibody binding to de-sialylated islet cells can be inhibited with lactose. Also E5C2 was found to bind to the purified glycolipid paragloboside but not to lactosylceramide, ceramide trihexoside, lacto-N-fucopentaosyl III ceramide, H1 blood group-active glycolipid, asialo GM1, and (alpha 2-3 or alpha 2-6) sialosylparagloboside. However the E5C2 antigen was revealed after removal of sialic acid from (alpha

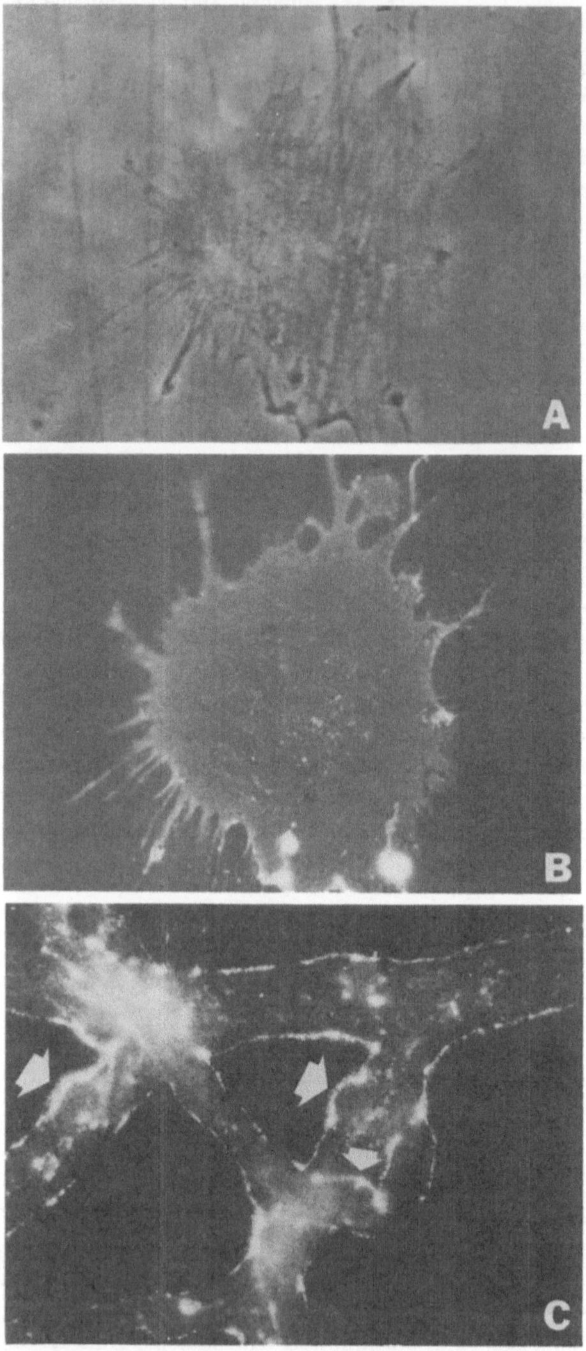

FIGURE 6.13. Immunofluorescent staining of microvascular pericytes by mAb 3G5. (**A**) Phase-contrast photomicrograph of bovine retinal capillary pericyte in vitro; (**B**) the same cell stained with mAb 3G5; (**C**) mAb 3G5 staining of primate retinal capillaries – arrows indicate membrane staining over characteristic protrusions of the pericyte from the vascular wall.

Table 6.4. Structures of some glycosphingolipids and oligosaccheride haptens.

Name	Structure[a]
Galactocerebroside	Galβ1-1′Cer
Sulfatide	Gal(3-SO_4)β1-1′Cer
GM4	NeuNAcα2-3Galβ1-1′Cer
Lactose	Galβ1-4Glc
N-acetyl-lactosamine	Galβ1-4GlcNAc
Lactosyl ceramide	Galβ1-4Glcβ1-1′Cer
Ceramide trihexoside	Galα1-4Galβ1-4Glcβ1-1′Cer
Globoside	GalNAcβ1-3Galα1-4Galβ1-4Glcβ1-1′Cer
Paragloboside	Galβ1-4GlcNAcβ1-3Galβ1-4Glcβ1-1′Cer
Forssman antigen	GalNAcα1-3GalNaAcβ1-3Galα1-4Galβ1-4Glcβ1-1′Cer
GM3	NeuNAcα2-3Galβ1-4Glcβ1-1′Cer
Asialo-GM2	GalNAcβ1-4Galβ1-4Glcβ1-1′Cer
GM2	GalNAcβ1-4Gal(NeuNAcα2-3)β1-4Glcβ1-1′Cer
Asialo-GM1	Galβ1-3GalNacβ1-4Galβ1-4Glcβ1-1′Cer
GM1	Galβ1-3GalNAcβ1-4Gal(NeuNAc α2-3)β1-4Glcβ1-1′Cer
GD3	NeuNAcα2-8NeuNAcα2-3Galβ1-4Glcβ1-1′Cer
GD1a	NeuNAcα2-3Galβ1-3GalNAcβ1-4Gal(α2-3NeuNAc)β1-4Glcβ1-1′Cer
GT1b	NeuNAc α2-3Galβ1-3GalNAcβ1-4(NeuNAc α2-3)Galβ1-4Glcβ1-1′Cer
GQ1c	NeuNAc α2-3Galβ1-3GalNAcβ1-4(NeuNAc α2-8NeuNAc α2-8NeuNAc α2-3)Galβ1-4Glcβ1-1′Cer

[a] Cer, Ceramide; Glc, glucose; Gal, galactose; GalNAc, N-acetylgalactosamine; GlcNAc, N-acetylglucosamine; NeuNAc, N-acetyl-neuraminic acid.

2-3 or alpha 2-6) sialosylparagloboside. Hapten inhibition studies indicated that N-acetyllactosamine (Gal beta 1-4 GlcNAc) was the best inhibitor of E5C2 binding (it is the nonreducing terminal disaccharide of paragloboside), although both lactose and lactulose could inhibit E5C2 binding but were 20 to 30 times less effective (see Table 6.4 for structures).

Western blotting analysis of neuraminidase-treated RINm5F cell extracts indicated that E5C2 specifically stains a glycoprotein doublet of relative molecular weights 60- and 68-kD. Interestingly, however, mAb E5C2 did not stain chromatographed glycolipid extracts of RINm5F cells, which suggests that this antigen may be found exclusively as a glycoprotein antigen on islet cells. Since E5C2 is an autoantibody, the authors of this study suggest that exposure of the normally cryptic antigen could initiate an autoimmune attack.

Monoclonal Autoantibody CNH8

Monoclonal autoantibody CNH8[37] was produced by transforming peripheral blood lymphocytes from type I diabetics with the Epstein-Barr virus. The CNH8 autoantibody is of the IgM(k) class. This antibody reacts with human pancreatic islet cells and the rat insulinoma line RINm5F. When an immunostaining method was used to detect glycolipids separated by TLC, CNH8 was found to recognize a glycolipid found in several human tissues. Lyophilized meconium (a rich source of this antigen) was used to purify the antigen, which was shown to have a sialic acid:sphingosine ratio of 1:1 and a galactose:N-acetylglucosamine:glucose ratio of 2:1:1. These results, combined with those of immunologic, chromatographic, and enzymatic digestion analyses, indicate that this glycolipid antigen is probably a monosialylated form of paragloboside (Gal beta 1-4GlcNAc beta 1-3 Gal beta 1-4 Glc-ceramide). Sialic acid is necessary for antibody binding and is probably in a terminal alpha 2-3 or alpha 2-6 linkage. Whether this antibody or antigen plays a role in the pathogenesis of type I diabetes is not known. However it is important to point out that this antigen is found in many tissues and, therefore, its involvement in autoimmune pathogenesis in diabetes would be at variance with the widely accepted notion that islet-cell autoantibodies are organ-specific. Of further interest, this antigen is similar to the putative ganglioside antigen we are studying in pancreatic sections and extracted pancreata. Both antigens have the properties of monosialogangliosides, although the sialosylparagloboside is not a member of the ganglio- series of glycosphingolipids.

Monoclonal Antibody R2D6

The monoclonal antibody R2D6[38] was raised by immunizing BALB/c mice with partially dispersed, cultured Wistar–Furth islet cells; it is of the IgM isotype. Within the pancreas, this monoclonal antibody is reported to be specifically reactive only with the plasma membranes of pancreatic beta-cells. Also R2D6 is reported to cross-react with beta-cells of mice and guinea pigs but not with those of humans or dogs, and with adrenal chromaffin cells, secretory cells in the anterior pituitary, and the myentric plexus of the gastrointestinal tract.

The R2D6 antigen is resistant to the effects of trypsin, chymotrypsin, papain, ficin, and Pronase®, since monoclonal antibody R2D6 could still bind to dissociated rat islets that had been pretreated with these proteases. Neuraminidase treatment of islet cells did, however, reduce R2D6 binding by 75%. The R2D6 antigen was quantitatively extracted from islets by dichloromethane/methanol (2:1) and, subsequently, was soluble in methanol or in phosphate-buffered saline. Antibody binding to the extracted antigen remained insensitive to Protease® treatment but was totally ablated by neuraminidase treatment, which indicated that sialic acid is either an integral part of, or is related sterically to, the epitope for R2D6. The antigen was found to migrate near the GD1a marker on TLC, and it bound to a weak anion exchange resin. Furthermore the R2D6 antigen was not denatured by mild base hydrolysis and was eluted in ganglioside-containing fractions on silicic acid chromatography. The R2D6 antibody, therefore, appears to bind to a beta-cell–specific ganglioside of rodent islets.

Coordinate Expression of mAb-Defined Glycolipid Antigens and Glycolipid Autoantigen(s) of Diabetes

Studies with mAbs have indicated that islet cells express a complex repertoire of glycosphingolipid antigens. Some of these monoclonal antibodies have been used in a recent study of antigen expression in the rat insulinoma (RIN) cell line RINm5F.[39] This study, which used viable RIN cells as the substrate for an ELISA assay for cell-surface antigens, showed that it was possible to detect cell surface ICA with this assay system. A more interesting finding in this study was that ICA surface antigen could not be detected on RIN cells after approximately 50 passages in vitro. Additionally a number of RIN-cell subclones were tested as substrates for antibody binding; a very interesting correlation emerged from these tests. All the subclones were seen to express the islet-cell surface protein antigen defined by mAb AID2; however some subclones did not highly express the islet ganglioside(s) bound by mAb A2B5. Subclones or lines that expressed little or no A2B5 ganglioside antigen similarly expressed little or no ICA-binding capacity. This relationship suggests that the differentiated state of the islet cells may be important to the expression of islet-cell gangliosides, as defined by mAb A2B5, and islet-cell autoantigen(s). The nature of the autoantigen in the rat cell line is unknown. It is known that ganglioside expression is a sensitive marker of both differentiation and transformation and, when considered with the correlation between A2B5 antigen expression and ICA Ag expression in RIN subclones, the findings suggest that the ICA surface antigen of RIN cells is a ganglioside. Furthermore the ICA antigen in frozen sections of rat pancreas has recently been shown to have immunochemical properties similar to those of frozen sections of human pancreas.[40]

Physiological Roles of Islet-Cell Gangliosides

Although we have evidence that islet-cell gangliosides may be autoantigens in type I diabetes, the physiologic role(s) of gangliosides in normal islets is essentially unknown. It has been shown that removal of sialic acid from the beta-cell surface moderately inhibits glucose oxidation and decreases insulin secretion in response to 20 mM *D*-glucose, 5 mM *D*-glyceraldehyde, 20 mM *L*-leucine, 10 mM *L*-arginine, and 5 μg/ml of glucagon.[41] Furthermore cholera toxin (which binds to GM1) potentiates the insulin-releasing action of 20 mM glucose, and this potentiating effect is abolished by neuraminidase treatment of islets. Neuraminidase treatment of islets also abolished the cholera toxin-induced rise in cyclic AMP (cAMP) but not the theophylline induced rise of cAMP. These results suggest that cell surface sialoglycoconjugates of pancreatic beta-cells may be involved in signal transduction, activation of adenylate cyclase, and insulin secretion. It has also recently been shown that cholera toxin acts as a secretagogue, to increase both total RNA and insulin gene-specific transcription in the insulin-producing RIN cell line.[42]

Antibodies to Glycolipids in Diseases Other than Type I Diabetes Mellitus

As recently as 1984, it was pointed out that relatively little attention has been paid to the human immune response to glycosphingolipids, with the notable exception of studies on autoantibodies that cause cold agglutination or lysis of

TABLE 6.5. Glycoconjugate reactivity of serum antibodies in various diseases.

Disease	Antigen designation	Glycoprotein	Glycolipid
Type I diabetes	Ganglioside(s)? Neutral glycolipid(s)?	?	+?
IgM paraproteinemia associated with polyneuropathy	Myelin-associated glyco-protein and gangliosides	+	+
Graves' disease	Thyrotropin receptor complex	+	+
Autoimmune hemolytic anemia (cold agglutinin disease)	Blood group Ii antigens	+	+
Systemic lupus erythematosus	GM1, asialoGM1	?	+
Multiple sclerosis	GM4, GM1, GM2, Galactosylcerebroside Digalactosyl diglyceride	?	+
Hanganutiziu–Deicher serum sickness	Erythrocyte glycoprotein		
heterophile antibodies	Hematoside (GM3) Sialosylparagloboside	+	+
Graves ophthalmopathy	Terminal alpha-galactosyl residues	?	+
Behcets' disease	AsialoGM1, globoside paragloboside, galactosylceramide	?	+
Acquired immunodeficiency syndrome	AsialoGM1	?	+
Chronic relapsing experimental allergic encephalomyelitis	Galactocerebroside Sulfatide GM1	?	+
Cranial trauma and cerebrovascular accident	Various glycolipids	?	+
Normal serum	Various glycolipids	?	+

human erythrocytes.[43] The literature does, however, indicate that a humoral immune response to glycolipid antigens can be demonstrated in a number of diseases (Table 6.5), which are briefly reviewed below.

IgM Paraproteinemia Associated with Polyneuropathy

In the majority of cases, the serum autoantibody in question is monoclonal and of the IgM class, although IgG and IgA autoantibodies have also been seen.[44] These autoantibodies have been shown to bind to frozen sections of peripheral nerve. Binding can be ablated by periodate oxidation and restored by borohydride reduction, but it is resistant to Protease® treatment of these sections.[45] Subsequently Ilyas and coworkers[46] showed that these autoantibodies react with both a glycoprotein and a glycolipid in myelin extracts. These antigens have been identified as myelin-associated glycoprotein and myelin gangliosides. It is interesting to note that the ganglioside antigen(s) were 10- to 50-fold more abundant than the myelin-associated glycoprotein, which probably accounts for the observed Protease®-insensitive immunostaining on frozen sections of peripheral nerve. The autoantibodies in patients' sera react with many different acidic glycosphingolipids, and in a few cases, specific myelin-associated gangliosides have been identified. In addition to autoantibodies that cross-react with both the myelin-associated glycoprotein and myelin gangliosides, some patients have been found to have autoantibodies that react only with the glycoprotein or only with the glycolipid. Other studies have shown that an acidic glycolipid (although not a ganglioside) that is bound by the monoclonal antibody HNK-1 is also a target for autoantibodies from these patients and is found in small amounts in myelin ganglioside extracts.[47] It has been postulated that these antibodies may be involved in the pathogenesis of the demyelinating disease in patients with polyneuropathy.

Graves' Disease

Graves' disease is an organ-specific autoimmune disease in which antibodies to the thyrotropin receptor (thyroid-stimulating hormone, TSH) are believed to cause a hyperthyroid state.[48] Many autoantibodies that have thyroid-stimulating and/or TSH-blocking activity have been found in this disease.[49] The thyroid-stimulating antibodies apparently mimic the effects of TSH on thyrocytes.

Thyroid-stimulating hormone has been shown to bind (at high affinity) to a glycoprotein of thyroid membranes and consequently to induce a physiologic response. Similarly TSH also can bind to a ganglioside through a lower-affinity interaction and can produce an analogous physiologic response.[50] A dual component model of the action of thyrotropin and antibodies to the thyrotropin receptor thus has been proposed.[50] Monoclonal antibodies to the thyrotropin receptor have been induced, and the physiologic action of mAb binding to thyrotropin receptor was found to mimic the effects of thyrotropin. Monoclonal antibodies that have two distinct effects on thyrocytes have been identified: (1) blocking antibodies and (2) stimulating antibodies.

Blocking antibodies completely inhibit TSH binding to the TSH receptor as well as TSH-stimulated adenylate cyclase activity and TSH stimulatable thyroidal iodine release or uptake. These blocking mAbs were also able to block TSH binding to the receptor glycoprotein. Stimulating antibodies were less potent at blocking TSH binding but were potent inhibitors of TSH-stimulated adenylate cyclase activity. These stimulating antibodies could also behave as agonists of adenylate cyclase at low concentrations of TSH. Serum IgG from Graves' disease patients and stimulatory monoclonal antibodies both bound to thyroid ganglioside preparations, but blocking antibodies did not. Neuraminidase predigestion of thyroid gangliosides abolished this antibody binding. Both the stimulatory mAbs and the Graves' disease IgG bound strongly to a thyroid disialoganglioside fraction.

Cold Agglutinin Disease

An autoimmune hemolytic anemia, also known as cold agglutinin disease, is occasionally found in association with chronic lymphoproliferative disease. Autoantibodies that react with erythrocyte glycoconjugates can be found as a monoclonal spike in the sera of many patients with this disease. The antigens bound by these autoantibodies have been found to be developmentally regulated, blood-group antigenic specificities borne on the oligosaccharide side chains of glycoconjugates. The glycoconjugates are the Ii blood-group antigens,[51] and are represented on both glycoproteins and glycolipids through shared oligosaccharide determinants. Erythrocyte plasma membrane gangliosides can inhibit autoantibody binding to the Ii-active glycoprotein.

Systemic Lupus Erythematosus

A high incidence of autoantibody against the neutral glycolipid asialo-GM1 in the serum of systemic lupus erythematosus (SLE) patients with neurologic disorders has been reported.[52] The sera of 14 of 17 cases of SLE with neurologic disorders had autoantibody to asialo-GM1 but not to GM2, GM1, or galactocerebroside. Antibody was not detected in 87 cases of SLE without any history of seizures. Antibody titer was high several months and even years before and/or after the seizure, although (surprisingly) the titer was low at the time patients showed definite neurologic symptoms; it was proposed that these antibodies may play a role in the pathogenesis of neurologic disorders accompanying SLE. However one study[53] found no correlation between the presence of anti-asialo–GM1 antibodies and neurologic symptoms in patients with SLE, although 19 of 31 SLE patients in this study had such autoantibodies.

Multiple Sclerosis and Experimental Allergic Encephalomyelitis

Multiple sclerosis (MS) is a neurologic disease involving the destruction of myelin in the central nervous system. This destruction is generally thought to be an autoimmune reaction. Antigen(s) that may provoke the autoimmune reaction are largely uncharacterized, but myelin basic protein has been found to be encephalitogenic in experimental animal models.[54] It has been suggested that myelin glycolipids may be targets of autoimmune attack in MS and experimental encephalomyelitis.[55-58] In 1974 it was shown that rabbits immunized with cerebroside produced antibodies to cerebroside that were able to inhibit sulfatide synthesis and myelination when applied to mouse embryo spinal cord explant cultures.[57] These antibodies also inhibited sulfatide synthesis and caused demyelination when added to nerve cell cultures.[57] Furthermore antibodies to cerebroside, produced by injecting rabbits with white matter, can fix complement. Absorption of this serum with purified cerebroside abolished inhibition and demyelination. The rabbits in which these antibodies were raised later developed encephalomyelitis. In 1984 Schwerer and coworkers[58] studied spinal cord-induced, chronic relapsing, experimental allergic encephalomyelitis (crEAE) in guinea pigs. The sera of these animals were tested for the presence of IgG antibodies against glycosphingolipids (galactocerebroside, GM1, sulfatide) by the enzyme-linked immunosorbent assay and for in vivo demyelinating ability by infusion into the lumbosacral subarachnoid space of normal rats. Sera from animals in the chronic stage of crEAE had a high incidence (21/26) and high titers (up to 1:2560) of antibodies against one or more of the glycosphingolipids tested. The high incidence and titers of antibody correlated with a high incidence (22/26) of in vivo demyelination, which suggests that these antibodies react with various glycosphingolipids in the process of demyelination.

In the human disease (MS), the situation is less clear. Using an assay to measure complement-dependent release of a liposome-entrapped fluorochrome, Hirsh and Parks[55] found that serologic reactions against glycolipid-sensitized liposomes could be observed in MS. In this assay, the antibodies in some of the MS sera tested reacted against cerebrosides, gangliosides, and white matter lipids, but reactions against digalactosyl diglyceride (an alpha-galactoside) were more frequent and were found more often in MS patients than in normal controls or in samples from patients with other neurologic diseases. These experiments do not provide much support for the hypothesis that galactocerebrosides or other known galactolipids of human white matter act as antigens to provoke an autoimmune attack on myelin in MS. Digalactosyl diglyceride appears to be a possible, although curious exception. Galactosyl diglycerides, which were once thought to be exclusively plant lipids, are found in brain tissue, apparently specifically in myelin. Arnon and coworkers,[56] however, have demonstrated in the liposome lysis assay, serologic activity against several purified brain gangliosides in some MS but not in normal individuals. Furthermore there was an apparent correlation between the severity of disease and the extent of liposome lysis. In a third study,[53] in which the liposome lysis assay was used, 34/46 MS patients were found to have antibodies to one or more glycolipids. The latter two studies did not screen for digalactosyl diglyceride antibodies.

Graves' Ophthalmopathy

It has recently been suggested that elevated anti-alpha galactosyl antibody titers may indicate progression in autoimmune thyroid disorders and in endocrine ophthalmopathy.[59] In 50 normal subjects, the titers of such antibodies ranged from 1/10 to 1/80, regardless of age and blood group. Elevated titers (> 1/80) were seen in 6/6 patients with progressive exophthalmos, in 5/5 patients with untreated Graves' disease, and in 11/12 patients with progressive nontoxic goiter. In contrast titers within the normal range were seen in primary myxedema (17 patients) and in residual exophthalmos (11 patients), whereas they nonreproducibly increased, upon a series of determinations, in only 1/31 patients with treated or cured Graves' disease (at least two years remission) and reproducibly increased in 5/36 patients with nonprogressive nontoxic goiter. Elevated titers were also found in 3/7 patients presenting with autoimmune thyroiditis. No correlations were found between elevated anti-alpha galactosyl antibody titers and antithyroglobulin or microsomal antibody titers. The antibodies were of the IgG class in both patient and normal control groups. These antibodies could bind to trypsinized human and porcine thyroid cells but not to untreated thyroid cells, which suggests that the antigenic determinants implicated in the enhanced production of anti-alpha–galactosyl antibodies are present but cryptic on the cell surface of thyroid cells. These antibodies also bind to immobilized melibiose (alpha-galactosyl glucoside), which confirms the alpha-galactosyl–specificity. Human tissues are generally thought not to express alpha-galactosyl residues under normal conditions. For example both murine and bovine thyroglobulin have been reported to have terminal alpha-galactose residues[60] as has the bovine TSH receptor,[61] but human thyroglobulin does not have this terminal structure. It is clear, however, that Protease® treatment can reveal this terminal saccharide structure on human cells, which suggests that proteolysis may expose galactolipids that are normally covered by proteins in the plasma membrane. Indeed it has recently been shown that human erythrocytes express small amounts of complex Gal-alpha 1-3 Gal glycosphingolipids.[62]

Acquired Immune Deficiency Syndrome

It has been reported that homosexual men with acquired immune deficiency syndrome (AIDS) express significantly high levels of IgG antibody to the neutral glycolipid asialo-GM1 (ganglio-N-tetraosylceramide) compared with healthy homosexual and heterosexual men.[63] The asialo-GM1 antibody could be removed from sera by absorption with spermatozoa. Antibody to asialo GM1 appears in the serum after rectal insemination of male rabbits with rabbit semen, which suggests that asialo-GM1 is a component of ejaculated spermatozoa. Rectal insemination by itself can lead to the production of antibodies to this glycolipid in the rabbit. It is interesting to note that NK cells express asialo GM1 and antibodies to asialo-GM1 generated through rectal immunization may contribute to the immunodeficiency by killing NK cells. Additionally, such antibodies may be of use in the early detection of individuals with AIDS.

Behcets' Disease

Behcet's disease is a recurrent generalized inflammatory disease, which manifests its major symptoms in the oral mucosa, eyes, external genitalia and skin. Although its etiology is still disputed, there is evidence that it involves some abnormality of the immune system. This evidence includes the presence of autoantibodies against oral mucosa, an increase in complement components, and the presence of immune complexes.[64]

Antibodies to various glycolipids have been found in sera of Behcet's disease patients,[64] and 2 of 32 patients had moderate titers of antibody to ceramide trihexoside. All controls and patients had demonstrable antibody against Forssman antigen, but the patients' titers were higher than the controls'. Additionally anti-asialo–GM2, asialo GM1, GM1, and GM3 titers were higher in patients. Some patients also had antibodies to globoside and paragloboside; these antibodies were not observed in healthy adults.

A major complication of Behcet's disease is a central nervous system lesion (the neuro-Behcet's syndrome). Anti-glycolipid antibodies have been described in the neuro-Behcet's syndrome and the spectrum of anti-glycolipid antibodies in neuro-Behcet's syndrome is distinct from but overlaps with the anti-glycolipid reactivity spectrum of Behcet's disease.[65] Appreciable antibody titers against GM1, globoside, Forssman glycolipid, asialo-GM1, and asialo-GM2 were found in some patients. Antibodies to asialo-GM1, found in the majority of patients with neuro-Behcet's syndrome, were less common in Behcet's disease without neurologic involvement and were rarely found in patients with miscellaneous neurologic disorders or collagen disease and in healthy adults.

Hanganutziu and Deicher (H-D) Heterophile "Serum Sickness"

The appearance of heterophile antibodies in sera of patients receiving therapeutic injections of foreign serum was independently described by Hanganutziu and Deicher. These antibodies differ in specificity from such heterophile antibodies as Forssman antibodies and the Paul–Bunnell antibodies of infectious mononucleosis. These H-D antibodies react with erythrocytes and sera of various animal species and can be absorbed by the sediment of guinea pig kidney homogenate. They also have been detected in sera from patients receiving gamma-globulin fractions of goat anti-human thymocyte serum; in sera of patients who have never received therapeutic injections of foreign serum; and in sera from patients with autoimmune disease, renal disease, and liver disease.[66] The H-D antigen is heat stable, extractable with hot ethanol, and precipitable by a 75% ethanol solution. Isolation and characterization of this antigen from bovine and equine erythrocytes have demonstrated that it is a ganglioside containing N-glycolyl-neuraminic acid (GcNeu). The antigen of equine erythrocytes was found to be GcNeu-hematoside (GcNeu-GM3) and of bovine erythrocytes, GcNeu-sialosylparagloboside.[67] More recently H-D antibodies also have been shown to react with a major glycoprotein of bovine erythrocytes,[68] glycoproteins from the submaxilliary gland,[69] and glycoproteins from fetal calf serum.[69] The H-D antigen also has recently been found to be expressed on Marek's disease lymphoma cell surface membranes; it is a Marek's disease, tumor-associated surface antigen.[70]

Cranial Trauma and Cerebrovascular Accident

Sera from patients suffering from cranial trauma (comatose patients in the Neurosurgical Intensive Care Unit) and from patients with major cerebrovascular accidents were evaluated for the presence of antibodies to glycolipids in a complement-dependent lysis test on liposomes.[53] In the sera from patients with cranial trauma, 8/11 reacted with one or more of the glycolipids tested. Of the eight positive sera, five sera reacted with GM1, five sera reacted with asialo-GM1, one serum reacted with GM4, two sera reacted with GD3, and one serum reacted with GD1b. None of the 11 sera reacted with galactosyl ceramide or GD1a. Six of six sera from patients with stroke reacted with one or more glycolipids. Anti-GM1 antibodies were found in four sera, anti-asialo GM1 in six sera, anti-GM4 in two sera, anti-galactosyl ceramide in one serum anti-GD3 in one serum, and anti-GD1b in one serum. None of these sera reacted with GD1a. These results indicate that, in certain instances, anti-glycolipid antibodies may reflect tissue damage rather than a pathogenic process.

Normal Sera

Antibodies to glycolipids have been detected in normal human sera. In one such study, antibodies from 30 healthy adults were screened against a panel of ten glycolipids.[71] Little or no anti-ceramide monohexoside (galactosyl ceramide), ceramide dihexoside, ceramide trihexoside, globoside, or paragloboside activity was detected. Several had weak anti-GM3 binding, and the majority had anti-asialo GM2, anti-asialo-GM1, and GM1 activity. All sera had anti-Forssman activity. Similar analysis of cord blood samples for antibodies against these 10 glycolipids showed little if any antibody reactivity, which suggests that "natural" antibodies to glycolipids are generated by postnatal environmental stimulation. Furthermore variations in the anti-glycolipid antibody "spectrum" of control groups may reflect the different environments experienced by the individuals in these various studies. It is clear, however, that various pathologic states can alter the pattern or spectrum of serum antibodies to glycolipid antigens.

Immunogenicity of Glycolipids

Attempts to produce antibodies to purified glycolipids in Freund's adjuvant have demonstrated that these molecules are very weak immunogens.[71,72] In cases in which antibodies to glycolipids were successfully produced, many anti-sera were found to cross-react with other glycolipids,[72,73] but the outcome of most immunization schemes generally depends on the nature of the immunogen and the immunized animal species.[73]

It is also clear that the mode of antigen administration is an important factor in the humoral immune response to glycolipid antigens. Although immunization with purified glycolipids is the most direct route to obtain antibodies to glycolipids, immunization with cells, cell fragments, and membrane fractions have yielded antibodies to novel glycolipids and have generally been more successful.[74] Monoclonal antibodies 3G5 and A2B5 (discussed above) were produced in this way. The success of immunization with tissues and tissue fractions over immunization with purified glycolipids strongly suggests that glycolipids presented/oriented in their native form (i.e., integrated into cellular membranes) are more immunogenic than purified molecules. A recent study in which it was shown that immunization with glycolipids incorporated into phospholipid/cholesterol liposomes produces specific antibodies to the glycolipid antigen,[75] supports this suggestion. In this study, asialo-GM1 was incorporated into liposomes that were then used to immunize mice intraperitoneally. Six monoclonal antibodies were produced and all reacted specifically with asialo-GM1. None of these monoclonal antibodies reacted with a panel of nine other structurally related glycolipids. In another recent study, a mAb that specifically reacts with sulfatide (3′-sulfogalactosylceramide) was produced by immunizing Balb/c mice with the purified glycolipid coated onto *Salmonella minnesota* bacterial membranes.[76] The antibody (Sulf I) showed similar binding avidity to three sulfated glycolipids: sulfatide, sulfolactosylceramide, and seminolipid. Lysosulfatide (sulfatide lacking the fatty acyl chain of ceramide) also bound the antibody, but with lower avidity, which suggests that terminal galactose-3-O-sulfate and part of the hydrophobic region of the glycolipid are recognized by the sulf I mAb. Thus spatial orientation of glycolipids in plasma membranes may be an important factor in the generation of humoral autoimmune responses to specific glycolipids and may also reflect a mechanism whereby the immune system may be recruited to clear senescent cells and cellular debris from tissues.

T-Cell Reactivity to Glycolipid Antigens

It is currently believed that autoreactive T-lymphocytes contribute to the pathogenesis of type I diabetes, at least partially. The evidence for this is strongest in animal models of type I diabetes (see Chapters five, six, seven, nine in this volume). The mononuclear infiltrate of pancreatic islets in a case of human type I diabetes consisted primarily of T-cells.[77] Furthermore the cytotoxic/suppressor T-cell subpopulation predominated, although T-helper cells and natural killer/killer cells were also present. Activated T-cells were present in this infiltrate, and they have been shown to be significantly elevated in the peripheral blood of type I diabetics.[78] These observations do not prove that these cells are the effector cells of islet-cell destruction. Recent reports of clinical trials of cyclosporin A in type I diabetics indicate that this drug can induce remission of disease, with decreased or no requirement for insulin therapy.[79] Since cyclosporin A inhibits T-cell function by inhibiting IL-2 production, a cyclosporin A-induced remission in diabetics supports the view that T-cells are involved in the effector mechanisms of anti-islet autoimmunity. In the NOD mouse and BB rat, the situation is much clearer in that diabetes can be transferred to nondiabetic animals by transfusion of diabetic lymphocytes.[80,81] Furthermore it has recently been shown that both the cytotoxic/suppressor lymphocyte (LYT-2$^+$) and the helper lymphocyte (L3T4$^+$) are required for adoptive transfer of diabetes in NOD mice.[82] The islet-cell antigens that T-cells react with to cause diabetes have not been identified, and consequently, it is not known whether the humoral autoimmune response is directed against the same antigens as is the cellular autoimmune response in rodent or human diabetes.

It may be instructive, however, to consider whether T-cells can respond to glycolipid antigens. This issue has not been seriously addressed in the literature, and it appears that many investigators may doubt that such a response is possible. Indirect evidence for autoreactive T-lymphocytes in type I diabetes may be found if the isotype distribution of ICA is considered. A number of studies indicate that ICA are of the IgG class, and predominantly IgG1.[7-9] The predominance of IgG subclasses of ICA may be taken to indicate that the antigen is a T-dependent antigen. The maturation of T-dependent antibody responses usually produce an IgG response, and T-independent antigens usually give rise to a response in which IgM predominates. The IgG–ICA response, therefore, implies a T-lymphocyte/B-lymphocyte collaboration. The T-cell collaboration in the T-dependent antibody response to protein antigens is usually antigen-specific and has a hapten carrier relationship (i.e., the T-cell recognizes a carrier determinant, the B-cell recognizes a haptenic determinant of the same antigen). For glycolipid antigens, however, it is not known whether such a hapten–carrier structure exists or whether an antigen-specific T-cell collaboration occurs. There is some evidence that lipids can enhance the immunogenicity of both oligosaccharide and protein antigens through a carrier effect.[25,36,83] It has been shown that the immune response to synthetic peptides of the beta subunit of cholera toxin is greatly enhanced by conjugation to dipalmityl side chains. A similar enhancement has been seen for peptides conjugated to tetanus toxoid (a macromolecular carrier) and for polymerized peptides in the absence of an external carrier. These results indicate that fatty acids covalently conjugated to a hapten can enhance immunogenicity of the hapten by acting as a carrier. Situations that can be considered to be the reverse of the above experiment (i.e., removal of conjugated lipid) seem to indicate that the lipid must act to increase the avidity of the antibodies generated. For example the monoclonal antibody E5C2 binds with high avidity to paragloboside, but it binds with 20- to 30-fold less avidity to N-acetyllactosamine (a non-reducing terminal disaccharide of paragloboside).[36] Additionally the binding of monoclonal antibody A2B5 is blocked with 2×10^{-7} M GQ1c, but 10,000-fold more N-acetyl-neuraminic acid is required to block antibody binding similarly.[25]

Autoreactive T-cell lines that react with ganglioside antigens have been produced from cerebrospinal fluid lymphocytes of MS patients.[84] Two T-cell lines were produced; these lines proliferated specifically in response to the gangliosides GM1, GD1a, GD1b, and GQ1b. A proliferative response to other brain or viral antigens was not seen. The T-cell lines from peripheral blood lymphocytes of normal individuals, MS patients, and patients with other neurologic diseases did not proliferate in response to these gangliosides. One line exhibited predominantly helper/inducer (CD4+) phenotype, whereas the other line was of the suppressor/cytotoxic (CD8+) phenotype: Further analysis suggested that the latter line may have a suppressor function since it did not react with an antibody marker of cytotoxic T-cells.

Additionally cell-mediated hypersensitivity to brain gangliosides in peripheral blood lymphocytes of MS patients has been demonstrated by the active E-rosette

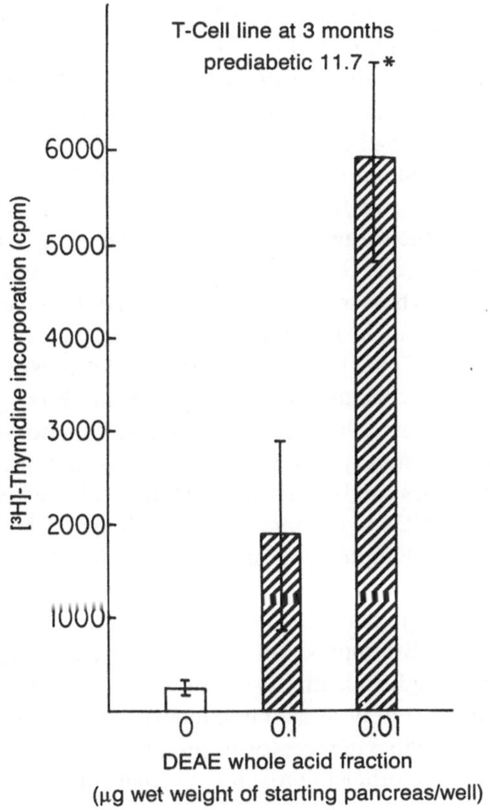

FIGURE 6.14. Stimulation of thymidine incorporation into a prediabetic T-cell line by a pancreatic acidic glycolipid fraction.

assay,[85-87] and T-lymphocytes may react specifically with a parasite glycolipid in Leishmaniasis.[88-93] We are currently attempting to isolate ganglioside-reactive T-cell lines from the peripheral blood of "prediabetics". As shown in Figure 6.14, a T-cell line that proliferates in response to exogenously added pancreatic acidic glycolipids was obtained. The T-cell line responded to pancreatic acidic glycolipids in a dose-dependent fashion and did not proliferate in the absence of acidic glycolipid extract (R. Keller, unpublished observation). Although T-cell lines from normal individuals are not yet available in our laboratory, it has been shown that peripheral blood lymphocytes from normal individuals do not proliferate in response to pancreatic acidic glycolipids at the dose levels that caused prediabetic T-cell lines to respond (R. Keller, personal communication). These results suggest that T-cells from prediabetics may react with islet gangliosides. The major problem in assessing cell-mediated immunity to gangliosides is the ability of these molecules to modulate the proliferative response of mononuclear cells to mitogens and antigens (for a review, see Ref. 43).

Recent findings suggest that gangliosides may also be used in type I diabetes as a novel therapeutic agent.[94] When gangliosides (5mg/100 g body weight) were given daily to young NOD mice from day 33 until the beginning of overt diabetes, or day 230, one of seven of the ganglioside-treated animal developed diabetes, compared with five of nine saline-treated control animals. In BB/W rats, the incidence of diabetes was 5 of 19 and 6 of 16 after treatment with gangliosides and saline, respectively (spontaneous diabetes incidence less than 40%). In the BB/Ph strain of rats (spontaneous diabetes incidence *ca* 86%) the incidence of diabetes was reported to be 17 of 21 in the ganglioside-treated group and 18 of 22 in the saline-treated group. Thus ganglioside treatment had no effect on the incidence of diabetes in BB rats but decreased the incidence of diabetes in NOD mice. These results suggest that NOD mice may differ from BB rats in their immunologic mechanism of diabetogenesis. The mechanism of the protective effect of gangliosides against diabetes in NOD mice has not been explained, but several considerations are pertinent. First gangliosides can depress antibody responses, inhibit IL-2–dependent lymphocyte proliferation,[43] and inhibit IL-1 secretion; any or all of these effects may protect against diabetes. Second gangliosides can bind to the CD4 molecule on helper T-lymphocytes and selectively cause CD4 to disappear from the helper T-lymphocyte surface.[95] It is known that CD4+ lymphocytes are necessary for the adoptive transfer of diabetes; consequently modulation of the CD4 antigen by gangliosides may inhibit the function of diabetogenic CD4+ T-cells. Finally if one assumes that NOD mice have T- and/or B-lymphocytes that are autoreactive against islet-cell ganglioside, then if the ganglioside mixture contains the autoantigenic ganglioside (either as a major component or as a contaminant), this ganglioside may be able to block the T- or B-lymphocyte receptor to produce an unreactive state through antigen blockade or some other tolerogenic mechanism.

Final Comments

Our studies over the last five years indicate that islet-cell autoantibodies may react with an acidic glycosphingolipid autoantigen(s) of pancreatic islet cells, which is most probably a ganglioside. There is biochemical and immunochemical evidence that islet-cell ganglioside(s) are autoantigens, but it is indirect in nature. Definitive evidence, that is, isolation and structural analysis of an autoantigenic ganglioside, has not been obtained. Once this evidence is obtained, it should be a relatively simple matter to construct a quantitative assay for ICA using the purified antigen. The availability of a defined substrate for ICA assay should solve the problem of interlaboratory variation and should also allow worldwide comparisons of ICA screening results on at risk populations to be made.

The production of autoantibodies by glycosphingolipid antigens is not unique to type I diabetes. We have reviewed some of the literature on glycosphingolipid-directed autoantibodies in this chapter, and it is clear that such autoantibodies are not found exclusively with autoimmune diseases. Such antibodies have been found in various malignancies and in cases of severe trauma. This suggests that anti-glycosphingolipid antibodies can be the result of, rather than the cause of, tissue damage and that they may be involved in clearing away "debris." We do not know what role these antibodies might play in type I diabetes; consequently the possibility that these antibodies are generated secondary to tissue destruction cannot be ignored. It is, however, worth pointing out that the vast majority of gangliosides are expressed at the cell surface, which means that these molecules might well be the primary targets of autoimmunity, although this does not disprove the secondary generation argument. Autoantibodies to glycosphingolipids clearing exist in other autoimmune states, as well as in type I diabetes, although the reports in the literature are, to some extent, conflicting with respect to the nature of the individual glycosphingolipid antigens detected; and there is also some overlap in reactivity range. This is probably due to the fact that normal sera contain anti-glycosphingolipid antibodies and that the absolute nature of the "natural" anti-glycosphingolipid antibody pattern type is affected by different environmental agents. This means that it may be very important to select control populations from the same environment as the at risk or disease study populations.

In conclusion very little is known about the role of glycolipid-directed autoimmunity in diabetes. Nevertheless recent pilot studies indicate that this area of research promises to be a rich field of investigation and innovation, for type I diabetes and for autoimmunity in general.

Acknowledgments. R. C. Nayak was supported by fellowship grants from the Juvenile Diabetes Foundation International (386248 and 387261) and is currently supported by a National Institutes of Health grant (DK-39783-01). G. S. Eisenbarth is supported by a National Institutes of Health grant (DK 32083-06). This work was also supported by a National Institutes of Health Diabetes and Endocrinology Research Center Grant (DK-38836).

We thank the National Diabetes Research Interchange for provision of human pancreas. The authors also wish to thank Ms. Patricia Griffiths for her excellent secretarial assistance.

References

1. Bottazzo GF, Florin-Christensen A, Doniach D. Islet cell antibodies in diabetes mellitus with autoimmune polyendocrine diseases. *Lancet* 1974;ii:1279–1283.
2. MacCuish AC, Barnes EW, Irvine WJ, et al. Antibodies to pancreatic islet cells in insulin dependent diabetics with coexistent autoimmune disease. *Lancet* 1974;ii:1529–1531.
3. Betterle C, Zanette F, Tiengo A, et al. Five year follow up of non-diabetes with islet cell antibodies. *Lancet* 1982;i:284–285.
4. Asplin CM, Cooney MK, Crossley JR, et al. Coxsackie B4 infection and islet cell antibodies three years before overt diabetes. *J Pediatr* 1982;101:398–400.
5. Gorsuch AN, Spencer KM, Lister J, et al. Evidence for a long prediabetic period in Type I (insulin dependent) diabetes mellitus. *Lancet* 1981;ii:1363–1365.
6. Srikanta S, Ganda OP, Eisenbarth GS, et al. Islet cell antibodies and beta cell function in monozygotic triplets and twins initially discordant for Type I diabetes mellitus. *N Engl J Med* 1983;308:322–325.

7. Dean BM, Bottazzo GF, Cudworth AG. IgG subclass distribution in organ specific autoantibodies. The relationship to complement fixing ability. *Clin Exp Immunol* 1983;52:61–66.
8. Omar MAK, Srikanta S, Eisenbarth GS. Human islet cell antibodies: Immunoglobulin class and subclass distribution defined by monoclonal antibodies. *Diabetes Res* 1987;4:155–157.
9. Schatz DA, Barrett DJ, MacLaren NK, et al. Polyclonal nature of islet cell antibodies in insulin-dependent diabetes. *Autoimmunity* 1988;1:45–50.
10. Nayak RC, Omar MAK, Rabizadeh A, et al. Cytoplasmic islet cell antibodies: Evidence that the target antigen is a sialoglycoconjugate. *Diabetes* 1985;34:617–619.
11. Nayak RC, Colman PG, Eisenbarth GS. How are monoclonal antibodies related to autoimmune serology? in Doniach D, Bottazzo GF (eds.): *Baillièrès Clinical Immunology and Allergy: Endocrine and other Organ-Oriented Autoimmune Disorders*. London: Baillière Tindall Ltd., 1987, pp. 81–99.
12. Srikanta S, Eisenbarth GS. Islet cell antigens: Initial studies of their biology and function. *Mol Biol Med* 1986;3:113–127.
13. Hakomori, S-I. Glycosphingolipids in cellular interaction, differentiation, and oncogenesis. *Ann Rev Biochem* 1981;50:733–764.
14. Curatolo W. The physical properties of glycolipids. *Biochim Biophys Acta* 1987;906:111–136.
15. Curatolo W. Glycolipid function. *Biochim Biophys Acta* 1987;906:137–160.
16. IUPAC–IUB Commission on Biochemical Nomenclature (CBN). Nomenclature of lipids recommendations 1976. *Eur J Biochem* 1977;79:11–21.
17. Svennerholm L. Chromatographic separation of human brain gangliosides. *J Neurochem* 1963;10:613–623.
18. Muresan V, Sarras MP Jr, Jamieson JD. Distribution of sialoglycoconjugates on acinar cells of the mammalian pancreas. *J Histochem Cytochem* 1987;30:947–955.
19. Colman PG, Nayak RC, Campbell IL, et al. Binding of cytoplasmic islet cell antibodies is blocked by human pancreatic glycolipid extracts. *Diabetes* 1988;37:645–652.
20. Svennerholm L, Fredman PA. A procedure for the quantitative isolation of brain gangliosides. *Biochim Biophys Acta* 1980;617:97–109.
21. Dotta F, Colman PG, Nayak RC, et al. Differential ganglioside expression between whole human pancreas and isolated human islets of Langerhans. Diabetes (in press).
22. Shienvold FL, Guinguis-Petasne S, Rabinovitch A, et al. Glycolipids as autoantigens in BB/Worcester rats. International Research Symposium, "The Immunology of Diabetes" American Diabetes Association, Woods Hole, MA, Oct 27–30, 1987, Abstr. 115.
23. Hofstetter W, Heusser CH, Blaser K. Nonspecific binding of mouse IgM antibodies to lipid antigens. *J Neuroimmunol* 1985;7:207–214.
24. Dotta F, Bonner-Weir S, Cahill C, et al. Immunoreactivity of islet tumor cells: A new diabetes-associated antigen. *Clin Res* 1988;36:480A.
25. Eisenbarth GS, Walsh FS, Nirenberg M. Monoclonal antibody to a plasma membrane antigen of neurons. *Proc Natl Acad Sci USA* 1979;76:4913–4917.
26. Eisenbarth GS, Shimizu K, Bowring MA, et al. Expression of receptors for tetanus toxin and monoclonal antibody A2B5 by pancreatic islet cells. *Proc Natl Acad Sci USA* 1982;79:5066–5070.
27. Kasai N, Yu RK. The monoclonal antibody A2B5 is specific to ganglioside GQ1c. *Brain Res* 1983;277:155–158.

28. Kundu SK, Pleatman MA, Redwin WA, et al. Binding of monoclonal antibody A2B5 to gangliosides. *Biochem Biophys Res Commun* 1983;116:836–842.
29. Fredman P, Magnani JL, Nirenberg M, et al. Monoclonal antibody A2B5 reacts with many gangliosides in neuronal tissue. *Arch Biochem Biophys* 1984;233:661–666.
30. Powers AC, Rabizadeh A, Akeson R, et al. Characterization of monoclonal antibody 3G5 and utilization of this antibody to immobilize pancreatic islet cell gangliosides in a solid phase radioassay. *Endocrinology* 1984;114:1338–1343.
31. Nayak RC, Colman PG, Halban PA, et al. Identification and purification of islet ganglioside antigen 3G5 and islet cell autoantigen from human pancreas. *Diabetes* 1987; 36(Suppl 1):64A.
32. Rabinowe SL, Nayak RC, Krisch K, et al. Aging in man: Linear increase of a novel T-cell subset defined by antiganglioside monoclonal antibody 3G5. *J Exp Med* 1987; 165:1436–1441.
33. Rabinowe SL, Larsen PR, Antman EM, et al. Autoimmune thyroid disease: Evaluation of a new monoclonal antibody defined T-cell subset. *Am J Med* 1986;81:53–57.
34. Nayak RC, Berman AB, George KL, et al. A monoclonal antibody (3G5)-defined ganglioside antigen is expressed on the cell surface of microvascular pericytes. *J Exp Med* 1988;167:1003–1015.
35. Nayak RC, Berman AB, King GL. Identification of a common antigen shared by retinal capillary pericytes and pancreatic beta cells using a monoclonal antibody (3G5). *Invest Ophthalmol Vis Sci* 1987;28(ARVO suppl):56 abstr.
36. Uchigata Y, Spitalnik SL, Tachiwaki O, et al. Pancreatic islet cell surface glycoproteins containing Galβ1-4GlcNac-R identified by a cytotoxic monoclonal autoantibody. *J Exp Med* 1987;165:124–139.
37. Spitalnik SL, Danley JM, Uchigata Y, et al. An islet cell specific human monoclonal autoantibody from a patient with Type I diabetes recognizes a monosialoganglioside. International Research Symposium "The Immunology of Diabetes" American Diabetes Association, Woods Hole, MA, Oct. 27–30, 1987, Abstr. 113.
38. Alejandro R, Shienvold FL, Vaerewyck-Hajek SA, Pierce M, et al. A ganglioside antigen on the rat pancreatic B-cell surface identified by monoclonal antibody R2D6. *J Clin Invest* 1984;74:25–38.
39. Thomas JW, Virta VJ, Nell LJ. Cell surface antigens on rat islet tumors. *J Immunol* 1987;138:2896–2901.
40. Dib SA, Colman PG, Dotta F, et al. Expression of cytoplasmic islet cell antigens by rat pancreas. *Diabetes* 1987;36:982–985.
41. Hahn HJ, Hellman B, Lernmark A, et al. The pancreatic beta cell recognition of insulin secretagogues. *J Biol Chem* 1974;249:5275–5284.
42. Nielsen DA, Welsh M, Casadaban MJ, et al. Control of insulin gene expression in pancreatic beta cells and in an insulin-producing cell line, RIN-5F cells. *J Biol Chem* 1985;260:13585–13589.
43. Marcus DM. A review of the immunogenic and immunomodulatory properties of glycosphingolipids. *Mol Immunol* 1984;21:1083–1091.
44. Ilyas AA, Quarles RH, Dalakas MC, et al. Polyneuropathy with monoclonal gammopathy: Glycolipids are frequently antigens for IgM paraproteins. *Proc Natl Acad Sci USA* 1985;82:6697–6700.
45. Leibowitz A, Gregson NA, Kennedy M, et al. IgM paraproteins with immunological specificity for a Schwann cell component and peripheral nerve myelin in patients with polyneuropathy. *J Neurol Sci* 1983;53:153–165.

46. Ilyas AA, Quarles RH, MacIntosh TD, et al. IgM in a human neuropathy related to paraproteinemia binds to a carbohydrate determinant in the myelin-associated glycoprotein and to a ganglioside. *Proc Natl Acad Sci USA* 1984;81:1225–1229.
47. Chou KH, Ilyas AA, Evans JE, et al. Structure of a glycolipid reacting with monoclonal IgM in neuropathy and with HNK-1. *Biochem Biophys Res Commun* 1985;128:383–388.
48. Adams DD. Thyroid stimulatory autoantibodies. *Vitam Horm* 1980;38:119–203.
49. Drexhage HA, van der Gaag RD, Wulffraat NM. The multiplicity of stimulating and blocking autoantibodies in relation to thyroid and extrathyroidal tissues, in Doniach D, Bottazzo GF (eds): *Baillière's Clinical Immunology and Allergy: Endocrine and Other Organ-Oriented Autoimmune Disorders*. London: Baillière Tindall Ltd., 1987, pp. 125–140.
50. Kohn LD, Yavin E, Yavin Z, et al. Autoimmune thyroid disease studied with monoclonal antibodies to the thyrotropin receptor, in Haynes BF, Eisenbarth GS (eds): *Monoclonal Antibodies-Probes for the Study of Autoimmunity and Immunodeficiency.* Orlando, FA: Academic Press, 1983, pp 221–258.
51. Feizi T, Hadler N. Autoantibodies and disease, in Elkeles RS, Tavill AS (eds): *Biochemical Aspects of Human Disease.* Oxford, England: Blackwell Scientific, 1983, pp 656–692.
52. Hirano T, Hashimoto H, Shiokawa Y. Antiglycolipid autoantibody detected in the sera from systemic lupus erythematosus patients. *J Clin Invest* 1980;60:1437–1440.
53. Endo T, Scott DD, Stewart SS, et al. Antibodies to glycosphingolipids in patients with multiple sclerosis and SLE. *J Immunol* 1984;132:1793–1797.
54. Kibler RF, Fritz RB, Chou FC-H, et al. Immune response of Lewis rats to peptide C1 (residues 68–88) of guinea pig and rat myelin basic proteins. *J Exp Med* 1977;146: 1323–1331.
55. Hirsch HE, Parks ME. Serological reactions against glycolipid-sensitized liposomes in multiple sclerosis. *Nature* 1976;264:785–787.
56. Arnon R, Crisp E, Kelley R, et al. Anti-ganglioside antibodies in multiple sclerosis. *J Neurol Sci* 1980;46:179–186.
57. Fry JM, Weissbarth S, Lehrer GM. Cerebroside antibody inhibits sulfatide synthesis and myelination and demyelinates in cord tissue cultures. *Science* 1974;183:540–542.
58. Schwerer B, Kitz K, Lassmann H, et al. Serum antibodies against glycosphingolipids in chronic relapsing experimental allergic encephalomyelitis. *J Neuroimmunol* 1984; 7:107–119.
59. Decerf-Etienne J, Malaise M, Mahieu P, et al. Elevated anti-galactosyl antibody titres: A marker of progression in autoimmune thyroid disorders and in endocrine ophthalmopathy? *Acta Endocrinol* 1987;115:67–74.
60. Spiro RC, Bhoyroo VD. Occurrence of α-D-galactosyl residues in thyroglobulin from several species. Localization in the saccharide chains of the complex carbohydrate units. *J Biol Chem* 1984;259:9858–9866.
61. Kress BC, Spiro RG. Studies on the glycoprotein nature of the thyrotropin receptor: Interaction with lectins and purification of the bovine protein with the use of Bandeiraea (Griffonia) simplicifolia I affinity chromatography. *Endocrinology* 1986; 118:974–979.
62. Galili U, Basbaum CB, Shohet SB, et al. Identification of erythrocyte Galα1-3Gal glycosphingolipids with a mouse monoclonal antibody, Gal-13. *J Biol Chem* 1987; 262:4683–4688.

63. Witkin SS, Sonnabend J, Richards JM, et al. Induction of antibody to asialo GM1 by spermatozoa and its occurrence in the sera of homosexual men with the acquired immune deficiency syndrome (AIDS). *Clin Exp Immunol* 1983;54:346–350.
64. Yasuda T, Ueno J, Matuhasu T. Antiglycolipid antibodies in Behcet's disease, in Inaba G (ed): *Behcet's Disease: Pathogenetic Mechanism and Clinical Future.* Tokyo: University of Tokyo Press, 1981, pp 413–420.
65. Inaba G, Aoyama J. Anti-glycolipid antibodies in neuro-Behcet's syndrome, in Inaba G (ed): *Behcet's Disease: Pathogenetic Mechanism and Clinical Future.* Tokyo: University of Tokyo Press, 1981, pp 145–152.
66. Higashi H, Naiki M. Antigen of "serum sickness" type of heterophile antibodies in human sera: Identification as gangliosides with N-glycolylneuraminic acid. *Biochem Biophys Res Commun* 1977;79:388–395.
67. Mukuria CJ, Fujii Y, Kato S, et al. Specificities of human heterophile Hanganutziu and Deicher (H-D) antibodies to glycosphingolipids and a glycoprotein. *J Biochem* 1986;100:469–475.
68. Naiki M, Higashi H. Detection of antibodies to gangliosides in pathologic human sera. Serum sickness type heterophile antibodies. *Adv Exp Med Biol* 1979;125:359–367.
69. Kasukawa R, Kyoichi K, Bloom ML, et al. Heterophile antibodies in pathologic human sera resembling antibodies stimulated by foreign species sera. *Clin Exp Immunol* 1976;25:122–132.
70. Naiki M, Fujii Y, Ikuta K, et al. Expression of Hanganutziu and Deicher type heterophile antigen on the cell surface of Marek's disease lymphoma, in Makita A, Handa S, Taketomi T, Nagai Y (eds): *New Vistas in Glycolipid Research.* New York/London: Plenum Publishing, 1982, pp 445–456.
71. Yasuda T, Ueno J, Naito Y, et al. Antiglycolipid antibodies in human sera, in Makita A, Handa S, Taketomi T, Nagai Y (eds): *New Vistas in Glycolipid Research.* New York/London: Plenum Publishing 1982, p 457–465.
72. Rapport MM, Graf L, Huang Y-YL, et al. Antibodies to total brain gangliosides: Titer and specificity of antisera. *Adv Exp Med Biol* 1979;125:327–334.
73. Marcus DM, Kundu SK. Preparation and properties of antibodies to gangliosides. *Adv Exp Med Biol* 1979;125:321–326.
74. Rapport MM, Huang Y-Y. Present status of the immunology of gangliosides. *Adv Exp Med Biol* 1984;174:15–25.
75. Watarai S, Handa S, Tadakuma T, et al. Application of liposomes to generation of monoclonal antibody to glycosphingolipid: Production of monoclonal antibody to $GgOse_4Cer$. *J Biochem* 1987;102:59–67.
76. Fredman P, Mattsson L, Andersson K, et al. Characterisation of the binding epitope of a monoclonal antibody to sulphatide. *Biochem J* 1988;251:17–22.
77. Bottazzo GF, Dean BM, McNally JM, et al. In situ characterization of autoimmune phenomena and expression of HLA molecules in the pancreas in diabetic insulitis. *N Engl J Med* 1985;313:353–360.
78. Jackson RA, Morris MA, Haynes BF, et al. Increased circulating Ia-antigen-bearing T cells in Type I diabetes mellitus. *N Engl J Med* 1982;306:785–788.
79. Bougneres PF, Carel JC, Castano L, et al. Factors associated with early remission of Type I diabetes in children treated with cyclosporine. *N Engl J Med* 1988;318: 663–670.
80. Wicker LS, Miller BJ, Mullen Y. Transfer of autoimmune diabetes mellitus with splenocytes from non-obese diabetic (NOD) mice. *Diabetes* 1986;35:855–860.

81. Mordes JP, Rossini AA. Keys to understanding autoimmune diabetes mellitus: The animal models of insulin dependent diabetes mellitus, in Doniach D, Bottazzo GF (eds): *Baillière's CLinical Immunology and Allergy: Endocrine and Other Organ-Oriented Autoimmune Disorders*. London: Baillière Tindall Ltd., 1987, pp 29–52.
82. Miller BJ, Appel MC, O'Neil JJ, et al. Both the LYT-2 and L3T4 T cell subsets are required for the transfer of diabetes in nonobese diabetic mice. *J Immunol* 1988;140: 52–58.
83. Jacob CO, Arnon R, Sela M. Effect of carrier on the immunogenic capacity of synthetic cholera vaccine. *Mol Immunol* 1985;22:1333–1339.
84. Bellamy A, Davison AN, Feldmann M. Derivation of ganglioside-specific T-cell lines of suppressor or helper phenotype from cerebrospinal fluid of multiple sclerosis patients. *J Neuroimmunol* 1986;12:107–120.
85. Offner H, Konat G. Stimulation of active E-rosette forming lymphocytes from multiple sclerosis patients by gangliosides and cerebrosides. *J Neurol Sci* 1980;46:101–104.
86. Offner H, Konat G, Della BA. Multi-sialo brain gangliosides are powerful stimulators of active E-rosetting lymphocytes from multiple sclerosis patients. *J Neurol Sci* 1981a;52:279–287.
87. Ilyas AA, Davison AN. Cellular hypersensitivity to gangliosides and myelin basic protein in multiple sclerosis. *J Neurol Sci* 1983;59:85–95.
88. Handman E, Greenblatt CL, Goding JW. An amphipathic sulphated glycoconjugate of Leishmania: Characterization with monoclonal antibodies. *The European Molecular Biology Organisation J* 1984;3:2301–2306.
89. Handman E, Goding JW. The Leishmania receptor for macrophages is a lipid containing glycoconjugate. *The European Molecular Biology Organisation J* 1985;4:329–336.
90. Mitchell GF, Handman E. The glycoconjugate derived from a Leishmania major receptor for macrophages is a suppressogenic, disease-promoting antigen in murine cutaneous leishmaniasis. *Parasite Immunol* 1986;8:255–263.
91. Scott P, Pearce E, Natovitz P, et al. Vaccination against cutaneous leishmaniasis in a murine model: II. Immunologic properties of protective and non-protective subfractions of a soluble promastigote extract. *J Immunol* 1987;139:3118–3125.
92. Handman E, Mitchell GF. Immunization with leishmania receptor for macrophages protects mice against cutaneous leishmaniasis. *Proc Natl Acad Sci USA* 1985;82: 5910–5914.
93. Titus RG, Lima GC, Engers HD, et al. Exacerbation of murine cutaneous leishmaniasis by adoptive transfer of parasite-specific helper T-cell populations capable of mediating leishmania major-specific delayed-type hypersensitivity. *J Immunol* 1984;133: 1594–1600.
94. Wilberz S, Herberg L, Renold AE, et al. Ganglioside (Cronassial®) treatment depresses diabetes incidence in NOD mice, not, however, in BB rats. Lessons from Animal Diabetes II, Second International Workshop, Geneva Switzerland. Sept. 9–13, 1987, Abstr 252.
95. Offner H, Thieme T, Vandenbart AA. Gangliosides induce selective modulation of CD4 from helper T lymphocytes. *J Immunol* 1987;139:3295–3305.

7 Insulin Autoantibodies

Jerry P. Palmer

Although the precise immunologic mechanisms that destroy beta-cells are not completely understood, it is currently accepted that insulin-dependent diabetes (IDDM) in man is an autoimmune disease. The immunologic attack on beta-cells usually begins several years before the patient presents with clinical IDDM and occurs during an asymptomatic "prediabetic" period. Antibodies directed against islet antigens and evidence of cell-mediated immunity against the islet were described in newly diagnosed IDDM patients over 10 years ago. Some investigators interpreted these findings as evidence that beta-cell destruction of IDDM was immunologically mediated, but others argued that these markers could result from beta-cell destruction by another mechanism. The more recent findings of Sutherland and colleagues,[1] that the pancreas from one nondiabetic twin in a set of discordant identical twins is rapidly destroyed by the diabetogenic process when transplanted into the diabetic twin, and the ability of such immunosuppressive drugs as azothioprine[2] and cyclosporin A[3] to preserve beta-cell function and increase remission rates in newly diagnosed, insulin-dependent diabetics, provide the strongest evidence that IDDM is the result of an immunologic attack on the beta-cells.

The antibodies against islet antigens that are found in the blood of newly diagnosed, insulin-dependent diabetics include islet-cell antibodies [both cytoplasmic (ICA) and cell-surface directed (ICSA)], antibodies to a 64-kD islet protein (A–64-kD), and insulin autoantibodies (IAA). Upon realizing that the beta-cell destructive process in IDDM is not acute, investigators looked for and found these antibodies in prediabetics years before the onset of clinical IDDM. As an immunologic marker of IDDM, IAA offers several advantages. The antigen was known and available in highly purified form, whereas this was not true for ICA, ICSA, and A–64-kD. Assays for IAA were quantifiable, compared to the subjective and/or difficult to quantify other assays, and finally, the antigen was also beta-cell–specific. A clear understanding of IAA and IDDM, however, is still not available. In this chapter, I will review the information currently available regarding IAA and discuss the possible importance of IAA in IDDM autoimmunity.

Insulin autoantibodies are defined by their presence in an individual or animal that has not received exogenous insulin. Before it was realized that IAA were

TABLE 7.1. Conditions Associated with Insulin Autoantibodies (IAA).

Insulin autoimmune syndrome	
Viral Infections	
	Chicken pox, mumps, measles
Autoimmune diseases other than diabetes	
	Graves' disease, systemic lupus erythematosus
Drugs	
	Penicillamine, methimazole, α-mercaptopropionyl glycine
Insulin-dependent diabetes	

found in IDDM, IAA had been described in rare patients with hypoglycemia. This was called the insulin autoimmune syndrome (IAS) and was first described by Hirata and colleagues in 1970.[4] Følling and Norman reported a thoroughly investigated case of IAS in *Diabetes* in 1972.[5] Although IAS is claimed to be the third leading cause of hypoglycemia in Japanese,[6] it is quite rare in Caucasians. Whether genetic or environmental factors account for this marked difference is unknown. As demonstrated in the Norwegian patient described by Følling and Norman,[5] and subsequently confirmed by other investigators studying other patients,[7] hypoglycemia is caused by a dissociation of biologically active insulin from the insulin antibody complex. In fact Hirata and colleagues reported hypoglycemia in some animals immunized with insulin after termination of insulin treatment.[8] The syndrome can occur spontaneously or after administration of methimazole, penicillamine, and α-mercaptopropionyl glycine.[9–11] Whether any of these drugs specifically induces the syndrome and, if so, how, is unknown. Penicillamine can cause other autoimmune reactions, and each of these drugs contains a sulfhydryl group that theoretically could interact with the disulfide bonds of insulin to make it immunogenic by forming haptens or by structurally altering the insulin molecule. Methimazole has been shown to enhance the lymphocyte proliferative response to nonspecific stimulation.[12] Possibly, it may also stimulate IAA production to detectable and sometimes clinically significant levels in individuals who were previously producing IAA at undetectable levels. The possible presence of very low levels of IAA in "normal" individuals is discussed in more detail subsequently.

In a few insulin autoimmune syndrome patients, IAA have been characterized and found to be IgG, with binding constants and capacities similar to those of insulin antibodies from insulin-treated diabetics.[7,13] Unlike insulin antibodies induced by insulin therapy, which are polyclonal and have both lambda and kappa light-chains, IAA from some patients with the insulin autoimmune syndrome are exclusively kappa light-chains,[14,15] which suggests a more restricted clonality.

IAA have also been identified in asymptomatic individuals. Common viral infections can induce IAA, although they then appear to be predominantly IgM.[16]

TABLE 7.2. Prevalence of IAA in Newly Diagnosed IDDM.

	Unextracted sera	Acid charcoal extraction	Acid charcoal extraction/ Displacement by cold insulin
Cut-point			
Upper range	17%	26%	39%
$\bar{X}$ + 2 SD	33%	36%	39%

Diaz and Wilkin screened over 2000 sera samples from patients with a provisional diagnosis of an autoimmune disorder and found IAA in 3.2%.[17] Nell and coworkers have also found IAA in asymptomatic patients with Graves' disease and with systemic lupus erythematosis.[18] We also occasionally find IAA in such individuals. Some of the sera from Wilkin's patients were specific for human versus porcine or bovine insulin.[19] Similar to the IAA from IAS patients, IAA from these autoimmune patients were either kappa or lambda light-chains, rather than both kappa and lambda.[20] We have not compared the various conditions in which IAA occur. These conditions, however, are listed in Table 7.1.

IAA were first reported in association with diabetes in 1963. Páv and colleagues studied 75 diabetics, only three of whom had been treated with insulin in the distant past. Using a complement consumption technique, they found insulin antibodies in 34% of the patients compared to 4% of the healthy controls. This was before IDDM and non-insulin-dependent diabetes mellitus (NIDDM) were recognized as separate diseases, and these diabetics, unfortunately, were not further characterized. Far ahead of their time, these authors suggested that the IAA may have been associated with a silent inflammation of the pancreas or that the antibodies were involved in the pathogenesis of diabetes. Their observation was not confirmed by other investigators who used a variety of techniques, and most authorities in the field had concluded that, in humans, insulin antibodies were not found in the absence of exogenous insulin treatment; nonetheless scattered reports continued to support the presence of IAA. Sebriakova and Little described two untreated diabetics who had insulin antibody indices above their assay sensitivity.[22] Insulin-induced lymphocyte transformation, which suggested sensitization to insulin, was reported in newly diagnosed untreated IDDM in 1975.[23]

Recognition of IAA as part of the autoimmune picture of IDDM started to become accepted with our finding of IAA in 18% of newly diagnosed, untreated, insulin-dependent diabetics,[24] although some workers continued to doubt the validity of this observation.[25] Improvements in assay methodology that decreased nonspecific binding, especially the availability of high-purity, mono-iodinated A-14 insulin, are probably responsible for our ability to detect IAA in untreated IDDM patients. With further refinements in the insulin antibody assay, we now find that approximately 40% (Table 7.2) of newly diagnosed, untreated IDDM patients have IAA.[26] This agrees with data from several other laboratories that have reported frequencies ranging from 20 to 50%.[27–32] Some of the variability between laboratories clearly relates to differences in assay technique and in patient selection, as well as

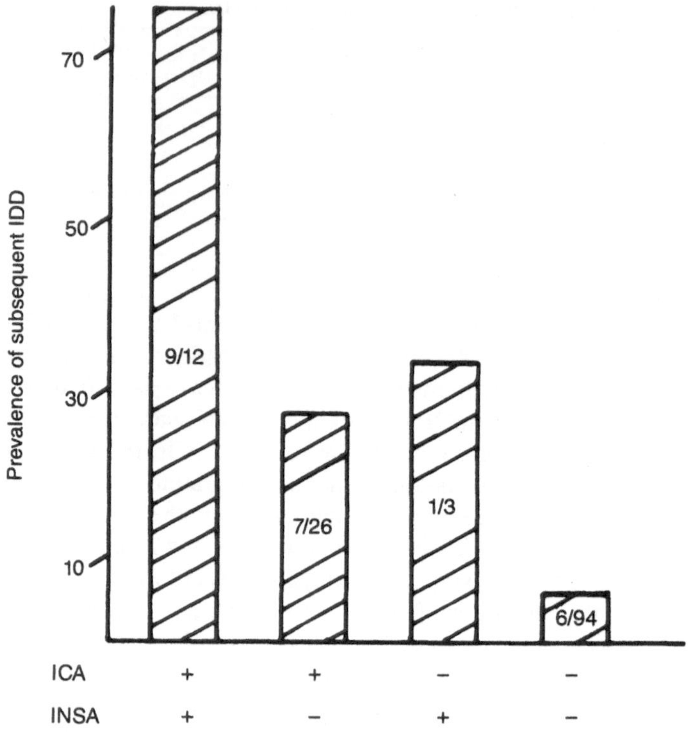

FIGURE 7.1. The prevalence of subsequent IDDM in high-risk, nondiabetic subjects positive (+) or negative (−) for islet-cell antibodies (ICA) and/or insulin autoantibodies (INSA). Numbers within the bars represent the actual number who developed IDDM over the number tested (from Ref. 34).

the cut-point used to define IAA, which has varied from mean plus 2 SD to mean plus 6 SD or an upper range of values in the normal population.[26-32]

As mentioned above, IAA have also been found in nondiabetic individuals who subsequently developed clinical IDDM, prompting the hypothesis that the presence of IAA has predictive value for subsequent development of IDDM. McEvoy and colleagues found IAA in sera obtained in the prediabetic period from four individuals who subsequently developed diabetes,[29] and Soeldner and colleagues reported a single patient who was IAA- and ICA-positive five years before the diagnosis of clinical IDDM was made.[33]

In collaboration with the above investigators, we measured IAA in sera that had been collected over many years from a heterogenous group of high-risk individuals, including discordant monozygotic twins; islet-cell–antibody (ICA)-positive, first-degree relatives of patients with IDDM; and patients with a history of transient glycosuria.[34] We found that IAA were present in 32% of ICA-positive individuals and that the presence of IAA plus ICA was associated with a greater

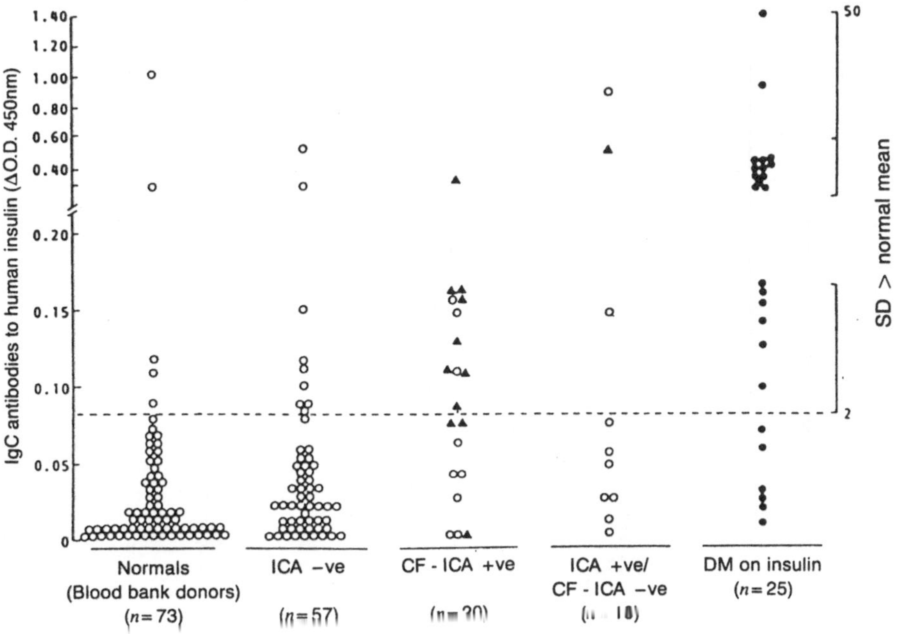

FIGURE 7.2. ELISA measurements of insulin antibodies in normal subjects, first-degree relatives of IDDM patients grouped according to ICA status, and diabetics on insulin. Subsequent development of IDDM (▲) (from Ref. 35 with permission of Springer-Verlag Heidelberg).

frequency of these individuals subsequently developing IDDM than the presence of either marker alone (Fig. 7.1). Similar results have recently been reported from the Barts–Windsor–Middlesex prospective family study for IDDM.[35] In a subset of individuals selected on the basis of ICA status, IAA were significantly associated with ICA-positive, and especially complement-fixing (CF-ICA)-positive subjects as opposed to ICA-negative, first-degree relatives. Of the subjects evaluated, 12 subsequently developed IDDM; of 11 individuals who were positive for both complement-fixing, islet-cell antibodies (CF-ICA) and IAA, 8 had developed IDDM as opposed to 3 of 9 of those subjects who were CF-ICA positive but IAA negative. Only 1 of 12 subjects was IAA positive but CF-ICA negative (Fig. 7.2). Further confirming the potential value of IAA as a marker of beta-cell destruction in IDDM, Atkinson and colleagues compared the beta-cell response to intravenous glucose in ICA-positive nondiabetics who were either IAA-positive or -negative. Of the ICA- and IAA-positive subjects, 70% were insulinopenic, whereas only 26% of the ICA-positive but IAA-negative subjects were insulinopenic.[27]

Although the above data strongly support the concept that IAA in individuals at high risk of IDDM may be a marker of beta-cell destruction, and therefore predictive of subsequent clinical IDDM, other data challenge this conclusion. Wilkin and colleagues found IAA in 47% of discordant identical twins of patients

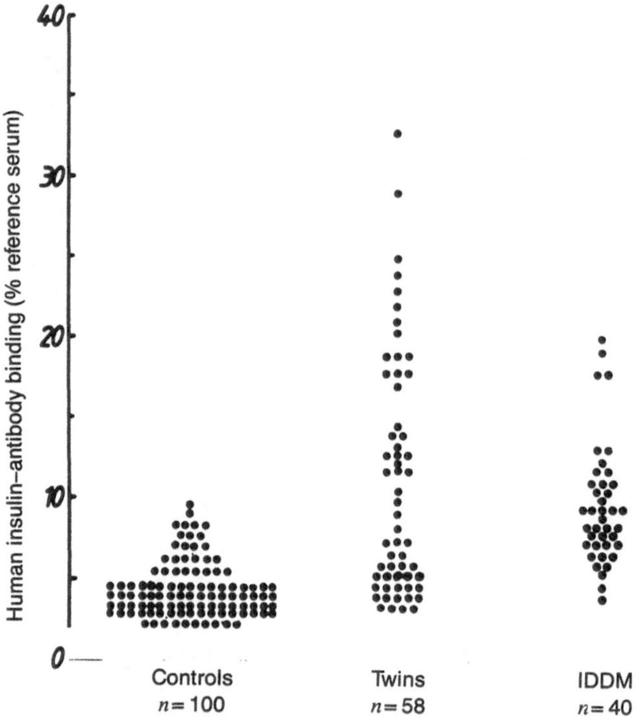

FIGURE 7.3. Insulin antibody binding in sera from controls, nondiabetic twins of insulin-dependent diabetics, and untreated, newly diagnosed patients with IDDM (from Ref. 31).

with insulin-dependent diabetes (Fig. 7.3). The frequency of IAA, however, was not related to duration of discordance nor to ICA status, and in those twins in whom more than one measurement was available, the IAA titer fluctuated over time.[31] Since twins who had been discordant for a long period of time were unlikely to develop diabetes, these authors argued that IAA were not highly predictive of subsequent IDDM. The BB rat is a strain of Wistar rat, in which approximately 50% of animals develop IDDM that is similar in many ways to the human disease. IAA were found in many of the BB rats that later developed IDDM, but in agreement with the above twin data, IAA were also found in many of the BB rats that did not develop diabetes.[36] Similarly in the NOD mouse, another animal model of IDDM, IAA have been found at an early age in the majority of animals but are not predictive of which animals will subsequently develop diabetes.[37] Adding to this controversy are our findings regarding IAA in sera collected from BB rats before and after the onset of overt diabetes. We are unable to detect IAA, even though we used high purity rat I and II insulin tracers in our assay and even though we can easily detect insulin antibodies in diabetic BB rats treated with exogenous insulin. Possibly these discrepant results in human IDDM and animal models are due to differences in assay methodology wherein IAA measured by some techniques have greater predictive value for subsequent IDDM than IAA determined by other methods.

It must also be emphasized that the data cited in the above discussion of the predictive value of IAA were all obtained from individuals who were at high risk of IDDM, genetically, from first-degree relatives of a proband with IDDM, or from rat and mouse animal models in which IDDM is genetically determined. The predictive value of IAA in the general population is unknown, but it is presumably quite low, considering the finding of IAA in the insulin autoimmune syndrome, in Graves' disease and systemic lupus erythematosus (SLE), after treatment with certain drugs, such as penicillamine and methimazole, and after certain viral infections.

Considerations regarding the predictive value of IAA (or other immunologic markers) for subsequent clinical IDDM are further compromised by our ignorance of the course of events during the preclinical period of IDDM. There are two major hypotheses. The first states that beta-cell destruction, once initiated, relentlessly and progressively destroys enough beta-cells so that clinical IDDM occurs.[33,38] Immunologic markers of this destructive process would then be highly predictive. The alternative hypothesis states that the beta-cell destructive process waxes and wanes, and even remits in some people, without causing end-stage, beta-cell secretory failure and clinical IDDM.[39-41] If IAA and/or other immunologic markers are present only when the disease process is active and disappear when the disease process remits, their predictive value for subsequent IDDM will be low. A definite answer to this question requires large-scale, population-based prospective studies employing careful evaluation of beta-cell function and immunologic markers. Such studies are underway.

Associations in IDDM patients of IAA with other defining characteristics such as HLA type, age, sex, presence of ICA, and C-peptide (CP) status have also been investigated. Unfortunately there is little agreement among the published reports. We found that IAA were more prevalent in ICA-positive, high-risk nondiabetics than in ICA-negative nondiabetics.[34] Some investigators have also noted this association,[27,35] but others have not.[28,30,31] Since IDDM is associated with DR3 and DR4 and since many investigators have found associations of HLA type with insulin antibodies induced by insulin therapy,[42,43] namely, higher antibody titers in DR4 and lower antibody titers in DR3 individuals, an association of IAA with HLA has been sought. Although some investigators have found an association between IAA and HLA-DR3 or HLA-DR4 or both,[27,28] we, and other investigators, have not found such an association.[30,31] Reports of apparent associations of IAA with sex[29] or CP status[28] are also disputed.[30,44] In contrast a number of groups agree that IAA tend to be more prevalent and that binding is higher in younger individuals.[28-30,34]

A completely satisfactory explanation for the apparently conflicting data regarding IAA is not available, but most assuredly some of the differences are the result of differences in the populations studied and the cut-points these investigators have used to define IAA. In addition several different IAA assays have been employed and some characteristics of IAA that vary between individuals could affect measurement of IAA differently in different assays. For instance some IAA are selective for human insulin and do not bind to porcine or bovine insulin.[17,19] In assays that employ porcine ligand, the results would be different from the results from assays employing human ligand. IAA have been reported to be IgG and IgM. Common viral infections can induce IAA of the IgM type,[16] but IAA of the IgG type appear to be more closely associated with IDDM.[35] Assays that measure binding due to both IgG and IgM would be expected to give different results from assays that measure only IgG binding. In insulin-treated diabetics, there are marked differences between individuals in terms of the binding capacity and affinity of the insulin antibodies formed.[43] If the binding characteristics of IAA also vary between individuals, assays which are more sensitive to either affinity or capacity would be expected to give different results. Insulin autoantibodies probably vary in terms of which epitopes on the insulin molecule are recognized. In fluid phase assays, all the epitopes on the ligand would be available, whereas the binding of insulin to plates in ELISA assays might make certain epitopes inaccessible. Consequently fluid- versus solid-phase assays might give different results depending upon the epitope-specificity of a patient's IAA. To date inadequate attention has been paid to the different assays used to measure IAA. In the First Workshop on Standardization of IAA, sponsored by the International Immunology and Diabetes Workshops, 21 laboratories assayed a panel of sera. Large variations between laboratories were observed, and some of the potential sources of variability mentioned above were identified.[45] Since the assays currently in use are markedly different, and each has advantages and disadvantages, a single "gold standard" assay is probably not a realistic or appropriate goal. Rather we must completely understand the complexities and limitations of each assay and define precisely what each assay is measuring. Until this is accomplished, observations from different laboratories will continue to be disparate and confusing.

The pathogenic mechanism(s) whereby the immune system loses tolerance and becomes directed against self-antigens in autoimmune diseases is an area of intense research albeit poorly understood. Similarly the mechanism responsible for IAA development is unknown, but several hypotheses can be proposed (Fig. 7.4). Changes in insulin structure that make it immunogenic may occur during beta-cell destruction in IDDM, with the consequent development of IAA. This mechanism was originally proposed by Páv and coworkers in 1963.[21] The fact that diabetic subjects develop insulin antibodies even when they are treated with human insulin suggests that very minor changes in the human insulin molecule can make it immunogenic.

Another hypothesis suggests that insulin per se may not be the antigen. The observation of Kaplan and colleagues that insulin immuno-reactivity, possibly pre-proinsulin, is present in the plasma membrane of beta-cells,[46] raises the possibility that an insulin precursor not normally circulating is released during beta-

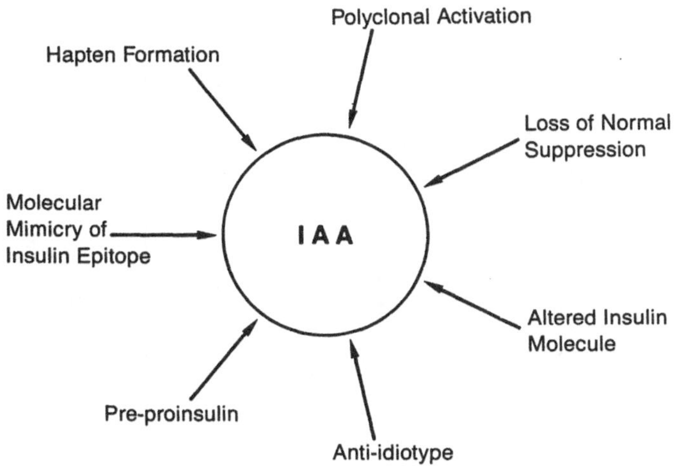

FIGURE 7.4. Possible mechanisms of IAA formation.

cell destruction and becomes antigenic. Alternatively the antigen may be unrelated to insulin (e.g., a virus), but the viral antibody may cross-react with insulin. It is well known that antibodies are frequently not specific for a single antigen but rather cross-react with a variety of different antigens. Molecular mimicry between epitopes on microorganisms and host proteins has been postulated for several diseases, including allergic encephalitis wherein antibodies to hepatitis B viral peptide cross-react with myelin basic protein, and rheumatic fever wherein antistreptococcal M-protein antibodies cross-react with cardiac myosin.[47] Even monoclonal antibodies that are derived from animals and humans with autoimmune diseases frequently react with a variety of endocrine tissues, including islet cells.[48–50] That is the monoclonal antibodies recognize epitopes on more than one protein and some of the antigens that have been identified as hormones include insulin. The finding of Notkins and colleagues that diabetes induced in mice by reovirus type 1 is associated with many autoantibodies, including IAA,[51] lends credence to this. Additional data supporting this concept of molecular mimicry come from the recent work of Leiter and colleagues. In two mouse models of diabetes, they found that transcription and translation of endogenous retrovirus within beta-cells are stimulated by glucose. The animals develop autoantibodies to p^{73}, a specific antigen of the retrovirus and autoantibodies to insulin. It appears that p^{73} and insulin share a common epitope, since the time-course of the autoantibodies measured against p^{73} and against insulin are very similar and binding to one antigen is reduced by absorption with the other antigen.[52]

The variable regions of antibody molecules (idiotypes) are themselves antigenic and induce anti-idiotypic antibodies. It has been postulated that in the

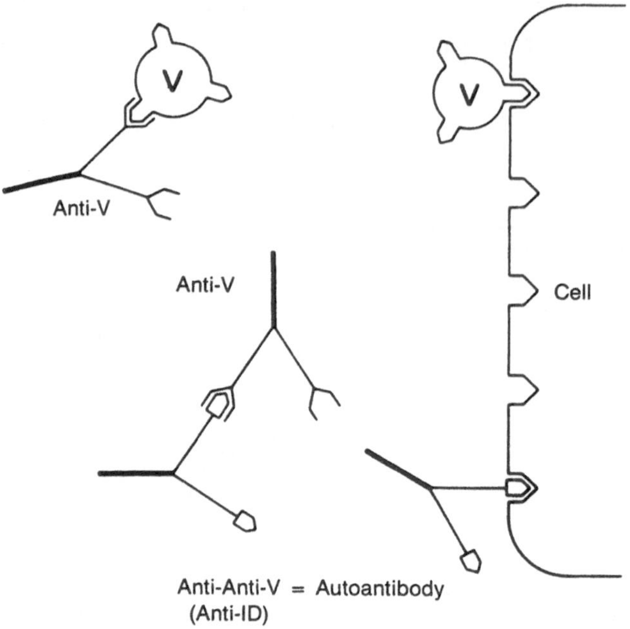

FIGURE 7.5. Schematic illustration of how an anti-idiotypic antibody might mimic the original antigen (v) and bind to its cell surface receptor (from Ref. 53).

cascade of anti-idiotypic antibodies, which are produced in response to a virus or some other foreign antigen (Fig. 7.5), anti-idiotypic antibodies that react with self-antigens are generated; they could thus account for the development of IAA.[53]

An additional hypothesis is that IAA arise as a result of polyclonal immunocyte activation. The finding of IAA (IgM in type) after several different common viral infections, including mumps, rubella, chicken pox, and measles, is most compatible with this mechanism.[16] Further support for polyclonal activation comes from a murine model of SLE. In certain strains of mice, chronic B-cell stimulation with a nonantigenic but mitogenic lipid A fraction of lipopolysaccharide resulted in autoantibody production, including single-strand DNA antibodies and a fatal immune complex-mediated glomerulonephritis.[54]

And finally it has been proposed that autoantibodies are part of the normal B-cell repertoire but that production is normally kept at very low levels, through immunologic suppression. Loss of this immunologic suppression could then enhance autoantibody production. Gaither and colleagues recently found low levels of autoantibodies characteristic of SLE and Sjögren's syndrome in normal individuals and suggested that the transition from a normal state to the expression of disease occurs through the loss of T-cell suppressor function, which activates and expands autoreactive clones.[55] Further support for this hypothesis comes from some recent work on autoimmune chronic active hepatitis. Sensitization of helper T-cells to a liver-cell membrane protein in these patients has been found to be associated with a defect in the T-cells that specifically induce suppressor lymphocytes.[56] At present it is not known whether one or more of these mechanisms accounts for IAA in humans. The mechanisms proposed above are not mutually exclusive and, given the heterogeneity of situations in which IAA have been described, it is likely that more than one mechanism accounts for the spectrum of insulin autoantibodies.

Although IAA are clearly associated with IDDM and are present during the prediabetic period, the pathophysiologic role, if any, of IAA in the beta-cell destructive process of IDDM is unknown. I will discuss four major possibilities, but first it is important to reemphasize that IAA vary and that it is likely that the pathophysiologic roles of different types of IAA also vary. In the Barts–Windsor–Middlesex study mentioned earlier, both IgM and IgG IAA were measured but only the IgG IAA were associated with CF-ICA and of predictive value for subsequent IDDM.[35] Wilkin recently reported preliminary observations that the IAA associated with IDDM versus the IAA found in nondiabetic patients with other autoimmune diseases are different in terms of ligand-specificity and in terms of reactivity with a panel of anti-IAA monoclonal antibodies.[57] In any discussion of the possible pathophysiologic roles of IAA, the likelihood that different types of IAA have different immunologic roles should be remembered.

First IAA may not even be markers of the beta-cell lesion of IDDM; but they might just tend to occur in individuals genetically predisposed to autoimmunity, and a small proportion of these people might, in turn, be predisposed to IDDM. Or IAA could occur in individuals genetically susceptible to IDDM and reflect a propensity for beta-cell autoimmunity; here, however, the immunologic mechanism destroying the beta-cells could be separate and not activated in many of the individuals. The finding of IAA in individuals with other autoimmune disorders such as Graves' disease and SLE, in the insulin autoimmune syndrome, and in individuals treated with such drugs as penicillamine suggests that IAA can occur without being a marker of beta-cell damage. Second IAA may not be pathogenetically involved but rather may be a marker of beta-cell destruction. One or more of the mechanisms previously discussed for the development of IAA may be activated during beta-cell destruction to produce IAA. IAA formation would be considered a result rather than a cause of beta-cell damage and could be likened to the bystanders who gather at the scene of an accident.

A third hypothesis suggests that IAA are directly involved in autoimmune beta-cell destruction of IDDM. Localization of immunoreactive insulin to the beta-cell membrane by Kaplan and colleagues[46] provides a mechanism whereby IAA, by detecting this marker, could give beta-cell–specificity to the autoimmune destructive process of IDDM. Potential mechanisms include complement-dependent, antibody-mediated cytotoxicity or antibody-dependent cellular cytotoxicity, but there is no experimental evidence that IAA are involved in either of these processes. Reports, in 1966, by two separate groups that immunization of animals with exogenous insulin produced lymphocytic infiltration of the pancreas, beta-cell destruction, and, in some animals, hyperglycemia or overt diabetes support this hypothesis[58,59] Alpha-cells were unaffected. The authors appropriately noted the similarities of these islet lesions to those observed in newly diagnosed human IDDM.[60] More recently Ludvigsson and colleagues have reported that residual beta-cell function deteriorates more rapidly in diabetics who have higher insulin antibody titers and suggest that diabetics should be treated with low antigenicity insulin to minimize beta-cell loss by this mechanism.[61] Whether the observations discussed above, which pertain to IA produced by exogenous insulin injection, are relevant to IAA is unknown and can be justifiably questioned. Opposing the hypothesis that IAA either directly or indirectly mediate beta-cell destruction are other reports that fail to find a relationship between residual beta-cell function and insulin antibody titer,[62] the lack of an association between IAA and CP levels at diagnosis of IDDM,[30,44] and the observation that insulin treatment of NIDDM with the subsequent development of insulin antibodies does not appear to affect underlying beta-cell function.

As discussed earlier, the variable regions of antibody molecules (idiotypes) can induce the formation of anti-idiotypic antibodies. In 1974 Jerne proposed that a network of idiotypic and anti-idiotypic antibodies served to regulate immune responses.[63] The antibodies formed in response to an antigen induce anti-idiotypic antibodies, which inhibit the idiotypic response to the original antigen. This proposed inhibitory effect of anti-idiotypic antibodies has recently been confirmed. Immunization of animals against an idiotype on the autoantibody to thyroglobulin suppressed their autoantibody response to thyroglobulin,[64] and a monoclonal anti-idiotypic antibody to anti-DNA antibodies has been shown to inhibit anti-DNA antibody production in vitro.[65] Studying the antibody response to bovine serum albumin (BSA), Eddy and colleagues have shown that the mechanism of this inhibition was anti-idiotypic antibody induction of suppressor T-cells that specifically inhibit antibody production to BSA.[66] Maron and coworkers reported the presence of anti-insulin receptor antibodies in the serum of newly diagnosed, insulin-dependent diabetics.[67] These could represent anti-idiotypic antibodies to IAA, since mice immunized with insulin develop not only antibodies to insulin but also antibodies to the insulin receptor. These antibodies to the insulin receptor were anti-idiotypic, since they were bound to and could be blocked by the antibodies to insulin.[68,69] If IAA were, in part, responsible for the beta-cell destruction of IDDM, and insulin receptor antibodies are anti-idiotypic antibodies to IAA, then the insulin receptor antibodies might have a protective function by downregulating IAA production.

The last hypothesis is that IAA do not mediate beta-cell damage but rather are part of the body's defense against immunologic destruction of beta-cells. Data have accumulated that suggest that autoantibodies are normal in healthy people, rather than always associated with disease.[70] For example, antibodies characteristic of SLE and Sjögren syndrome have been detected in normal individuals.[55] Avrameas and colleagues have demonstrated that autoantibodies to such common antigens as tubulin, actin, thyroglobulin, myoglobin, fetuin, transferrin, albumin, cytochrome C, and collagen are also found in normal individuals.[71] More recently using hybridoma and other techniques, activated B-lymphocytes that produce autoantibodies were found to be common in both adult and newborn mice.[72–74] In fact about 50% of all IgM-secreting cells that appear after both mitogen and immunogen injection of mice produce IgM, which although not specific for the immunogen is specific for self-IgG (rheumatoid factor).[75] Reviewing this topic, Cohen and Cooke recently suggested that between 10 and 30% of B-lymphocytes in normal healthy individuals were engaged in producing autoantibodies.[76] In fact the lymphocytes responsible for autoantibody production may be a distinct and separate population of lymphocytes.[77] A large percentage of the autoreactive monoclonal antibodies secreted by hybridomas derived from normal mice have multiple specificities; that is, they react with many structurally unrelated antigens.[73] Cohen and Cooke have proposed that autoantibodies may help protect against autoimmune disease by blinding the immune system to environmental epitopes cross-reactive with self. They argue that self-mimicking epitopes on microorganisms are not rare and that, as part of the coordinated immune response against foreign structures, autoantibodies block responses to dangerously self-similar epitopes while allowing a vigorous response to foreign epitopes on the microorganism.[76] This blinding of the immune system to self-mimicking epitopes could be accomplished by passively blocking the antigen by autoantibodies. The use of anti-Rh antibodies to prevent the maternal antibody response to an Rh positive fetus is an example of this ability of antibodies to blind the immune system to an antigen.

Whether one of the above mechanisms, or another mechanism, accounts for the association of IAA with IDDM is presently unknown. In fact the mechanisms discussed above may not be mutually exclusive; possibly one or the other mechanism may obtain, depending upon the specific antibody. Most natural protective autoantibodies may be of the IgM class, whereas autoantibodies associated with disease may be IgG. It is tempting to speculate that the IgM-IAA that occurs after common viral infections[16] may be part of the protective autoimmune network. Perhaps when we are better able to measure IAA and to characterize different populations of IAA, we will find that some types of IAA are present in normal individuals and serve a protective function whereas other types of IAA are associated with IDDM. The simple concept that IAA are found only in individuals with beta-cell disease and that all these individuals will eventually develop clinical IDDM is certainly not valid.

In reviewing the topic of IAA and IDDM, I have drawn few firm conclusions and raised many questions. The predictive value of IAA in nondiabetic individuals for the diabetogenic disease process or subsequent clinical diabetes, the mechanisms underlying the production of IAA, and the precise pathophysiologic role of IAA in beta-cell destruction of IDDM remain undefined. Although at first one might find this distressing, I would argue that this is the result of a better understanding of the complexities of the immune system in autoimmune disease and of the availability of highly specific, sensitive, and quantitative assays for IAA. I would predict that the situation will become similarly confusing for the other immunologic markers associated with IDDM when the antigens for ICA, ICSA, and A–64-kD are identified and similar assays become available. With the knowledge that the relationship between autoantibodies and disease is extremely complex, sensitive, quantitative, and specific assays for IAA that will also differentiate the different types of IAA should provide the tools we need to answer more questions pertaining to IAA and IDDM.

Acknowledgment. The author thanks A. Rafn for secretarial assistance. This work was supported in part by National Institutes of Health grants AM-17047 and AM-30780 and a Fogarty International Fellowship.

References

1. Sibley RK, Sutherland ER, Goetz F, Michael AF. Recurrent diabetes mellitus in the pancreas iso- and allograft. *Lab Invest* 1985;53:132–144.
2. Harrison CL, Colman PG, Dean B, Baxter R, Martin FIR. Increase in remission rate in newly diagnosed type I diabetic subjects treated with azathioprine. *Diabetes* 1985; 34:1306–1308.
3. Feutren G, Papoz L, Vialettes B, Karsenty G, Vexiau P, Rostu HD, Rodier M, Sirmai J, Lallemand A, Bach JF. Cyclosporin increases the rate and length of remission in insulin dependent diabetes of recent onset. Results of a multicentre double-blind trial. *Lancet* 1986;ii:119–123.
4. Hirata Y, Ishizu H, Ouchi N, Motumura S, Abe M, Hara Y, Wakasugi H, Takahashi I, Sakano H, Tanaka M, Kawano H, Kanesaki T. Insulin autoimmunity in a case with spontaneous hypoglycaemia. *Japn J Diabetes* 1970;13:312–319.
5. Føllng I, Norman N. Hyperglycemia, hypoglycemic attacks, and production of anti-insulin antibodies without previous known immunization. *Diabetes* 1972;21:814–826.
6. Takayama S, Hirata Y. Incidence of insulin autoimmune hypoglycemia in Japan during the three-year period from 1979 to 1981. *Diabetes* 1983;32:150A.
7. Goldman J, Baldwin D, Rubenstein AH, Klink DD, Blackard WG, Fisher LK, Roe TF, Schnure JJ. Characterization of circulating insulin and proinsulin-binding antibodies in autoimmune hypoglycemia. *J Clin Invest* 1979;63:1050–1059.
8. Hirata Y, Arimachi M. Insulin autoimmune syndrome: The second case. *Japn J Diabetes* 1972;15:187–192.
9. Ichihara K, Shima K, Saito Y, Nonaka K, Tarui S, Nishikawa M. Mechanism of hypoglycemia observed in a patient with insulin autoimmune syndrome. *Diabetes* 1977; 26:500–506.
10. Benson EA, Healey LA, Barron EJ. Insulin antibodies in patients receiving penicillamine. *Am J Med* 1985;78:857–960.
11. Hirata Y. Methimazole and insulin autoimmune syndrome with hypoglycaemia. *Lancet* 1983;ii:1037–1038.
12. Okabe N, Inoue K, Mori R. Effects of antithyroid drugs on lymphocyte proliferative responses to lectins: Relationship between insulin autoimmune syndrome and methicazole. *J Clin Lab Immunol* 1983;11:167–171.
13. Seino S, Fu ZZ, Marks W, Seino Y, Imura H, Vinik A. Characterization of circulating insulin in insulin in autoimmune syndrome. *J Clin Endocrinol Metab* 1986;62:64–69.
14. Nakagawa S, Suda N, Kudo M, Kawasaki M. A new type of hypoglycaemia in newborn infant. *Diabetologia* 1973;9:367–375.

15. Kahn CR, Rosenthal AS. Immunologic reactions to insulin: Insulin allergy, insulin resistance, and the autoimmune insulin syndrome. *Diabetes Care* 1979;2:283–295.
16. Bodansky HJ, Dean BM, Bottazzo GF, Grant PJ, McNally J, Hambling MH, Wales JK. Islet-cell antibodies and insulin autoantibodies in association with common viral infections. *Lancet* 1986;II:1351–1353.
17. Diaz J-L, Wilkin T. Differences in epitope restriction of autoantibodies to native human insulin (IAA) and antibodies to heterologous insulin (IA). *Diabetes* 1987;36: 66–72.
18. Nell LJ (personal communication).
19. Wilkin TJ, Nicholson S. Autoantibodies against human insulin. *Br Med J* 1984;288: 349–352.
20. Wilkin TJ, Casey CR. The specificity of autoantibodies to human insulin. British Diabetes Association Program, Edinburgh, March, 1984.
21. Páv J, Prague MD, Jezková Z, Skrha F. Insulin antibodies. *Lancet* 1963;ii:221–222.
22. Sebriakova M, Little JA. A method for the determination of plasma insulin antibodies and its application in normal and diabetic subjects. *Diabetes* 1973;22:30–40.
23. MacCuish AC, Jordan Jennifer, Cambell CJ, Duncan LJP, Irvine WJ. Cell-mediated immunity in diabetes mellitus. Lymphocyte transformation by insulin and insulin fragments in insulin-treated and newly-diagnosed diabetics. *Diabetes* 1975;24:36–43.
24. Palmer JP, Asplin CM, Clemons P, Lyen K, Tatpati O, Raghu PK, Paquette TL. Insulin antibodies in insulin-dependent diabetics before insulin treatment. *Science* 1983;222: 1337–1339.
25. Reeves WG. The immune response to insulin: Characterisation and clinical consequences. *Diabetes Ann* 1986;2:81–93.
26. Palmer JP, Asplin CM, Raghu PK, Clemons P, Lyen K, Tatpati O, McKnight B, Paquette T, Sperling M, Baker L, Guthrie R. Anti-insulin antibodies in insulin-dependent diabetics before insulin treatment—a new marker for autoimmune beta-cell damage? *Pediatr Adolesc Endocrinol* 1986;15:111–116.
27. Atkinson MA, Maclaren NK, Riley WJ, Winter WE, Fisk DD, Spillar RP. Are insulin autoantibodies markers for insulin-dependent diabetes mellitus? *Diabetes* 1986;35: 894–898.
28. Karjalainen J, Knip M, Mustonen A, Ilonen J, Åkerblom HK. Relation between insulin antibody and complement-fixing islet cell antibody at clinical diagnosis of IDDM. *Diabetes* 1986;35:620–622.
29. McEvoy, Witt ME, Ginsberg-Fellner F, Rubinstein P. Anti-insulin antibodies in children with type I diabetes mellitus. Genetic regulation of production and presence at diagnosis before insulin replacement. *Diabetes* 1986;35:634–641.
30. Arslanian SA, Becker DJ, Rabin B, Atchison R, Eberhardt M, Cavender D, Dorman J, Drash AL. Correlates of insulin antibodies in newly diagnosed children with insulin-dependent diabetes before insulin therapy. *Diabetes* 1985;34:926–930.
31. Wilkin T, Armitage M, Casey C, Pyke DA, Hoskins PJ, Rodier M, Diaz JL, Leslie RDG. Value of insulin autoantibodies as serum markers for insulin-dependent diabetes mellitus. *Lancet* 1985;ii:480–482.
32. Bergman S, Ludvigsson J, Binder C, Mandrup-Poulsen T. Insulin antibodies before treatment in ICA-positive children with IDDM. *Diabetes Res Clin Prac* 1985 pg 545, abstract 112 volume 1 (Suppl 1).
33. Soeldner JS, Tuttleman M, Srikanta S, Ganda OmP, Eisenbarth SG. Insulin-dependent diabetes mellitus and autoimmunity: Islet-cell autoantibodies, insulin autoantibodies, and beta-cell failure. *N Engl J Med* 1985;313:893–894.

34. Srikanta S, Ricker AT, McCulloch DK, Soeldner JS, Eisenbarth GS, Palmer JP. Autoimmunity to insulin, beta-cell dysfunction, and development of insulin-dependent diabetes mellitus. *Diabetes* 1986;35:139–142.
35. Dean BM, Becker F, McNally JM, Tarn AC, Schwartz G, Gale EAM, Bottazzo GF. Insulin autoantibodies in the pre-diabetic period: Correlation with islet cell antibodies and development of diabetes. *Diabetologia* 1986;29:339–342.
36. Wilkin T, Kiesel U, Diaz JL, Burkart V, Kolb H. Autoantibodies to insulin as serum markers for autoimmune insulitis. *Diabetes Res* 1986;3:173–174.
37. Elliott RB (personal communication).
38. Srikanta S, Ganda OP, Gleason RE, Jackson RA, Soeldner JS, Eisenbarth GS. Pre-type I diabetes. Linear loss of beta-cell. Response to intravenous glucose, *Diabetes* 1984;33:717–720.
39. McCulloch DK, Palmer JP, Benson EA. Beta-cell function in the preclinical period of insulin-dependent diabetes. *Diabetes Metab Rev* 1987;3:1–17.
40. Millward BA, Alviggi L, Hoskins PJ, Johnston C, Heaton D, Bottazzo GF, Vergani D, Leslie RDG, Pyke DA. Immune changes associated with insulin-dependent diabetes may remit without causing the disease: A study in identical twins. *Br Med J* 1986; 292:793–796.
41. Heaton DA, Millward BA, Gray P, Tun Y, Hales CN, Pyke DA, Leslie RDG. Evidence of β cell dysfunction which does not leas on to diabetes: A study of identical twins of insulin dependent diabetics. *Br Med J* 1987;294:145–146.
42. Reeves WG, Barr D, Douglas CA, Gelsthorpe K, Hanning I, Skene A, Wells L, Wilson RM, Tattersall RB. Factors governing the human immune response to injected insulin. *Diabetologia* 1984;26:266–271.
43. Asplin CM, Dornan TL, Raghu PK, Hansen JA, Palmer JP. The antibody response to insulin therapy. A prospective study in HLA-typed insulin-dependent diabetic subjects. *Diabetes* 1984;33:966–969.
44. Palmer JP (unpublished observations).
45. Wilkin T, Palmer JP, Bonifacio E, Diaz JL, Kruse V. First International Workshop on the Standardization of Insulin Autoantibodies (IAA). *Diabetologia* 1987;30:676–677.
46. Kaplan DR. Possible target antigens in autoimmune endocrine disease. *Immunol Today* 1984;5:130–131.
47. Dyrberg T, Oldstone MBA. Peptides as probes to study molecular mimicry and virus-induced autoimmunity. *Curr Top Microbiol Immunol* 1986;130:25–37.
48. Saegusa J, Onodera T, Ray UR, Prabhakar BS, Notkins AL. Monoclonal autoantibodies that react with hormones also react with cells not containing those hormones. *Endocrinology* 1985;116:761–764.
49. Haspel MV, Onodera T, Prabhakar BS, McClintock PR, Essani K, Ray UR, Yagihashi S, Notkins AL. Multiple organ-reactive monoclonal autoantibodies. *Nature* 1983; 304:73–76.
50. Satoh J, Prabhakar BS, Haspel MV, Ginsberg-Fellner F, Notkins AL. Human monoclonal autoantibodies that react with multiple endocrine organs. *N Engl J Med* 1983;309:217–220.
51. Haspel MV, Onodera T, Prabhakar BS, Horita M, Suzuki H, Notkins AL. Virus-induced autoimmunity: Monoclonal antibodies that react with endocrine tissues. *Science* 1983;220:304–306.
52. Serreze DV, Leiter EH, Kuff EL, Jardieu P, Ishizaka K. Molecular mimicry between insulin and retrovirus p^{73}. *Diabetes* 1988;37:351–358.

53. Plotz PH. Autoantibodies are anti-idiotype antibodies to antiviral antibodies. *Lancet* 1983;ii:824–826.
54. Leming H, Slack JH, Amundson C, Izui S, Theofilopoulos AN, Dixon FJ. Induction of murine autoimmune disease by chronic polyclonal B cell activation. *J Exp Med* 1983;157:874–883.
55. Gaither KK, Owen FF, Yamagata H, Mamula MJ, Reichlin M, Harley JB. Implications of anti-RO/Sjögren's syndrome. A antigen autoantibody in normal sera for autoimmunity. *J Clin Invest* 1987;79:841–846.
56. Vento S, O'Brien CJ, McFarlane IG, Williams R, Eddleston ALWF. T-cell inducers of suppressor lymphocytes control liver-directed autoreactivity. *Lancet* 1987;i:886–888.
57. Wilkin TJ, Mirza I, Armitage M, Casey C. IAA alloautotypes: Insulin autoantibody subsets with greater discrimination for diabetes relatedness. British Diabetes Association Program, Warwick, 1987, p 70.
58. LeCompte PM, Steinke J, Soeldner JS, Renold AE. Changes in the islets of Langerhans in cows injected with heterologous and homologous insulin. *Diabetes* 1966;15: 586–596.
59. Grodsky GM, Feldman R, Toreson WE, Lee JC. Diabetes mellitus in rabbits immunized with insulin. *Diabetes* 1966;15:579–585.
60. Gepts W. Pathologic anatomy of the pancreas in juvenile diabetes mellitus. *Diabetes* 1965;14:619–633.
61. Ludvigsson J. Insulin antibodies in diabetic children treated with monoçomponent porcine insulin from the onset: Relationship to B-cell function and partial remission. *Diabetologia* 1984;26:138–141.
62. Hoogwerf BJ, Rich SS, Barbosa J. Meal-stimulated C-peptide and insulin antibodies in type I diabetic subjects and their nondiabetic siblings characterized by HLA-DR antigens. *Diabetes* 1985;34:440–445.
63. Jerne NK. Towards a network theory of the immune system. *Ann Immunol (Paris)* 1974;125C:373–389.
64. Zanetti M, Glotz D, Rogers J. Perturbation of the autoimmune network. II. Immunization with isologous idiotype induces auto-anti-idiotypic antibodies and suppresses the autoantibody response elicited by antigen: A serologic and cellular analysis. *J Immunol* 1986;137:3140–3146.
65. Epstein A, Greenberg M, Diamond B, Grayzel AI. Suppression of anti-DNA antibody synthesis in vitro by a cross-reactive antiidiotypic antibody. *J Clin Invest* 1987;79: 997–1000.
66. Eddy KN, Raymond JA, Michael JG. Idiotype regulation of the anti-bovine serum albumin response. I. Generation of an idiotype-specific Lyt-1^-2^+. *J Immunol* 1987; 138:1693–1698.
67. Maron R, Elias D, de Jongh BM, Bruining GJ, van Rood JJ, Shecter Y, Cohen IR. Autoantibodies to the insulin receptor in juvenile onset insulin-dependent diabetes. *Nature* 1983;303:817–818.
68. Schechter Y, Maron R, Elias D, Cohen IR. Autoantibodies to insulin receptor spontaneously develop as anti-idiotypes in mice immunized with insulin. *Science* 1982; 216:542–544.
69. Reilly TM, Root RT. Production of idiotypic and anti-idiotypic antibodies by balb/c mice in response to immunizations with glucagon, vasopressin, og insulin: Supporting evidence for the network concept. *J Immunol* 1986;137:597–602.

70. Holmberg D, Coutinho A. Natural antibodies and autoimmunity. *Immunol Today* 1985;6:356–357.
71. Guilbert B, Dighiero G, Avrameas S. Naturally occurring antibodies against nine common antigens in human sera. *J Immunol* 1982;128:2779–2786.
72. Dighiero G, Lymberi P, Holmberg D, Lundquist I, Coutinho A, Avrameas S. High frequency of natural autoantibodies in normal newborn mice. *J Immunol* 1985;134: 765–771.
73. Prabhakar BS, Saegusa J, Onodera T, Notkins AL. Lymphocytes capable of making monoclonal autoantibodies that react with multiple organs are a common feature of the normal B cell repertoire. *J Immunol* 1984;133:2815–2817.
74. Steele EJ, Cunningham AJ. High proportion of Ig-producing cells making autoantibody in normal mice. *Nature* 1978;274:483–484.
75. Dresser DW, Popham AM. Most IgM-producing cells in the mouse secrete autoantibodies (rheumatoid factor). *Nature* 1976;264:552–554.
76. Cohen IR, Cooke A. Natural autoantibodies might prevent autoimmune disease. *Immunol Today* 1986;7:363–364.
77. Stall AM, Lalor PA, Herzenberg LA, Herzenberg. LY-1 B cells and autoantibodies. *Ann Immunol (Paris)* 1986;137D(1):173–177.

8
Cell-Mediated Anti-Islet–Cell Immune Response: Clinical Experience and Lessons from Animal Models

ROBERT C. MCEVOY AND NANCY M. THOMAS

Introduction

The hypothesis that type I, insulin-dependent diabetes mellitus (IDDM) results, at least in part, from the autoimmune destruction of pancreatic beta-cells has been discussed in several reviews,[1-3] as well as in other chapters in this volume. Relatively early in the analysis of autoimmunity and diabetes, a small number of papers presented evidence of a direct anti-beta–cell (or pancreatic islet-cell) cellular immune response in individuals with diabetes. It has become clear over the last decade, however, that the results of such studies in humans are limited, at least to the extent that they may be considered beta-cell–specific, by the so-called major histocompatibility complex (MHC)-restriction of the cell-mediated arm of the immune system. The specific effector cell, typically a thymus-derived cell, the T-cell, and the target cell must share identity at one of the class I (cytotoxic/suppressor cells, CD8⁺) or class II (helper/inducer cells CD4⁺). This issue of MHC-restriction in the immune system has been reviewed recently.[4-5] Although MHC-restriction is not now considered to be an absolute requirement for cell-mediated cytolysis, further research in the description and analysis of the role of the cell-mediated immune response in human diabetes has virtually stopped because of the lack of a biologically relevant, MHC-restricted beta-cell for use as a target. Fortunately studies of the cell-mediated immune response against pancreatic beta-cells are possible in experimental models of diabetes using both in vitro and in vivo techniques. Consequently after a brief review of the limited work in humans, the remainder of this chapter will be devoted to a discussion of data that suggest a cell-mediated, anti-beta–cell immune response in each of the animal models of diabetes that demonstrate the pathognomonic lesion of autoimmune diabetes insulitis. These include BB rats, NOD mice, and mice made diabetic by multiple, subdiabetogenic doses of the beta-cell toxin streptozotocin.

Cell-Mediated, Anti-Beta–Cell Autoimmunity in Human Diabetes

The first studies implicating cell-mediated, anti-beta–cell immunity in diabetes were those of Nerup et al.,[6] which demonstrated that peripheral blood mononuclear cells (PBMNC) from diabetic patients could inhibit leukocyte migration when extract from rat islets was used as an antigen. Of the diabetic subjects, 28% had migration indices below the normal range calculated from the results of 45 healthy controls.

The direct examination of a cell-mediated, anti-beta–cell immune response in human diabetes owes its beginning to a classic study by Huang and McLaren[7] in which PBMNC from diabetic and control patients were incubated in culture dishes in which a human insulinoma had been grown. A much stronger specific attachment to the insulinoma cells occurred in the dishes incubated with diabetic lymphocytes. Similar results were reported recently, using $CD4^+$ lymphocytes and a rat insulinoma (RIN) cell line as the target.[8] Neither of these papers reported any evidence of a specific beta-cell cytotoxicity induced by these adherent cells. Suzuki and colleagues[9] studied various immune mechanisms of cytotoxicity, including direct cell-mediated cytotoxicity, antibody-dependent, cell-mediated cytotoxicity, and complement-augmented, cell-mediated cytotoxicity, using lymphocytes from diabetic patients and from control subjects. The targets in this study were dispersed rat islet cells that had been labeled with chromium-51 (^{51}Cr). The results indicated that non-T–cell cytotoxicity against rat islet cells could be demonstrated more commonly in diabetics than in controls, although patients and controls were not matched for MHC type and the diabetic group may have included individuals with such high immune response HLA types as DR3 and DR4. Scott and her colleagues[10] were able to show that cell-mediated cytolysis could be demonstrated against intact rat islets using PBMNC obtained from humans with IDDM. Another group[11,12] demonstrated a number of different cellular and antibody-dependent mechanisms by which islet cells could be killed in the presence of PBMNC from human diabetes. All these studies suffered from an inability to demonstrate the cell type(s) responsible for the cytotoxicity, although the description of such presumptively nonspecific immune mechanisms by which beta-cells could be killed was provocative. Unfortunately most of the studies were uncontrolled for cell type, since the investigators did not have parallel studies using nonislet cells to demonstrate that there was no difference between controls and diabetic patients with regard to nonislet cell cytotoxicity. In addition, patients and controls were not matched for MHC (HLA) type in any of these studies.

Although one could argue that nonspecific immune mechanisms are, in fact, responsible for the final destruction of pancreatic beta-cells in IDDM, the specificity of the destruction and its localization to the pancreatic islets argue that a cell- or organ-specific immune response must be necessary, at least as the inducer of such a nonspecific response.

Diabetes Induced by Multiple, Subdiabetogenic Doses of Streptozotocin

Description

In 1976 Like and Rossini described the effects of multiple injections of subdiabetogenic doses of streptozotocin (Multi-Strep).[13] The principal findings were delayed onset of hyperglycemia, development of insulitis preceding onset of hyperglycemia, and activation of C type virus in pancreatic beta-cells. Subsequently work in a number of laboratories elucidated a role of the immune system in the production of diabetes in the Multi-Strep model. Many of these studies have been summarized in a recent review.[14]

Several studies have reported a difference in the susceptibility of different mouse strains to Multi-Strep.[15-19] These studies indicate that in Multi-Strep diabetes, as in human IDDM, genes of (or linked to) the MHC have a role in immune pathogenesis, although the diverse findings in these inbred mice indicate that, as in humans, any effect of MHC is modified significantly by other factors. Multi-Strep IDDM differs remarkably from human IDDM in that, even in the most highly susceptible strains, only males develop diabetes.[20-23]

Autoimmune Basis of Multi-Strep–Induced Diabetes

It is now well established that induction of diabetes by Multi-Strep in mice depends upon activation of an autoimmune response by thymus-dependent lymphocytes. This hypothesis was supported by studies in which antimouse lymphocyte serum partially[24] or completely[25] prevented IDDM. Although it has been reported that athymic mice do not develop diabetes after Multi-Strep,[21] attempts to confirm this have not been entirely successful.[26-28] Some strains of mice (normal and athymic) are particularly sensitive to the direct beta cytotoxic effects of Multi-Strep. Thus in these strains, the standard Multi-Strep dose is not a "subdiabetogenic" dose. Buschard and Rygaard[26] reported that splenic lymphocytes from mice diabetic after Multi-Strep injection could transfer the diabetes to their syngeneic normal mice or nude mice. Irradiation of donor lymphocytes did not prevent transfer. The degree of hyperglycemia after adoptive transfer was mild but this observation has not been confirmed,[27] although insulitis has been noted in the absence of significant hyperglycemia.[29-31] In analogy to the requirement for (Con-A) activation of the donor lymphocytes for passive transfer of IDDM in the BB/W rat,[32] after transfer of Con-A–activated lymphocytes from Multi-Strep mice, hyperglycemia could be produced in syngeneic recipients by a single low dose of Multi-Strep.[33] That Multi-Strep alone was not responsible for beta-cell destruction recently was confirmed by a study showing that mice maintained on a diet deficient in essential fatty acids did not develop diabetes after Multi-Strep; when essential fatty acids were introduced into the diet three weeks after Multi-Strep injection, insulitis and hyperglycemia developed within several days.[34]

Immunologic manipulations have further established an immune component in Multi-Strep diabetes. Prevention of diabetes by the administration of anti-mouse lymphocyte serum has been noted.[24,25] Production of immune deficiency by irradiation of neonatal thymectomy[35-38] also prevented the development of Multi-Strep diabetes. Susceptibility was restored by a T-lymphocyte–enriched splenocyte preparation but not a B-lymphocyte–enriched splenocyte one.[38] Cyclophosphamide treatment of mildly susceptible mice increased the degree of hyperglycemia attained after Multi-Strep,[39,40] as did nonspecific activation of the immune system by lipopolysaccharide injection.[41] Injection of 10^8 syngeneic lymphocytes decreased the hyperglycemia of a highly susceptible strain,[40] which suggests that Multi-Strep diabetes may result from an inactivation/inhibition of suppressor T-lymphocytes. Similarly injection of monoclonal antibodies against immune response gene products (Ia and Ij) reduced the degree of hyperglycemia induced by Multi-Strep.[42] In contrast to its effects in some humans and in BB rats, cyclosporin immunosuppression enhanced the diabetogenic effect of Multi-Strep.[43] The dose used in this study was not excessive, but direct inhibition of insulin biosynthesis by cyclosporin has been reported.[44] Nicotinamide prevents the diabetogeneic effects of high doses of Multi-Strep, but does not prevent Multi-Strep IDDM.[45] Free radical scavengers can prevent high-dose Multi-Strep diabetes, but injections of supraoxide dismutase failed to prevent Multi-Strep IDDM.[46] The hydroxy-radical scavenger dimethyl urea reduced the degree of hyperglycemia, but not the insulitis induced by Multi-Strep.[47] These latter studies support the concept that Multi-Strep–induced IDDM is not due solely to direct toxic damage to the beta-cells by the streptozotocin.

The susceptibility of inbred strains to Multi-Strep IDDM has led to attempts to ameliorate the diabetes by syngeneic islet transplantation. Islet transplantation in C57BL/KsJ mice ameliorated Multi-Strep IDDM, and as in high-dose Multi-Strep diabetes, the recipients were permanently cured.[48-49] However in contrast to diabetes in mice after high-dose streptozotocin, insulitis and hyperglycemia recurred in the Multi-Strep, syngeneic-islet transplant recipients when three small additional doses of streptozotocin were given.[49-50]

Cell-Mediated, Anti-Beta–Cell Autoimmune Response in Multi-Strep Diabetes

To address the role of cell-mediated immunity in the pathogenesis of insulin-dependent diabetes in the Multi-Strep mouse directly, we examined such mice for the presence of splenic lymphocytes that were specifically cytotoxic for a RIN cell line.[51] Cytotoxicity was quantitated by the release of ^{51}Cr from damaged cells. A low, but statistically significant level of cytolysis (5%) by splenic lymphocytes was first detectable on day 8 after the first dose of Multi-Strep. The cytotoxicity reached a maximum of approximately 9% on day 10 and slowly decreased thereafter; it was undetectable 42 days after Multi-Strep injection. The time-course of the in vitro cytotoxic response correlated with the degree of insulitis demonstrable in the pancreata of the Multi-Strep mice. The degree of cytotoxicity after Multi-Strep injection was related to the number of splenic effector lymphocytes to which the target RIN cells were exposed and was comparable to that detectable after immunization by intraperitoneal injection of RIN cells in normal mice. The cytotoxicity was specific for insulin-producing cells; syngeneic, allogeneic, and xenogeneic lymphocytes and lymphoblasts, 3T3 cells, and a human keratinocyte cell line were not specifically lysed by splenic lymphocytes from Multi-Strep mice. Lymphocytes from mice made diabetic by a single high dose of streptozotocin exhibited a very low level of cytotoxicity against the RIN cells.

To establish that this cytotoxicity was mediated by thymus-derived lymphocytes, the response was quantitated in splenocytes from control and Multi-Strep mice (10 days after the first dose of streptozotocin) before and after culture with mitomycin-treated RIN cells in the presence of T-cell growth factor (TCGF). The specific cytotoxicity of the Multi-Strep lymphocytes was enhanced more than fourfold after such culture, indicating that cytotoxic T-lymphocyte proliferation could be stimulated and supported in vitro by TCGF.

In additional studies,[52] we excluded a major role for B-lymphocytes, macrophages, or their products in this cytotoxic response by passing splenocytes from the mice treated with Multi-Strep over a Nylon wool column. Removal of the adherent cells enhanced the cytotoxicity against a RIN cell line in vitro to the same extent as did enrichment with T-lymphocytes, that is, by approximately twofold. To examine the strain dependency of this cytotoxicity, mice of five strains were immunized with RIN cells. Only splenocytes from the two strains susceptible to Multi-Strep demonstrated a significant cytotoxic response against RIN cells in vitro. Mice preimmunized with either RIN cells or with syngeneic islets labeled in vitro with the hapten trinitrophenol developed hyperglycemia more rapidly than control mice after Multi-Strep. In the latter experiment, the control mice, immunized with complete Freund's adjuvant only, also became hyperglycemic after a modified Multi-Strep using only 3 doses of 40 mg/kg that did not cause diabetes in nonimmunized mice. In mice pretreated with either adjuvant or cyclophosphamide and then given modified Multi-Strep (5 doses of 35 mg/kg rather than 40 mg/kg), the degree of hyperglycemia was reduced, but there was no protective effect by cyclophosphamide. However mice pretreated with adjuvant developed hyperglycemia more rapidly and to a much higher level than did the mice given Multi-Strep only.

Summary

The data presented above support the hypothesis that the hyperglycemia that develops in mice after multiple subdiabetogenic doses of streptozotocin (Multi-Strep) results, at least in part, from the activation of an autoimmune response against pancreatic beta-cells. However further studies, using both in vivo and in vitro techniques, will be needed to identify the specific effector cell(s) responsible for the cytotoxicity against B-cell targets in vitro and the production of diabetes in vivo.

Diabetes in the Bio-Breeding Rat

Description

The Bio-Breeding (BB) rat was discovered in 1974 in a commercial colony of Wistar-derived rats at the Bio-Breeding Laboratories in Ottawa, Canada.[53,54] A complete, up-to-date summary of the basic characteristics of this rat is contained in a recent review.[55] It is clear that BB rats in colonies throughout the world vary with respect to both the frequency and severity of diabetes.[56-58] However the "classic" BB rat manifests frank, insulin-dependent diabetes with the classic signs of polyuria, polydipsia, polyphagia, weight loss, and insulin deficiency and progression to ketosis, ketoacidosis, and death unless treated.[59,60,55] The onset of diabetes occurs from about 60 to 120 days of age and is preceded for several days to weeks by an insulitis in the pancreatic islets.[54,55,59] Diabetes in these rats is associated with a gene linked to the MHC of the rat (*RT1^u^* haplotype).[61-63] Breeding studies have demonstrated an autosomal recessive inheritance with an approximately 50% penetrance of the recessive gene(s).[59,64] The diabetes in BB rats was also believed to be linked to a gene outside the MHC, which was responsible for a severe lymphopenia.[65] Although the majority of diabetic and diabetes-prone BB rats are severely lymphopenic, there is some evidence to suggest that this lymphopenia is not an obligate immunogeneic feature of the diabetes. Like and colleagues[66] recently have developed a line of diabetic rats that does not exhibit the usual lymphopenia of the BB rat and that, in fact, have normal T-lymphocyte subset percentages. A diabetes-resistant BB rat has also been bred at the Worcester colony.[55,66] These diabetes-resistant BB rats do not exhibit lymphopenia and have normal T-cell subsets and T-cell in vitro responses, whereas the classic, diabetes-prone animals are markedly lymphopenic with fewer T-cells and a deficiency of both cytotoxic ($CD8^+$) and helper ($CD4^+$) lymphocytes. In addition it has been reported that the rat lymphocyte maturation alloantigen RT-6 is absent in the diabetes-prone BB rat, whereas it is present on the normal lymphocytes of the diabetes-resistant strain.[67]

Autoimmune Basis of Diabetes in the BB Rat

The role of an autoimmune response in the production of diabetes in the BB rat was first suggested by the presence of insulitis in the pancreatic islets of rats at the onset of hyperglycemia.[54] Most of the infiltrating mononuclear cells stain with corresponding monoclonal antibodies, which indicate that they are Ia^+-"activated" cells,[68,69] although some $CD4^+$ and $CD8^+$ cells are also present. B-lymphocytes are uncommon. It has been reported that the endothelial cells of BB-rat islet appear to express class II MHC antigens (Ia) prior to the onset of the insulitis.[69] This observation coupled with the finding that there is an increased number of $CD4^+$ lymphocytes and some macrophages in the islet expressing Class II antigens, suggests that immune activation is occurring within the pancreatic islet. The $CD8^+$ lymphocytes appear at a later stage of insulitis.[69]

Although insulitis is the pathognomonic lesion of BB-rat diabetes, there is substantial evidence to suggest that BB-rat islets are not intrinsically defective. Several reports have indicated that the beta-cells of BB-rat islets do not express class II MHC antigens on their surface. More evidence to document the normality of the BB-rat islet comes from transplantation studies in which islets from diabetes-prone BB rats could reverse the diabetes of streptozocin–treated tolerant Wistar rats.[70-72] Conversely transplantation of islets from other rats and from diabetes-resistant BB rats to diabetic BB rats has always produced insulitis and islet destruction within the grafts.[73-74] These data suggest that a putative, BB-rat–specific, beta-cell antigen is not essential to initiate the insulitis and subsequent immune destruction of beta-cells. This interpretation is supported by other studies in which islets and pituitary tissue from the same MHC-incompatible donor were transplanted to diabetes-prone BB rats.[70] The grafted islets, but not the grafted pituitary tissue, was destroyed; this suggests that, at least in the immune-incompetent BB rat, MHC-incompatible tissue can survive, yet MHC-compatible islet cells are destroyed. The interpretation of these findings has been that the autoimmune mechanisms producing diabetes in the BB rat differ from those involved in immune rejection of a tissue graft (see Chapter nine).

Cell-Mediated Anti-Beta–Cell, Autoimmunity Response in BB-Rat Diabetes

More specific information is available to suggest that the diabetes of the BB rat depends upon a cell-mediated response. First several investigators have reported that cells from the blood, spleen, or thoracic duct of the BB rat can specifically destroy beta-cells when assessed by an in vitro chromium release assay.[75] Further studies demonstrated that the effector cells responsible for islet-cell killing were natural killer (NK)-cells.[76] In addition it has been reported that lymphocytes and T-cell hybrids from BB rats can proliferate and secrete interleukin-2 specifically in the presence of islet antigen.[77-78] Immunosuppression with cyclosporin has been shown to prevent diabetes in diabetes-prone BB rats if it is given before the onset of hyperglycemia,[79,80] but other immunosuppressive agents, such as cyclophosphamide[81] or low-dose irradiation,[82] can actually produce diabetes in diabetes-resistant BB rats.

Unfortunately the widespread effects of such agents on the immune system do not allow us to draw specific conclusions as to which immunoregulatory or effector cells may have been altered. One series of experiments has addressed this problem with regard to the absence of RT-6$^+$ cells in the diabetes-prone BB rat. When young, diabetes-resistant BB rats were treated twice weekly with an antibody against RT-6$^+$ cells, these cells were depleted and almost half the treated rats became diabetic by 60 days of age, as compared to less than 3% of the untreated animals.[83] Rossini and his colleagues[55] have postulated that the autoimmune pathogenesis of diabetes in the BB rat involves RT-6$^+$ lymphocytes, which would play an immunoregulatory role in suppressing an anti-beta–effector cell or perhaps an NK-cell (see below). Effector cells can produce diabetes if such suppressor cells are absent, as in diabetes-prone BB rats, or reduced, as in immunosuppressed, diabetes-resistant BB rats.

Direct evidence that cells from acutely diabetic BB rats can produce diabetes in other animals was first published by Koevary and colleagues,[32] who treated spleen cells from acutely diabetic BB rats with the mitogen Con-A. Such activated cells produced both insulitis and diabetes within 14 days in over 50% of young, diabetes-prone animals versus less than 1% of their untreated littermates. Subsequently investigators were able to transfer diabetes with BB-rat splenocytes or lymphocytes if the recipients were pretreated with cyclophosphamide, radiation, or some other immunosuppressive agent.[84,85] Interestingly it has been shown that frequent administration of cell-free conditioned medium, after incubation of splenocytes from acutely diabetic BB rats with Con-A, can induce diabetes in young diabetes-prone and even in diabetes-resistant BB rats.[86] These data suggest that a soluble substance, possibly lymphokine(s) produced by the Con-A–activated splenocytes, can induce diabetes or perhaps accelerate the onset of diabetes. However, other studies have demonstrated that irradiation of Con-A–activated splenocytes[87] or pretreatment of the cells with specific cytotoxic monoclonal antibodies to rat lymphocytes[88] prevented the adoptive transfer of diabetes in BB rats, which suggests that viable lymphocytes are required for successful transfer.

Immunotherapy has been found to modulate diabetes in the BB rat. Although anti-rat lymphocyte serum was the first tested immunologic intervention,[88] a number of other nonspecific interventions have also been shown to prevent diabetes. These include total-body irradiation,[82] lymphoid tissue irradiation,[89] silica injection,[90] and combinations of nonspecific immunosuppressive agents. As mentioned above, cyclosporin is extremely effective in preventing diabetes if given in the young (>70=day=old) BB rat.[79,80] In fact it has been established that even if cyclosporin treatment is discontinued after 120 days, the BB rat never develops diabetes.[91] If the normal complement of MHC-compatible RT-6$^+$ cells is restored in diabetes-prone BB rats, the animals do not develop diabetes.[83,92] This has been shown to be the specific mechanism through which bone-marrow transplants from nondiabetes-prone BB rats to diabetes-prone BB rats prevent diabetes.[93] In addition blood transfusions[94] and transfusions of CD3$^+$ lymphocytes[95] or CD4$^+$ lymphocytes,[96] can also prevent diabetes. Taken together, these data indicate that the CD4$^+$, RT-6$^+$, T-cell subset plays an important role in modulating the immune system of the BB rat and in preventing the autoimmune reactions that would have eventually destroyed the beta-cells.

Reduction of RT-6$^+$ cell populations in unrelated rat strains does not result in diabetes, which suggests that a specific autoimmune, anti-beta–cell effector cell is present in the BB rat and in its genetically similar substrains. Very recent data suggest that, although the NK-cell may be the cytotoxic effector cell in the diabetes-prone BB rat,[76,97] it is not the effector cell in the diabetes-resistant, RT-6$^+$-depleted, diabetes-resistant BB rat.[98] Depletion of NK-cells in vivo by monoclonal antibodies reduced the cytotoxicity against a NK-cell target (YAC-1) and against Wistar-Furth–rat, pancreatic islets. However the administration of such an antibody to diabetes-resistant rats that also received monoclonal antibody to deplete the RT-6⁺ cell population reduced the cytotoxicity against the YAC-1 cells but did not reduce the cytotoxicity demonstrated against the Wistar-Furth islet cells.

Summary

The BB rat has been shown to be an exciting model in which many aspects of human diabetes mellitus can be demonstrated. There are significant differences, however, between BB-rat diabetes and human diabetes, the most striking of which is the presence of a profound immunodeficiency in many substrains of diabetes-prone BB rats. Apparently deficiency of the RT-6⁺ regulatory lymphocytes permits the expression and/or activation of specific, anti-beta–cell, autoimmune effector cells. It seems likely at this point that the effector cell in the standard diabetes-prone rat is a NK-cell. Nonspecific killing by these cells suggests that the beta-cell of the BB rat, although functionally and immunologically normal, may be more susceptible to the cytotoxic effects of NK-cells than other, non-beta–cells of the islets or the exocrine pancreas. The mechanism for targeting these cells to the islets, and the relevance of such information to human diabetes, is unknown at this time, since neither a specific immunoregulatory cell deficiency nor the presence of NK-cell activity against beta-cells have been demonstrated in human diabetes.

Diabetes in Nonobese Diabetic Mice

Description

The newest model of spontaneous autoimmune diabetes was first described by researchers in Japan.[99,100] This model also has been the subject of recent reviews.[14,101-103] A spontaneously diabetic female mouse died before extensive testing could be done, but fortunately her offspring were available for inbreeding. Eventually two sublines were produced, one with a high incidence of diabetes [nonobese diabetes (NOD)] and one without diabetes (Nonobese, nondiabetic). Investigations using this model have been slow, since these mice have only been available outside of Japan in the last few years. The NOD mice begin to develop ketotic diabetes at 12 weeks of age,[99,100] but their diabetes differs from human IDDM in that it is predominantly a disease of females. Depending upon the colony, 60 to 90% of NOD females are diabetic by 30 weeks of age, while only 10 to 40% of the males are.[100,104] Insulitis is present in most mice of both sexes by 5 to 6 weeks of age, and in virtually every mouse examined after 30 weeks of age.[103] Classic signs of diabetes develop beginning at four months of age and progress to ketosis; without treatment, death occurs within three to four weeks. Castration of male mice at six weeks of age increases the incidence of diabetes to approximately that of the female, which suggests that androgens suppress the diabetes.[103] However castration of females before the onset of diabetes also reduces the incidence of IDDM (to that of intact males), which indicates a positive role for estrogens or progesterone in the pathogenesis of IDDM in these mice.

Although many initial reports indicated that NOD mice had a moderate-to-marked, T-cell immunodeficiency characterized by pan-T–cell lymphopenia; defective killing by both NK-cells and cytotoxic T-cells; a polyclonal antibody activation suggestive of a defect in helper/suppressor cell regulation; and a reduced ability to mount an immune response against viral antigens,[105] studies in laboratories outside of Japan have not been able to corroborate this immunodeficiency.

The NOD mouse is an ideal model in which to examine the genetics of diabetes in the mouse. It is now clear that three recessive genes are required for diabetes to develop in this model.[106,107] One gene is located near, or is identical with, MHC locus.[106-108] Since the NOD mouse has a unique Ia molecule that does not react with the majority of anti-Ia antibodies, this may actually be the gene involved. Recent DNA sequencing has revealed that the alpha chain of the Ia antigen is identical to that of Ia^d but that the beta chain differs from all other reported haplotypes. Thus it may be that the Ia beta gene is the unique diabetes susceptibility gene associated with the MHC. Genetic crosses have permitted mapping of a second gene to a locus near the *THY-1* gene. Finally genetic breeding experiments have identified the requirement for a third recessive gene the location of which has not been precisely mapped.[106,107]

Autoimmune Basis of Diabetes in the NOD Mouse

As noted above, insulitis occurs in pancreases of NOD mice of both sexes before the onset of diabetes, even if most of these mice will not develop diabetes. Thus insulitis by itself is necessary, but not sufficient, for the production of diabetes in the NOD mouse; the immune system, however, seems to be clearly involved. Athymic mice with a NOD genetic background do not develop either diabetes or insulitis.[109] Reconstitution of the immune system by allogeneic bone marrow transplantation prevents diabetes in the NOD mouse,[110] whereas transplantation of islets in tolerant mice only briefly ameliorates hyperglycemia; this is followed by a recurrence of the diabetes, but not by rejection of the transplanted islets (see Chapter 10).

Immune system involvement in the pathogenesis of diabetes in the NOD mouse also has been supported by the results of studies that use immunotherapy to prevent diabetes. An early paper demonstrated that nicotinamide injections, if given to female NOD mice from the time of weaning, prevented the onset of diabetes. Two-thirds of the mice in a control group developed diabetes by 40 days after weaning, whereas none of the nicotinamide-treated animals developed glucosuria; and histologically, these mice had only a mild insulitis. Four of the six NOD mice that developed diabetes in the control group, if given nicotinamide from the day of the first occurrence of glycosuria, were "cured" of their glycosuria; their glucose tolerance also improved during the course of therapy.

This observation is the rationale for trials of such treatment in humans. One such trial of nicotinamide was recently reported to be ineffective in preventing human diabetes.[112] Cyclosporin also prevents diabetes in the NOD mouse.[113]

Cell-Mediated Anti-Beta–Cell Autoimmunity in NOD Mice

Diabetes can be passively transferred from NOD mice at the onset of hyperglycemia to naive mice (who then develop diabetes before six weeks of age) either by injecting activated splenocytes[114] or T-cell–enriched splenocytes.[115] Recently two different laboratories demonstrated that the transfer of diabetes by T-cells requires that both CD4⁺ and CD8⁺ cells be transferred.[115,116] These are difficult experiments to interpret, since recipient animals are genetically prone to diabetes and their own immune systems must be significantly depressed by radiation before cell transfer to prevent rejection of these cells. More recently some of these issues have been clarified by the demonstration that mice of an F_1 generation of a cross between the NOD mouse and the nondiabetic mouse can develop diabetes if injected with NOD bone marrow.[117] A cytotoxic response against BALB/c islets has been demonstrated to be greater with splenocytes from NOD mice than with splenocytes from control NON mice, even though indices of both NK cell activity and antibody-dependent cellular cytotoxicity were reduced in the NOD-mouse cells.[118] Recently this observation has been extended by selection of the effector cells using monoclonal antibodies. The $CD8^+$ cells proved to be the cytotoxic effector cells.[119]

More specific therapy has been attempted using monoclonal antibodies against specific T-cell subsets. Monoclonal antibody GK1.5 (a rat cytotoxic monoclonal antibody that binds to the mouse lymphocyte L3T4 (CD4) receptor) can block the appearance of spontaneous diabetes if given to NOD mice before severe hyperglycemia occurs.[120,121] The antibody was only effective if given in a large dose for the first week (200 μg/day for three days) followed by 100 μg/week for the duration of the study. With such treatment, only 2 of 25 female mice developed diabetes, compared with 29 of 35 untreated mice or 18 of 21 mice who received only the initial 200-μg (× 3) loading dose. Treatment with this anti-CD4 antibody cleared the preexisting islet mononuclear cell infiltration; the islet cells appeared completely normal 90 days after therapy. Most importantly diabetes did not recur when the monoclonal antibody therapy was discontinued after 100 to 150 days. When the animals were followed for up to 225 days, it was found that only 1 of 18 mice had become diabetic. Infection of NOD mice with lymphocytic choriomeningitis virus (LCMV) completely prevented diabetes in the mice who survived the viral illness.[122] Lymphocytic choriomeningitis virus is known to infect and kill mouse $CD4^+$ lymphocytes, and thus these results are consistent with the above observations that specific monoclonal antibodies ablate these cells and prevent diabetes.

Finally to identify the subset of $CD4^+$ lymphocytes that may be involved in the pathogenesis of diabetes more specifically, Kelly and colleagues employed a cytotoxic monoclonal antibody that binds to the murine IL-2 receptor.[123] These receptors are present on very recently activated, proliferating T-cells and some activated B-cells. There is some evidence that such high-affinity, IL-2 receptors are only transiently expressed on lymphocytes during activation. Groups of NOD mice were treated intraperitonealy with 5 μg of purified monoclonal antibody N7-20 beginning at five weeks of age, before insulitis usually occurs. Female littermates were treated with an unrelated monoclonal antibody as a control. Treatment with the anti-IL2 receptor suppressed insulitis in the NOD mice; four of seven treated mice had no evidence of insulitis, whereas all six of the control animals treated with the nonspecific monoclonal antibody had an inflammatory-cell islet infiltrate.

Summary

The diabetes of NOD mice differs in many ways from human diabetes mellitus, although there are also marked similarities. Like human diabetes, there appear to be diabetes susceptibility genes in the mouse that seem to be linked to, if not identical with, their unique MHC antigens, although additional non-MHC–related genes have also been described. Diabetes in NOD mice seems to be cell-mediated, in that it can be transferred to naive or even genetically different mice using preparations that contain both cytotoxic and helper T-cell populations. Finally helper T-cells seem to play a major pathogenetic role, since they are required to transfer diabetes in NOD mice adoptively. Reduction in helper T-cell numbers by monoclonal antibody administration markedly reduces the incidence of diabetes, and infection with a virus known to reduce the helper T-cell population prevents diabetes in this animal model.

Summary and Conclusions

Although direct evidence of a cell-mediated immune response in human diabetes has not yet been obtained, the animal models of diabetes that manifest many of the characteristics of human diabetes have proved to be extremely valuable in determining the role of a cell-mediated autoimmune component in the pathogenesis of diabetes mellitus. These animal models have also been extremely valuable in demonstrating the efficacy of certain immunologic therapies that may prove beneficial in the treatment of human diabetes. The BB-rat model demonstrated the possible efficacy of cyclosporin treatment before such treatment was used in human diabetes. The demonstration that nicotinamide could prevent diabetes in NOD mice led to ongoing clinical trials to determine whether this therapy has any efficacy in human diabetes, although a short trial with a limited number of subjects yielded negative results. Even in this area, it is clear that the diabetes seen in animal models is not identical to human diabetes. Cyclosporin treatment of BB rats can be discontinued after approximately 120 days of age, with no subsequent development of diabetes. Discontinuance of cyclosporin treatment in human diabetes, however, always results in a recurrence of hyperglycemia and the need for insulin therapy. Nevertheless experimentation with these animal models of autoimmune, insulin-dependent diabetes mellitus should continue to provide important insights into the pathogenesis of this fascinating, clinically important disease.

Acknowledgments. The authors wish to acknowledge the generous support of their research by the National Institutes of Health (DK 19631, DK 39286), the American Diabetes Association, the Juvenile Diabetes Foundation, the Robert Wood Johnson Trust, the New York Diabetes Research and Education Foundation, and the Swedish Diabetes Association.

References

1. Powers AC, Eisenbarth GS. Autoimmunity to islet cells in diabetes mellitus. *Ann Rev Med* 1985;36:533–544.
2. Janeway C. The immune destruction of the pancreatic beta cells. *Immunol Today* 1985;6:229–232.
3. Handwerger B. The immunology of diabetes mellitus, in Volpe R (ed): *Autoimmunity and Endocrine Disease*. New York: Marcel Dekker, 1985, pp 287–344
4. McDevitt HO. The HLA system and its relation to disease. *Hosp Prac* 1985;15: 57–72.
5. Marrack P, Kappler J. The T cell and its receptor. *Sci Am* 1986;254:36–45.
6. Nerup J, Andersen O, Bendiken G, et al. Antipancreatic cellular hypersensitivity in diabetes mellitus. *Diabetes* 1971;20:424–427.
7. Huang S-W, MacLaren NK. Insulin-dependent diabetes: A disease of autoaggression. *Science* 1976;192:64–66.
8. Segain J, Valentin A, Bardet S, et al. T4 lymphocytes from type I diabetics specifically bind in vitro to xenogeneic beta cell antigens via the T cell receptor. *Diabetes* 1988;37(Suppl 1):56A, Abstr 223.
9. Suzuki M, Charles A, Ong K, et al. In vitro islet cell cytotoxicity assays for evaluation of potential cellular and antibody-mediated immunological mechanisms in diabetes mellitus. *Diabetes* 1980;30(Suppl 1):65A, Abstr 224.
10. Scott J, Daniels JC, Poffenbarger PL. In vitro cell-mediated cytolysis of intact islets of Langerhans in studies of IDDM. *Diabetologia* 1981;21:78, Absr. 12.
11. Charles MA, Suzuki M, Waldeck N, et al. Immune islet killing mechanisms associated with insulin-dependent diabetes: In vitro expression of cellular and antibody-mediated islet cell cytotoxicity in humans. *J Immunol* 1983;130:1189– 1194.
12. Gupta S, Charles MA, Waldeck N, et al. Multiparameter immunologic studies in patients with newly diagnosed type I insulin-dependent diabetes mellitus. *Diabetes Res* 1986;3:225–229.
13. Like AA, Rossini AA. Streptozotocin-induced pancreatic insulitis: A new model of diabetes mellitus. *Science* 1976;193:415–417.
14. Kolb H. Mouse models of insulin-dependent diabetes low-dose streptozotocin-induced diabetes and non-obese diabetic (NOD) mice. *Diabetes Metab Rev* 1987; 3:751–778.
15. Rossini AA, Appel MC, Williams RM, et al. Genetic influence on streptozotocin-induced insulitis and hyperglycemia. *Diabetes* 1977;26:916–920.

16. Kolb H, Kiesel U. Localization of genes controlling an experimental model of human IDDM within and without the major histocompatibility complex. *Immunobiology* 1981;160:58–62.
17. Wolf J, Lilly F, Shin S-I. The influence of genetic background on the susceptibility of inbred mice to streptozotocin-induced diabetes. *Diabetes* 1984;33:567–571.
18. Leiter EH. Genetic control of the pathogenesis of diabetes in C3H mice. Influence of the major histocompatibility complex. *Diabetes* 1984;33:1068–1072.
19. Le PH, Leiter EH, Leyendecker JR. Genetic control of susceptibility to streptozotocin diabetes in inbred mice. Effect of testosterone and H-2 haplotype. *Endocrinology* 1985;116:2450–2455.
20. Rossini AA, Williams RM, Appel MC, et al. Sex differences in the multiple, low dose streptozotocin model of diabetes. *Endocrinology* 1978;103:1518–1520.
21. Paik S-G, Fleischer N, Shin S-I. Insulin dependent diabetes mellitus induced by subdiabetogenic doses of streptozocin: Obligatory role of cell-mediated autoimmune processes. *Proc Natl Acad Sci USA* 1980;77:6129–6133.
22. MacLaren NK, Neufeld M, McLaughlin JV, et al. Androgen sensitization of streptozotocin-induced diabetes in mice. *Diabetes* 1980;29:710–716.
23. Kuttler B, Schneider E. Diabetes mellitus in mice induced by multiple subdiabetogenic doses of streptozotocin: Age and sex dependence. *Acta Biol Med Ger* 1982; 41:1199–1201.
24. Rossini AA, Like AA, Chick WL, et al. Studies of streptozotocin-induced insulitis and diabetes. *Proc Natl Acad Sci USA* 1977;74:2485–2489.
25. Rossini AA, Williams RM, Appel MC, et al. Complete protection from low-dose streptozotocin-induced diabetes in mice. *Nature* 1978;276:182–184.
26. Buchard K, Rygaard J. T-lymphocytes transfer streptozotocin-induced diabetes mellitus in mice. *Acta Pathol Microbiol Scand [C]* 1978;86:277–282.
27. Beattie G, Lannom R, Lipstick J, et al. Streptozotocin-induced insulitis model in athymic mice. *Diabetes* 1980;29:146–150.
28. Rossini AA, Like AA, Appel MC, et al. Streptozotocin-induced insulitis model in athymic mice. *Diabetes* 1980;29(Suppl 2):51A, Abstr. 204.
29. Kiesel U, Freytag G, Kolb H, et al. Transfer of experimental autoimmune insulitis by spleen cells in mice. *Diabetologia* 1980;19:516–520.
30. Kiesel U, Kolb H, Freytag G. Strain dependence of the transfer of autoimmune insulitis in mice. *Clin Exp Immunol* 1981;43:430–433.
31. Nakamura M, Nagafuchi S, Yamaguchi K, et al. The role of thymic immunity and insulitis in the development of streptozotocin-induced diabetes in mice. *Diabetes* 1984;33:894–900.
32. Koevary SB, Williams RM, Chick WL, et al. Passive transfer of diabetes in the BB rat. *Science* 1983;220:727–729.
33. Kim YT, Steinberg C. Immunological studies on the induction of diabetes in experimental mice. *Diabetes* 1984;33:771–777.
34. Schreiner G, Rossini A, Mordes J, et al. Essential fatty acid deficiency inhibits the development of diabetes. *Diabetes* 1988;37 (Suppl 1):80A, Abstr. 318.
35. Paik S-G, Blue ML, Fleischer N, et al. Diabetes susceptibility of BALC/cBom mice treated with streptozotocin: Inhibition by lethal irradiation and restoration by splenic lymphocytes. *Diabetes* 1982;31:808–815.
36. Hahn HJ, Barnsdorf K, Nadrowitz R, et al. The effect of irradiation on the development of low dose streptozotocin diabetes in mice. *Acta Biol Med Ger* 1982;41: 1191–1197.

37. Nedergaard M, Egeberg J, Kromann H. Irradiation protects against pancreatic islet degeneration and hyperglycemia following streptozotocin treatment of mice. *Diabetologia* 1983;24:382–386.
38. Blue ML, Shin S-I. Diabetes induced by sub-diabetogenic doses of streptozotocin in BALB/cBom mice: Noninvolvement of host B lymphocyte function. *Diabetes* 1984; 33:105–113.
39. Kiesel U, Grevlich B, Marx-Soho Moume C, et al. Induction of diabetes by low dose streptozotocin treatment in genetically resistant mice. *Immunol Lett* 1981;3:227–230.
40. Kolb H, Greulich B, Kiesel U, et al. Demonstration of suppressor lymphocytes which inhibit the manifestations of IDDM in mice. *Diabetologia* 1981;21:292, Abstr 282.
41. Flechner I, Muntefering H, Smadja Y, et al. Immunomodulation of streptozotocin-induced diabetes in mice by a lipopolysaccharide. *Diabetes Res* 1984;1:231–232.
42. Kiesel U, Kolb H. Suppressive effect of antibodies to immune response gene products on the development of low dose streptozotocin-induced diabetes. *Diabetes* 1983; 32:869–871.
43. Sestier C, Odent-Pugo S, Bonneville M, et al. Cyclosporin enhances diabetes induced by low-dose streptozotocin-treatment in mice. *Immunol Lett* 1985;10:57–60.
44. Andersson A, Hallberg A, Sandler S, et al. Direct toxicity of cyclosporin on beta cells in tissue culture. *Diabetologia* 1984;27:66–69.
45. Sandler S, Andersson A. Modulation of streptozotocin-induced insulitis and hyperglycemia in the mouse. *Acta Pathol Microbiol Immunol Scand [A]* 1985;93:93–98.
46. Gold G, Manning M, Heldt A, et al. Diabetes induced by multiple, subdiabetigenic doses of streptozotocin. Lack of protection by exogenous superoxide dismutase. *Diabetes* 1981;30:634–638.
47. Sandler S. Protection by dimethyl urea against hyperglycemia but not insulitis in low dose streptozotocin-induced diabetes in the mouse. *Diabetologia* 1984;26:386–388.
48. Andersson A. Islet transplantation normalizes hyperglycemia caused by streptozotocin-induced insulitis in the mouse. *Lancet* 1979;i:581–584.
49. Sandler S, Andersson A. Islet implantation into diabetic mice with pancreatic insulitis. *Acta Pathol Micro Scand [A]* 1981;89:107–112.
50. Sandler S, Andersson A. Survival of intrasplenically implanted islets in mice with experimental insulitis and hyperglycemia. *Diabetes* 1982;31(Suppl 4):78–82.
51. McEvoy RC, Andersson J, Sandler S, Hellerstrom C. Multiple, low dose streptozotocin induced diabetes in the mouse: Evidence for stimulation of a cytotoxic cellular immune response against an insulin producing beta cell line. *J Clin Invest* 1984; 74:715–722.
52. McEvoy RC, Thomas NM, Hellerstrom C, et al. Multiple, low dose streptozotocin diabetes in the mouse: Further evidence for involvement of an anti-B cell cytotoxic cellular autoimmune response. *Diabetologia* 1987;30:232–238.
53. Chappel CI, Chappel ER. The discovery and development of the BB rat colony. An animal model of spontaneous diabetes mellitus. *Metabolism* 1983;32(Suppl. 1): 8–10.
54. Nakhooda AF, Like AA, Chappel CI, et al. The spontaneously diabetic Wistar rat. Metabolic and morphologic studies. *Diabetes* 1977;26:100–112.
55. Mordes JP, Desemone J, Rossini AA. The BB rat. *Diabetes Metab Rev* 1987;7: 325–750.
56. Marliss EB. Recommended nomenclature for the spontaneously diabetic syndrome of the BB rat. *Metabolism* 1983;32(Suppl 1):6–7.

57. Nakhooda AF, Wei CN, Like AA, et al. The spontaneously diabetic Wistar rat (the BB rat): The significance of transient glucosuria. *Diabetes Metab* 1978;4:255–259.
58. Nakhooda AF, Poussier P, Marliss EB. Insulin and glucagon secretion in BB Wistar rats with impaired glucose tolerance. *Diabetologia* 1983;24:58–62.
59. Like AA, Rossini AA. Spontaneous autoimmune diabetes in the BioBreeding/Worcester rat. *Surv Synth Pathol Res* 1984;3:131–138.
60. Butler L, Guberski DL, Like AA. The effect of inbreeding on the BB/W diabetic rat. *Metabolism* 1983;32:51–53.
61. Colle E, Guttman RD, Seemayer TA. Spontaneous diabetes mellitus in the rat. I. Association with the major histocompatibility complex. *J Exp Med* 1981;154:1237–1242.
62. Colle E, Guttman RD, Seemayer TA, et al. Spontaneous diabetes mellitus syndrome in the rat. IV. Immunogenetic interactions of MHC and non-MHC components of the syndrome. *Metabolism* 1983;32(Suppl. 1):54–61.
63. Colle E, Guttman RD, Fuks A. Insulin dependent diabetes mellitus is associated with genes that map to the right of the class I RT1.A locus of the major histocompatibility complex of the rat. *Diabetes* 1986;35:454–458.
64. Butler L, Guberski DL, Like AA. Genetic analysis of the BB/W diabetic rat. *Can J Genet Cytol* 1983;25:7–15.
65. Jackson RA, Buse JB, Rifai P, et al. Two genes required for diabetes in BB rats. Evidence from cyclical intercrosses and backcrosses. *J Exp Med* 1984;159:1629–1636.
66. Like AA, Guberski DL, Butler L. Diabetic BioBreeding/Wistar (BB/Wor) rats need not be lymphopenic. *J Immunol* 1986;136:3254–3258.
67. Greiner DL, Handler ES, Nakano K, et al. Absence of the RT-6$^+$ T cell subset in diabetes prone BB/W rats. *J Immunol* 1986;136:148–151.
68. Like AA, Forster RM, Woda BA, et al. T cell subsets in islets and lymph nodes of Bio-Breeding/Worcester (BB/W) rats. *Diabetes* 1983;32(Suppl 1):51A, Abstr 201.
69. Dean BM, Walker R, Bone AJ, et al. Prediabetes in the spontaneously diabetic BB/E rat: Lymphocyte sub-populations in the pancreatic infiltrate and expression of rat MHC class II molecules in endocrine cells. *Diabetologia* 1985;28:464–466.
70. Weringer EJ, Like AA. Immune attack on pancreatic islet transplants in the spontaneously diabetes BioBreeding/Worcester (BB/W) rat is not MHC-restricted. *J Immunol* 1985;134:2383–2386.
71. Prowse SJ, Bellgrau D, Lafferty KJ. Islet allografts are destroyed by disease occurrence in the spontaneously diabetic BB rat. *Diabetes* 1986;35:110–114.
72. Naji A, Silvers WK, Barker CF. Cell-mediated immunity in type I (insulin-dependent) diabetes of man and the BB rat. *Concepts Immunopathol* 1985;2:32–46.
73. Naji A, Bellgrau D, Anderson A, et al. Transplantation of islets and bone marrow cells to animals with immune insulitis. *Diabetes* 1982;31(Suppl. 4):84–91.
74. Woehrle M, Markmann JF, Silvers WK, et al. Transplantation of cultured pancreatic islets to BB rats. *Surgery* 1986;100:334–341.
75. MacKay P, Boulton A, Rabinovitch A. Lymphoid cells of BB/W diabetic rats are cytotoxic to islet beta cells *in vitro. Diabetes* 1985;34:706–709.
76. MacKay P, Jacobson J, Rabinovitch A. Spontaneous diabetes mellitus in the BioBreeding/Worcester rat. Evidence *in vitro* for natural killer cell lysis of islet cells. *J Clin Invest* 1986;77:916–924.
77. Prud'Homme GJ, Fuks A, Colle E, et al. Isolation of T lymphocyte lines with specificity for islet cell antigens from spontaneously diabetic (insulin-dependent) rats. *Diabetes* 1984;33:801–803.

78. Prud'Homme GJ, Fuks A, Guttman RP, et al. T cell hybrids with specificity for islet cell antigens. *J Immunol* 1986;136:1535–1536.
79. Laupacis A, Gardell C, Dupre J, et al. Cyclosporin prevents diabetes in BB Wistar rats. *Lancet* 1983;i:10–12.
80. Like AA, Dirodi V, Thomas S, et al. Prevention of diabetes mellitus in the BB/W rat with cyclosporin A. *Am J Pathol* 1984;117:92–97.
81. Like AA, Weringer EJ, Holdash A, et al. Adoptive transfer of autoimmune diabetes mellitus in Bio-Breeding/Worcester (BB/W) inbred and hybrid rats. *J Immunol* 1985;134:1583–1587.
82. Handler ES, Mordes JP, Geisberg M, et al. Effect of ultraviolet (UV) and X-irradiation on diabetes-prone and resistant BB/W rats. *Diabetes* 1985;34(Suppl 1):69A, Abstr 275.
83. Greiner DL, Mordes JP, Handler ES, et al. Depletion of RT6.1$^+$ T lymphocytes induces diabetes in resistant Bio-Breeding/Worcester (BB/W) rats. *J Exp Med* 1987; 166:461–475.
84. Koevary SB, Williams DE, Williams RM, et al. Passive transfer of diabetes from BB/W to Wistar-Furth rats. *J Clin Invest* 1985;75:1904–1907.
85. Yale JF, Vigeant C, Ivanic D. Metabolism time course and immunological concomitants of passive transfer of type I (insulin dependent) diabetes in the BB rat. *Diabetologia* 1986;29:608A.
86. Handler ES, Mordes JP, Seals J, et al. Diabetes in the Bio-Breeding/Worcester (BB/W) rat. Induction and acceleration by spleen cell conditioned media. *J Clin Invest* 1985;76:1692–1694.
87. Mordes JP, Handler ES, Like AA, et al. Irradiated lymphocytes do not adoptively transfer diabetes or prevent spontaneous disease in the BB/W rat. *Metabolism* 1986; 35:552–554.
88. Like AA, Rossini AA. Guberski DL, et al. Spontaneous diabetes mellitus: Reversal and prevention in the BB/W rat with antiserum to rat lymphocytes. *Science* 1979; 206:1421–1423.
89. Rossini AA, Slavin S, Woda BA, et al. Total lymphoid irradiation prevents diabetes mellitus in the BioBreeding/Worcester (BB/W) rat. *Diabetes* 1984;33:543–547.
90. Oschilewski U, Kiesel U, Kolb H. Administration of silica prevents diabetes in BB rats. *Diabetes* 1985;34:197–199.
91. Brayman KL, Armstrong J, Shaw LM, et al. Prevention of diabetes in BB rats by intermittent administration of cyclosporine. *Surgery* 1985;102:235–241.
92. Brayman K, Markmann J, Barker C, et al. Prevention of diabetes in BB rats requires lymphocytes functionally restricted to MHC-compatible thymic determinants. *Diabetes* 1988;37(Suppl 1):55A, Abstr 218.
93. Naji A, Silvers WK, Bellgrau D, et al. Prevention of diabetes in rats by bone marrow transplant. *Ann Surg* 1981;194:328–330.
94. Rossini AA, Mordes JP, Pelletier AM, et al. Transfusions of whole blood prevent spontaneous diabetes in the BB/W rat. *Science* 1983;219:975–977.
95. Scott J, Engelhard VH, Curnow RT, et al. Prevention of diabetes in BB rats. I. Evidence suggesting a requirement for mature T cells in bone marrow inoculum of neonatally injected rats. *Diabetes* 1986;35:1034–1040.
96. Mordes JP, Gallina DL, Handler ES, et al. Transfusions enriched for W3/25$^+$ helper/inducer T lymphocytes prevent spontaneous diabetes in the BB/W rat. *Diabetologia* 1987;30:22–26.
97. Jacobson JD, Markmann JF, Brayman K, et al. Prevention of recurrent autoimmune diabetes in BB rats by anti-asialo-GM1 antibody. *Diabetes* 1988;37:838–841.

98. Nakamura N, Greiner DL, Reynolds CW, et al. Cytotoxic effector cells of diabetes-prone and diabetes-resistant BB/Wor rats are different. *Diabetes* 1988;37(Suppl 1):55A, Abstr 220.
99. Tochino Y, Kanaya T, Makino S. Studies on spontaneously diabetic non-obese mice. *J Japn Diabetes Soc* 1978;21:295.
100. Makino S, Kunimoto K, Maraoka Y, et al. Breeding of a non-obese, diabetic strain of mouse. *Exp Animal* 1980;29:1–13.
101. Makino S, Hayashi Y, Muraoka Y, et al. Establishment of the non-obese diabetic mouse, in Sakamoto N (ed): *Current Topics in Clinical and Experimental Aspects of Diabetes Mellitus*. Amsterdam: Elsevier Science Publishers, 1985, pp 25–32.
102. Tarui S, Tochino Y, Nonaka K (eds). *Insulitis and Type I Diabetes—Lessons from the NOD Mouse*. New York: Academic Press, 1986.
103. Leiter EH, Prochazka M, Coleman DL. The non-obese diabetic (NOD) mouse. *Am J Pathol* 1987;128:380–383.
104. Makino S, Kunimoto K, Muraoka Y, et al. Effect of castration on the appearance of diabetes in NOD mice. *Exp Anim* 1980;30:137–140.
105. Kataoka S, Satoh J, Fijiya H, et al. Immunologic aspects of the non-obese diabetic mouse. *Diabetes* 1983;32:247–253.
106. Wicker LS, Miller BJ, Coker LZ, et al. Genetic control of diabetes and insulitis in the non-obese diabetic (NOD) mouse. *J Exp Med* 1987;165:1639–1654.
107. Prochazka M, Leiter EH, Serreze DV, et al. Three recessive loci required for insulin-dependent diabetes in non-obese diabetic mice. *Science* 1987;237:286–289.
108. Hattori M, Buse JB, Jackson RA, et al. The NOD mouse: Recessive diabetogenic gene in the major histocompatibility complex. *Science* 1986;231:733–735.
109. Makino S, Harada M, Kishimoto Y, et al. Absence of insulitis and overt diabetes in athymic nude mice with NOD genetic background. *Jikken Dobutsu* 1986;33:495–498.
110. Ikehara S, Ohtsuki H, Good RA, et al. Prevention of type I diabetes in non-obese diabetic mice by allogeneic bone marrow transplantation. *Proc Natl Acad Sci USA* 1985;82:7743–7747.
111. Yamada K, Nonake K, Hanafusa T, et al. Preventive and therapeutic effects of large-dose nicotanamide injections on diabetes associated with insulitis. An observation in diabetic (NOD) mice. *Diabetes* 1982;31:749–753.
112. Herskowitz R, Jackson RA. Pilot trial of preventive therapy: Progression to overt hyperglycemia by 3/3 "prediabetics" despite oral nicotinamide. *Diabetes* 1988; 37(Suppl 1):59A, Abstr 233.
113. Mori Y, Suko M, Okudaira H, et al. Preventive effects of cyclosporin on diabetes in NOD mice. *Diabetologia* 1986;29:244–247.
114. Wicker LS, Miller BJ, Mullen Y. Transfer of autoimmune diabetes mellitus with splenocytes from non-obese diabetic (NOD) mice. *Diabetes* 1986;35:855–860.
115. Bendelac A, Carnaud C, Boitard C, et al. Syngeneic transfer of autoimmune diabetes from diabetic NOD mice to healthy neonates: Requirement and both L3T4$^+$ and Lyt-2$^+$ T cells. *J Exp Med* 1987;166:823–832.
116. Miller BJ, Appel MC, O'Neil JJ, et al. Both the Lyt-2$^+$ and L3T4$^+$ T cell subsets are required for the transfer of diabetes in non-obese diabetic mice. *J Immunol* 1988; 140:52–58.
117. Serreze DV, Leiter EH, Worthen SM, et al. NOD marrow stem cells adoptively transfer diabetes to resistant (NOD × NON) F_1 mice. *Diabetes* 1988;37:252–255.

118. Maruyama T, Takei I, Taniyama M, et al. Immunological aspect of non-obese diabetic mice: Immune islet cell killing mechanism and cell-mediated immunity. *Diabetologia* 1984;27:121–123.
119. Nagata M, Yokono K, Hayakawa M, et al. Destruction of pancreatic islets by cytotoxic T-lymphocytes in NOD mice. *Diabetes* 1988;37(Suppl 1):97A, Abstr 387.
120. Shizuru JA, Taylor-Edwards C, Banks BA, et al. Immunotherapy of the non-obese diabetic mouse: Treatment with an antibody to T-helper lymphocytes. *Science* 1988;240:659–662.
121. Charlton B, Mandel TE. Progression from insulitis to beta cell destruction in NOD mouse requires L3T4$^+$ T lymphocytes. *Diabetes* 1988;37:1108–1112.
122. Oldstone MBA. Prevention of type I diabetes in non-obese diabetic mice by virus infection. *Science* 1988;239:500–502.
123. Kelly VE, Gaulton GN, Hattori M, et al. Anti-interleukin 2 receptor antibody suppresses murine diabetic insulitis and lupus nephritis. *J Immunol* 1988;140:59–61.

9 The Pathogenesis of Immunologically Mediated Diabetes

YI WANG, LIMING HAO, AND KEVIN J. LAFFERTY

Introduction

Diabetes as a pathologic entity has been with us as long as recorded medical history. Effective treatment for this disease, however, has developed only over the last 60 years, and it is probably fair to say that we still do not understand the physiologic trigger for the process that leads to insulin-dependent (type I) diabetes nor the detailed mechanisms of beta-cell destruction. This lack of detailed knowledge is no reason to shy away from the subject. We are at a time when discussion and speculation is needed to focus our research approach. This is a stage of scientific development when new ideas and theories are hotly debated in the light of experimental data. In this chapter, we approach our subject as immunologists rather than endocrinologists; it is now clear that type I diabetes, both in man and in animal models, is a pathologic process with an immunologic etiology.[1-3]

Type I diabetes is sometimes said to be autoimmune.[1-3] This is a loose use of the term. Control of the disease process in man with immunosuppressive drugs points to an immunologic etiology[1-3]; not all immunopathology results from autoimmunity, however. The classic example is lymphocytic choriomeningitis (LCM), a virus infection in rodents.[4] In carrier colonies, the virus is transmitted from mother to offspring with no obvious ill effects; the virus replicates but is noncytopathic and the host is "tolerant" to the agent that has been with it throughout neonatal development. When animals from noncarrier colonies are infected with the virus, however, a severe encephalitis develops. Inflammatory cells are seen in the brain. The immune response to the nontolerated virus causes this disease. This is not autoimmunity, although clearly we are dealing with a form of immunopathology. The term *autoimmunity* must be retained to describe the generation in the animal of what Burnet called "forbidden clones"[5]; that is, clones of either T- or B-cells, which normally would not develop, and when they do develop, are specific for certain self-antigens. It is the interaction of the target organ or cells with these components of the immune system that is responsible for the pathologic process that is termed *autoimmunity*. In a strict sense, a disease process cannot be classified as autoimmune until we have isolated the T-cell or antibody responsible for the pathologic process and shown that one or both of these agents interact specifically with a true autoantigen. We have not reached this point in the study of type I diabetes. Only if and when this point has been reached can we say that the disease process, is autoimmune. In the interim we are dealing with an immunopathology of unknown etiology.

How do we study this disease process? First we need input from epidemiologists and geneticists. Their work has been the basis for the notion that we are dealing with an immunologic process in insulin-dependent diabetes mellitus (IDDM).[3,6] The next step is analysis – the testing of hypotheses in animal models. We have two relevant models for the study of type I diabetes – the BioBreeding (BB) rat and the Non-obese Diabetic (NOD) mouse. In both models, inflammation of the islets (insulitis) is necessary for diabetes to develop, but it is not always sufficient for this disease to be expressed.[7–9]

What do we know about the pathogenesis of diabetes from studies of the BB rat and the NOD mouse? First in both cases the disease process is thymus dependent and can be inhibited by appropriate immunosuppressive treatment before disease onset.[10–13] The disease process can be accelerated in diabetes-prone animals by the transfer of lymphocytes from diabetic donors.[14,15] In the BB rat, the lymphocytes must be activated by concanavalin A (Con-A) before transfer[14]; the situation is similar to that seen in the transfer of experimental allergic encephalitis (EAE).[16] In the BB rat, however, disease can also be accelerated by the transfer of the supernatant from Con-A–activated spleen cells.[17] At this stage it is not clear just what is responsible for the acceleration of the disease process in this animal model. The initial cell transfers were interpreted as transfer of reactive T-cells. However the fact that a similar acceleration of disease is obtained when the supernatant of Con-A–activated cells is injected into diabetes-prone animals suggests that one or more of the cytokines may be involved in triggering the disease process. This later notion may tie in with the concept proposed by Nerup and his colleagues that the proximate cause of islet damage is mediated by such cytokines as monokine interleukin-1 (IL-1) and tumor necrosis factor (TNF) (this volume).

In the NOD mouse disease transfer requires the cooperative interaction of CD4 and CD8 T-cells.[20] Monoclonal antibody treatment (anti-CD4) of prediabetic NOD mice delays (possibly indefinitely) the onset of disease, which suggests a role for the CD4 cell in the disease process.[21] Like's group, however, found evidence that CD8+/OX 19-cells (NK-cells?) were required for disease expression in the BB rat.[22]

Analysis of the disease process in animals models using classic cellular and antibody transfer techniques has shown that we are dealing with an immunologic disease. However questions concerning the detailed mechanisms of the disease process still require clarification. In this chapter, we discuss a different approach to the problem. Rather than studying the adoptive transfer of immunity, we have asked what information can be gained by studying islet-tissue grafts in spontaneously diabetic animals. Our approach has been to ask what happens when we "bring the mountain to Mohammed."

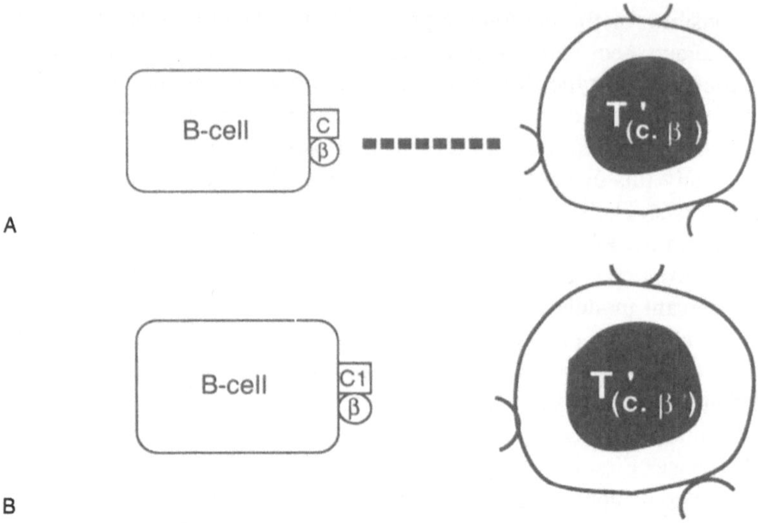

FIGURE 9.1. Diagrammatic representation of disease recurrence in grafted islet tissue mediated by a direct attack of islet-reactive T-cells, T′ (c.β). These cells are specific for MHC antigen, c, on the surface of the host beta-cell and some islet-associated antigen, β. **A**: This direct attack results in the recognition and initiation of disease in the MHC-compatible islet graft. **B**: The MHC-incompatible islet graft c1.β) is not recognized by the activated T-cell, T′ (c.β).

Mechanism of the Disease Process

MHC-Restriction–Role of Islet Tissue

Our laboratory has used islet grafting to analyze T-cell involvement in the generation of immunologically mediated islet damage. The immunogenicity of islet tissue can be reduced or eliminated by organ culture in an oxygen-rich atmosphere before grafting.[23,24] The grafting of such tissue in spontaneously diabetic animals provides a model for the study of the disease process; in this case, we study disease recurrence in the grafted tissue. Figure 9.1A,B shows the situation we might expect if the disease results from a direct T-cell attack on the target beta-cell. The specificity of the T-cell is denoted (c.β), where c represents the major histocompatibility complex (MHC) expressed on the islet beta-cell, and β is some islet-associated antigen. If such an interaction were the basis of islet damage, we would expect the disease process to be restricted by the MHC carried on the surface of the grafted islet tissue. Thus, an allogeneic islet graft that presents antigen c1.β) would resist direct attack by T-cells of specificity (c.β); see Figure 9.1B. Studies in the rat have produced somewhat conflicting results. Our own studies suggest that the disease process is not restricted by the MHC of the islet tissue grafted to spontaneously diabetic rats or diabetic NOD mice.[25,26] Thus when allogeneic PVG (RT1^{c}) islets and pituitary are first cultured and then transplanted to the kidney capsule of spontaneously diabetic BB rats ($RT1^{u}$), the islet tissue is destroyed; however the pituitary placed along side the islet tissue remains intact (Fig. 9.2, pp. 188–189). The disease process is characterized by an intense hemorrhagic reaction within the grafted islet tissue, with mononuclear and eosinophilic cellular accumulation in the grafted tissue prior to its destruction. These findings confirm those reported by Weringer et al.[27] Naji, however, reported that Lewis islets ($RT1^{l}$) resist the disease process when grafted to the liver of spontaneously diabetic BB recipients.[28] This finding does not agree with that of Weringer and colleagues[27] who reported damage to Lewis islets grafted to the kidney capsule of diabetic BB rats. Further studies from our laboratory helped clarify this situation (Table 9.1). When we examined the function of PVG and Lewis islets—precultured to reduce immunogenicity—in spontaneously diabetic BB rats, the PVG islets were acutely rejected whereas the Lewis islets continued to function (Table 9.1). Upon histologic examination, however, some of the Lewis islets showed definite signs of damage. Endothelium was stripped from blood vessels in the graft and many beta-cells were degranulated. Sufficient granulated beta-cells remained to maintain graft function, however. This finding raises questions concerning the sensitivity of different types of islet tissue to the disease process. We will return to this question when we discuss the mechanism of immunologically mediated islet destruction.

TABLE 9.1. Survival of PVG and Lewis islets following transplantation to spontaneous diabetic BB rats.[a]

Islet donor	Graft survival time (days)	Time of nephrectomy	Graft histology: H & E[b]	Graft histology: AF (% positive)[c]
PVG	7	Day 20	++++[d]	0[e]
	5	Day 15	++++	< 5
	8	Day 13	+++	10
	0	Day 41	ND	ND
	0	Day 40	+++	0
Lewis	> 50	Day 70	+	75
	25	Day 41	ND	ND
	> 50	Day 62	0	> 95
	0	Day 40	+++	0
	> 50	Day 50	0	> 95

[a] Islet tissue from PVG or Lewis rats, cultured in a 95% oxygen/5% CO_2 atmosphere for 14 days, was transplatned into the kidney capsule of diabetic BB rats.[24] Graft survival was monitored by weekly blood glucose measurements. The islet graft was removed for histologic examination at the times indicated.

[b] Section stained by hematoxylin and eosin H&E.

[c] Section stained by aldehyde-fuchsin (AF).

[d] Histologic scores: 0, no infiltrate; +, minor focal infiltrate; ++, major focal infiltrate; +++, diffuse infiltration with destruction of islet architecture; ++++, complete destruction of islet parenchyma; ND, not determined.

[e] Mean percentage of AF-positive cells per graft.

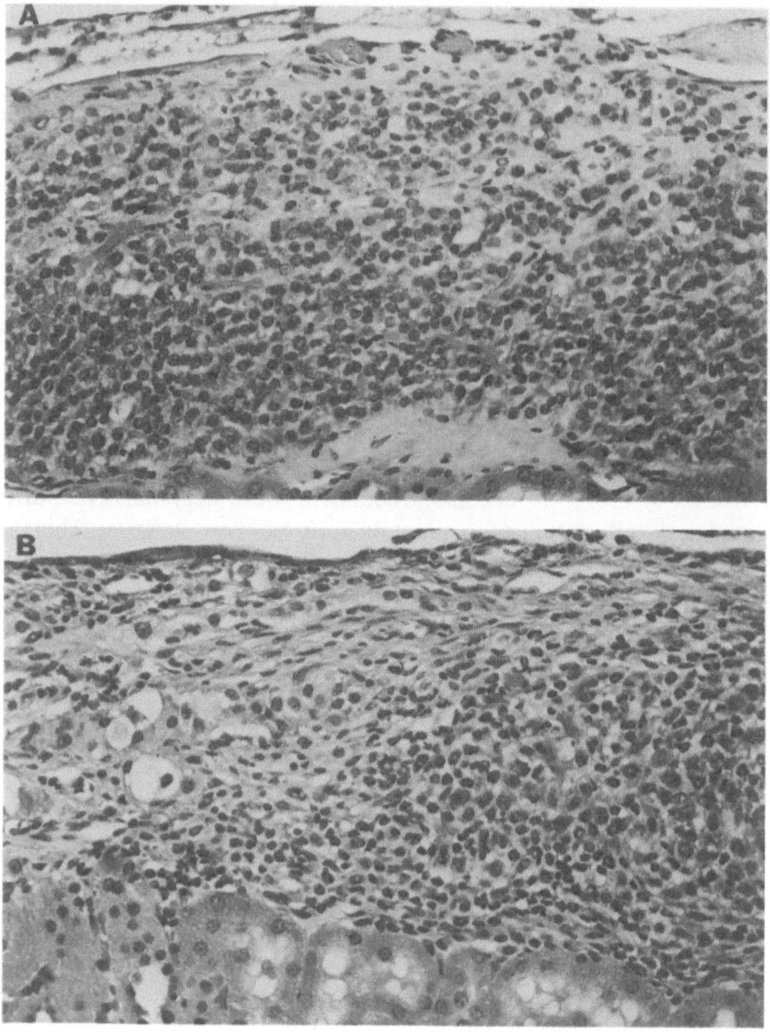

FIGURE 9.2. Histologic appearance 18 days after grafting cultured pituitary and islet allografts taken from PVG donors (RT1^c) in kidney capsules of diabetes-prone BB/Den recipients rats (RT1^u). **A**: Pituitary allograft shows no evidence of an allograft response. **B**: The islet graft adjacent to the pituitary is destroyed; eosinophilic granulocytes are prominent in the inflammatory infiltrate. Eosinophils are readily identified in the hematoxylin- and eosin-stained sections by their doughnut-shaped nuclei and eosinophilic granules in the cytoplasm.

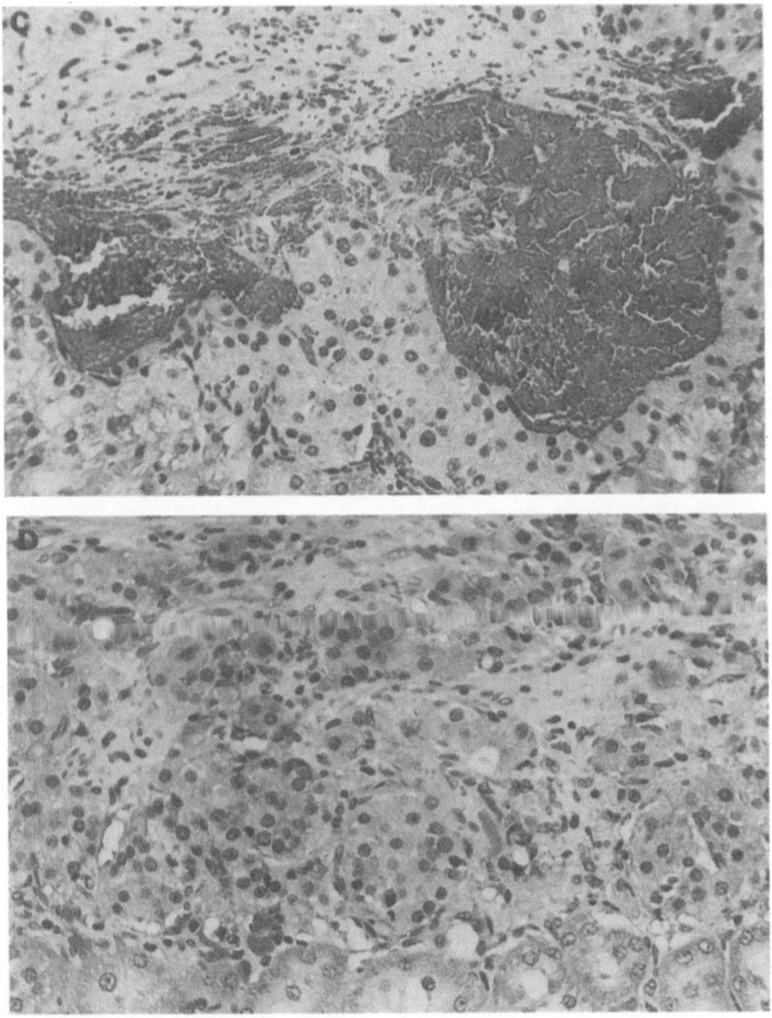

FIGURE 9.2. *Continued.* **C**: The intense hemorrhagic reaction within the grafted islet tissue. **D**: The normal histologic appearance, 28 days after grafting of cultured islet allografts taken from PVG donors ($RT1^c$) and grafted in kidney capsules of DA rats ($RT1^a$). The islet graft shows no evidence of an allograft response (H&E, × 200).

TABLE 9.2. Cultured BALB/c islet grafts are destroyed by disease recurrence in spontaneously diabetic NOD mice.

Islet			Grafts surviving (*N*)/Total (*N*)[a]	
Group no.	Donor	Recipient	Islet graft	Pituitary graft
1	BALB/c	Diabetic NOD	0/6	6/6
2	BALB/c	Nondiabetic NOD[b]	4/4	NT[c]
3	BALB/c	Nondiabetic NOD primed with BALB/c spleen cells[d]	NT	0/3

[a] NOD mice received cultured BALB/c islet and pituitary grafts. Graft survival was monitored by macroscopic and histologic examination in all groups at 28 days postgraft and by weekly blood glucose measurements in group 1.
[b] Five months-old male, nondiabetic NOD mice were the islet recipients. Animals were monitored by blood glucose measurements for diabetes.
[c] NT, not tested.
[d] Animals in group 3 were injected IP with 2×10^6 BALB/c spleen cells at the time of grafting.

Disease Recurrence Model

Our laboratory has used islet grafting in spontaneously diabetic NOD mice to investigate pathogenesis of the disease process. As mentioned earlier, the immunogenicity of islet tissue can be reduced or eliminated by organ culture in an oxygen-rich atmosphere prior to grafting.[23,24] Thus grafting of cultured islet tissue in diabetic NOD mice provides a model for the study of the disease process. Diabetic NOD mice grafted with cultured islet and pituitary tissue from BALB/c donors destroy the islet tissue graft but not the pituitary graft (Table 9.2). That is there is a tissue-specific destruction of islet tissue grafts in the spontaneously diabetic NOD mouse. This tissue specificity is characteristic of the disease process. When recipient NOD mice were challenged with spleen cells of donor origin at the time of grafting, however, the pituitary graft was promptly destroyed (Table 9.2); the allograft reaction is not tissue-specific.

Histologic examination of tissue grafted in diabetic NOD recipients at various times postgraft, showed well-granulated islets four days after grafting. However by seven days, an intense mononuclear cell infiltrate was seen in and around the islet tissue that, by this stage, was degranulated. Fourteen days postgraft, all the animals were hyperglycemic and histologic examination revealed that the islet tissue was totally destroyed. Little or no damage was seen in cultured pituitaries from the same donors that were grafted at the same time as the islet tissue (Table 9.2). When cultured BALB/c islet tissue was grafted in nondiabetic NOD mice, little or no infiltration was seen around the islet graft and the islets were still well granulated when examined 28 days postgraft (Table 9.2). The tissue-specificity of the pathologic process observed when cultured BALB/c islet tissue was grafted in diabetic NOD mice, and its association with diabetes in the recipient animal, is strong evidence that this destructive process represents disease recurrence in the grafted tissue and is not the result of an allograft response.

Disease Recurrence in Islet Grafts is CD4 T-Cell Dependent

The immune system may destroy islet tissue in a T-cell–dependent manner in several ways. First islet destruction could result from the direct interaction of cytotoxic T-cells with pancreatic beta-cells (Fig. 9.1). The results of the experiments described above that indicate that the disease process is not restricted by the MHC of the grafted islet tissue tend to argue against such a proposition. Miller et al. have recently reported that diabetes can be induced in young nondiabetic NOD mice by the intravenous injection of unfractionated spleen cells obtained from diabetic NOD donors – the adoptive transfer of immunity.[20] Spleen cells from donors depleted of either CD4 or CD8 T-cell subsets were unable to transfer disease in this system, which suggests that both CD4 T-cells and CD8 T-cells were required to initiate the disease process. Similar findings, that both CD4 and CD8 T-cells are required to transfer the disease to newborn NOD mice, have been reported by Bendelae and colleagues.[29] The interpretation of these studies is that initiation of the disease process in recipient mice requires an interaction between CD4 and CD8 T-cells and that the CD8 "cytotoxic" T-cell is the effector that causes islet damage.[20,29] Studies of disease recurrence in MHC-incompatible islets (see above) suggest that the disease process results from inflammatory tissue damage activated by CD4 T-cells interacting with antigen processed and presented by host antigen-presenting cells (APC) in and around the target islet tissue (Fig. 9.3). This apparent conflict between these two sets of data led us to reinvestigate the process.

In our studies, we examined both the T-cell requirement for transfer of disease to the NOD mouse and the T-cell requirement for expression of disease when islets were grafted in spontaneously diabetic NOD recipients (see Fig. 9.4). Young recipient NOD animals were irradiated with γ-rays (775 rads) and injected intraperitoneally with spleen cells from diabetic NOD donors. Spleen cells from diabetic animals were transferred to three groups of recipient animals. The control group received no further treatment. The two experimental groups were treated with anti-CD4 (GK1.5) or anti-CD8 (116–13.1) antibody in vivo to inactivate the function of one of the T-cell subsets (Fig. 9.4). The results of this study (Table 9.3) indicate that both CD4 and CD8 subsets are required to initiate the disease process following passive transfer of immunity to recipient NOD mice. This finding is consistent with earlier reports from other groups.[20,29]

We obtained very different results when we used the same antibodies to control disease recurrence in islets grafted in spontaneously diabetic NOD mice. Grafting of cultured BALB/c islet tissue in spontaneously diabetic NOD mice failed to bring the blood glucose into the normal range (Table 9.4). This inability of islet tissue to function is the result of disease recurrence in the graft (Table 9.2). By depleting CD4 or CD8 T-cells prior to grafting islet tissue in diabetic NOD recipients, we were able to determine which T-cell subset was required for disease recurrence in the grafted tissue.[30,31] When cultured islet tissue was grafted in animals treated in vivo with anti-CD4 antibody, all the grafts brought the blood sugar into the normal range within a week (Table 9.4). Normoglycemia was not maintained indefinitely; grafted animals became hyperglycemic 10 to 47 days

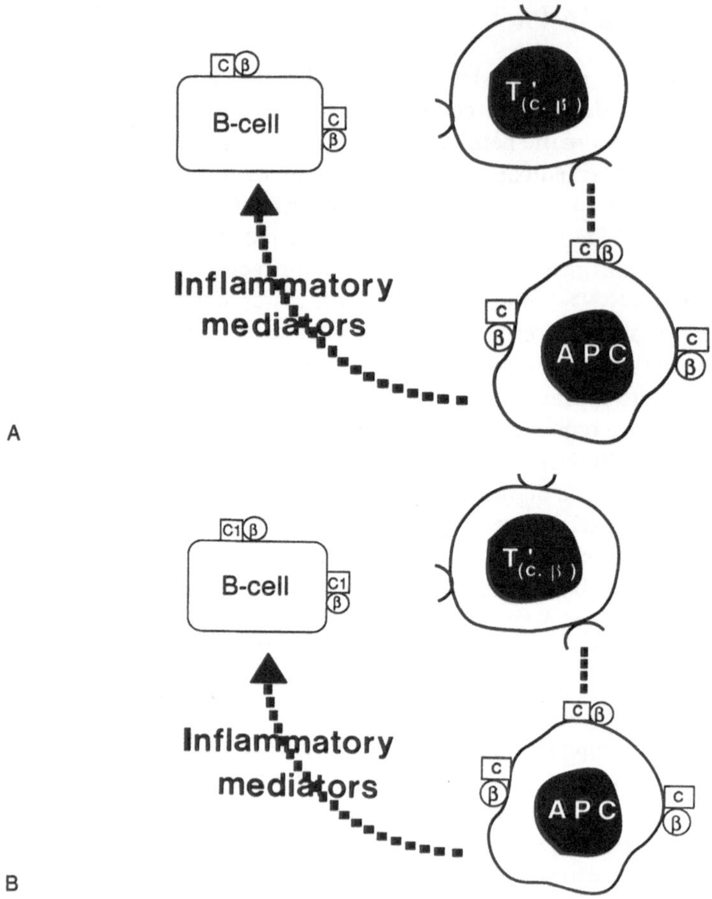

FIGURE 9.3. Diagrammatic representation of disease recurrence in grafted islet tissue mediated by an indirect attack of islet-reactive T-cells, T′ (c.β). These cells are specific for MHG antigen, c, on the surface of host beta-cells and some islet-associated antigen, β. This indirect attack results from the recognition of antigen β, which has been processed by an antigen-presenting cell (APC) of the host. Islet damage results from the effect of inflammatory mediators produced by the host APC. In this situation, MHC-compatible (**A**) and MHC-incompatible (**B**) islet tissue are equally sensitive to the disease process.

postgraft. This return of the disease correlated with the reappearance of CD4 cells in the peripheral blood; the percent of CD4 T-cells at the time of disease recurrence was approximately 18 (n=6). When cultured islet tissue was grafted in animals treated with anti-CD8 antibody, however, the grafts failed to bring the blood glucose into the normal range and all the animals remained hyperglycemic (Table 9.4). Histologic examination revealed that all grafts in the group treated with anti-CD8 antibody were destroyed within 14 days. These results lead us to conclude that expression of disease in this situation is CD4 T-cell dependent.

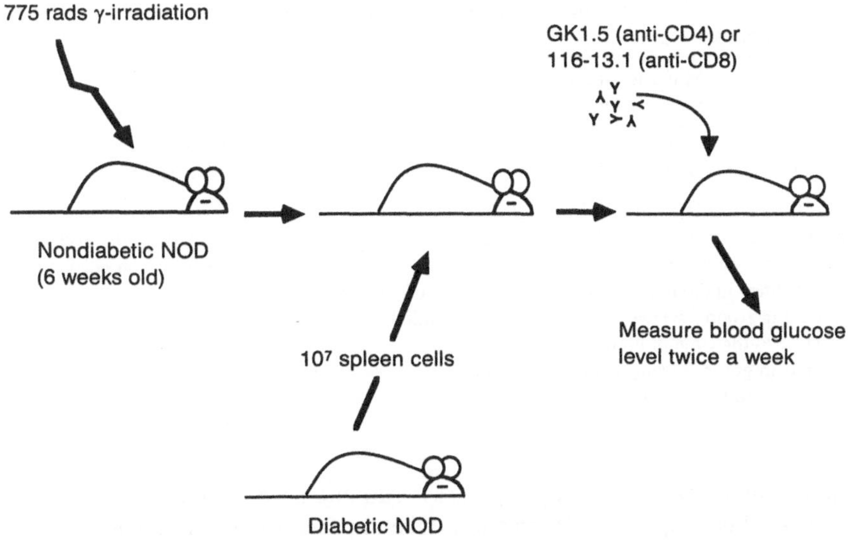

FIGURE 9.4. Protocol used to determine the involvement of CD4 and CD8 T-cells in the transfer of disease to diabetes-prone NOD mice with spleen cells from diabetic NOD donors. Six-week-old nondiabetic NOD animals were γ-irradiated (775 rads) and injected intraperitoneally with 10^7 spleen cells from diabetic NOD donors on day 0. Animals were then given anti-CD4 (GK1.5) or anti-CD8 (116–13.1) monoclonal antibody, 200 mg/kg, on days 0, 7, and 14. The control group received no further treatment. All animals were followed by urine and blood glucose measurements.

The finding that both CD4 and CD8 T-cells are required to initiate the disease process in diabetes-prone NOD mice, but that only CD4 T-cells are required for disease recurrence in tissue grafted in animals with active disease, leads to the conclusion that the CD4 T-cell is the immunologic effector in the process of islet destruction, and that some form of interaction between CD4 and CD8 T-cells is

TABLE 9.3. CD4 and CD8 T-cells are required to initiate the disease process.

Antibody treatment	Diabetic (*N*)/Total (*N*)[a]	Days to diabetes (mean ± SD)
None	9/10	22.6 (±5.9)
GK1.5 (anti-CD4)	0/5	> 120
116–13.1 (anti-CD8)	0/5	> 120

[a] Six-week-old, nondiabetic NOD mice, after γ-irradiation (775 rads), were injected with 10^7 spleen cells from diabetic donors. The presence of diabetes was determined by blood glucose measurements. Animals were considered diabetic when two consecutive blood sugar readings were greater than the normal blood range of NOD mice (6.2 mmol/l with a 99% interval of ±3.1 mmol/l.

TABLE 9.4. Disease recurrence in BALB/c islets grafted to diabetic NOD mice is CD4 T-cell dependent.

Group no.	Monoclonal antibody pretreatment	Graft survival time after grafting[a]
1 (n =9)	None	<14, <14, <14, <14, <14, <14, <14, <14, <14
2 (n=11)	GK1.5 (anti-CD4)[b]	<14, 30, 29, 14, 28, 28, 36, 18, 34, 47, 39
3 (n=7)	116-13.1(anti-CD8)[c]	<14, <14, <14, <14, <14, <14, <14

[a] Diabetic NOD mice received grafts of cultured BALB/c islet tissue on day 0; graft survival was monitored by weekly blood glucose measurements. Graft survival was scored as consecutive blood sugar readings in the normal blood glucose range of NOD mice.
[b] Animals in group 2 were treated with anti-CD4 antibody (GK1.5) 200 mg/kg on days −14, −7, and 0, relative to the time of grafting.
[c] Animals in group 3 were treated with anti-CD8 antibody (116-13.1) 200 mg/kg on days −3 and +4, relative to the time of grafting.

required to initiate the disease process. Although we do not understand the details of this cooperative interaction, taken at face value, these studies would suggest that the CD8 T-cell is required to activate the CD4 effector cell. This notion goes against generally immunologic dogma. However there is a situation in which CD8 cells have been shown to modulate the function of CD4 effector cells.[32] Further analysis of this problem must await the cloning of islet-reactive T-cells from the NOD mouse and an investigation of the pathogenesis of the disease process using such cloned T-cells.

Free Radical Involvement in the Disease Process

The finding in both the NOD mouse and the BB rat that the disease process is not restricted by the MHC antigens of the grafted tissue[25-27] suggests that disease results from inflammatory tissue damage (see Fig. 9.3). This proposition is also consistent with the finding that disease is a CD4 T-cell-dependent process in the NOD mouse. In the BB rat, both CD4 T-cells and NK-cells appear to be involved in the expression of disease.[22] If the disease is the result of an inflammatory response around the islet tissue, we must explain the relatively specific destruction of beta-cells in this nonspecific attack.

According to the Okamoto model for islet damage,[33,34] generation of such free radicals as superoxide at the inflammatory site could be a source of islet toxicity (Fig. 9.5). Superoxide can be converted to the reactive hydroxyl radical within the cell.[35] This radical is extremely toxic; it can cause DNA breaks and it attacks cellular membranes.[33,34] The breakage of DNA activates the repair enzyme poly-(ADP-ribose)-synthetase, which uses cellular adenosine diphosphate (ADP) as a source of adenine. Thus the repair process depletes the cells' nicotinamide adenine dinucleotide (NAD) and in this way shuts off proinsulin synthesis (Fig.

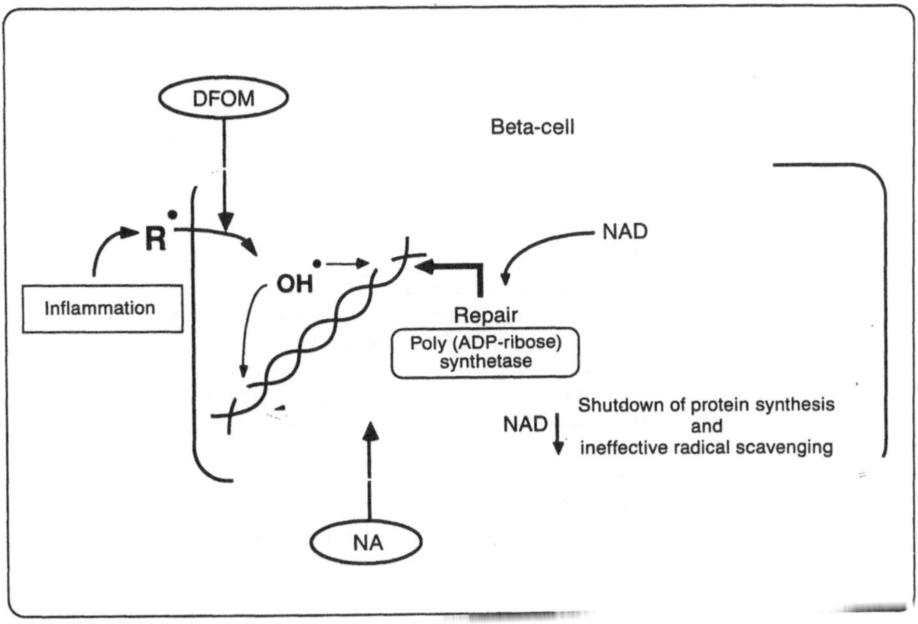

FIGURE 9.5. The Okamoto model sees the generation of hydroxyl radicals from such radicals as superoxide, produced in the inflammatory process, as a prime source of tissue damage. These hydroxyl radicals attack cellular membranes and cause DNA breaks; DNA strand breakage promotes a repair process in which the enzyme poly(ADP-ribose) synthetase is activated. Activation of this enzyme depletes NAD levels in the cell, inhibiting proinsulin synthesis and making the cell more sensitive to damage by free radicals. Desferrioxamine (DFOM) could interfere with the process by blocking hydroxyl radical production. Nicotinamide (NA) would be protective by inhibiting the DNA-repair enzyme and thus maintaining the cellular NAD levels required for efficient radical scavenging. The model predicts that these two agents will be synergistic. Modified from Okamoto[33] by permission of Kluwer Academic Publishers and from Okamoto[34] by permission of Akademie-Verlag, Berlin.

9.5).[33,34] Since NAD is also involved in the electron-transport system required for radical scavenging, the damaged cell becomes more susceptible to radical attack. The final result is death of the cell.

These steps in the process of inflammatory islet damage are illustrated in Fig. 9.5. According to this model, it is theoretically possible to interfere with this process. The chelating agent desferrioxamine could be used to remove iron from the system; the conversion of superoxide to the reactive hydroxyl radical is catalyzed by iron. It is also possible to block the DNA repair process with nicotinamide, which inhibits poly-(ADP-ribose)-synthetase. Such a blockade would reduce the degree of NAD depletion and so protect the cell from radical damage.

TABLE 9.5. Treatment with DFOM and NA prevents destruction of islet tissue grafted in spontaneously diabetic NOD mice.

Treatment[a]	Graft survival time (days)[b]
None (n=8)	0, 0, 0, 0, 0, 0, 0, 0
DFOM (n=4)	5, 5, 8, 12
NA (n=4)	7, 8, 8, 8
DFOM + NA (n=8)	9, 13, 20, 29, 38, >50, >50, >50

[a] Diabetic NOD mice received cultured BALB/c islet grafts and were treated for 24 days with DFOM (4 mg/day) or NA (15 mg/day) or a combined treatment with both DFOM and NA.
[b] Graft survival was monitored by weekly blood glucose measurements. Graft survival was scored as the consecutive blood sugar readings in the normal blood glucose range of NOD mice.

Nicotinamide has been shown to protect islet tissue from streptozotocin-induced damage.[33,34] Nicotinamide also has been shown to inhibit disease development in NOD mice.[38]

We have recently tested the validity of the Okamoto model using this approach to control the disease process.[39] Diabetic NOD mice were treated with either desferrioxamine (DFOM) or nicotinamide (NA) or a combination of the two and then received precultured BALB/c islet grafts. All grafted animals with no prior therapy showed no evidence of graft function (Table 9.5). However animals treated with either DFOM or NA became normoglycemic within a few days of grafting and remained so for approximately seven days. Further all animals treated with both DFOM and NA became normoglycemic and remained so for an extended period (Table 9.5). It is clear that these agents afford some protection from inflammatory tissue damage and that they act synergistically. We have also shown that enzyme superoxide dismutase (SOD) protects islet tissue grafted in spontaneously diabetic NOD mice from disease recurrence.[40] Such features are consistent with the notion that islet damage is the result of a free radical attack on the beta-cell.

Disease Recurrence in Grafted Tissue: A Summary

Grafting either syngeneic or cultured allogeneic tissue in spontaneously diabetic BB rats or NOD mice results in the destruction of the grafted tissue. There are some exceptions to this general rule. One such exception is the prolonged function of Lewis islets grafted to spontaneously diabetic BB rats (Table 9.1). In this situation, there is evidence that Lewis islets may be somewhat resistant to the disease process (Table 9.1). The lack of disease restriction by the MHC of the grafted islet tissue suggests that we are not looking at a destructive process resulting from a direct interaction of reactive T-cells with the beta-cells of the grafted islets. What appears more likely is that we have a situation in which a local inflammatory response—a CD4 T-cell-dependent process[30,31]—causes beta-cell destruction.

Damage in this case would result from inflammatory tissue damage (Fig. 9.3). The Okamoto model proposes that such damage results from a direct free radical attack on the target beta-cells.[33,34] Protection of grafted islets from disease recurrence by SOD, DFOM and NA, and the synergistic action of these agents, supports this hypothesis.[39,40] Another possibility that should not be ruled out at this stage is that beta-cell damage is CD4 T-cell dependent but is mediated by either natural killer (NK)-cells[22] or through antibody-dependent, cell-mediated cytotoxicity mechanisms.[41] The involvement of such cytokines as IL-1 and TNF in beta-cell damage must also be considered.[18,19] In this regard, we should note that both cytokine production and superoxide production may be regulated and amplified by each other at inflammatory foci,[36,42] and IL-1 may not damage islet beta cells directly in vivo.[37] More investigation into the disease process is required before we can outline a reasonable model that will include all the known phenomena.

We can, however, draw the following general conclusions: The disease process is T-cell dependent.[14,15] Activation of the "autoimmune" process requires an interaction between CD4 and CD8 T-cell subsets,[20,29] but the CD4 subset is required for expression of the disease.[30,31] Both macrophages and NK-cells appear to be involved in the development of disease,[22,43] probably because they are activated at the inflammatory focus. Free radicals[33,34,39,40] and cytokines[18,19,37] can be involved in the destructive process, but the exact mechanism of this destruction is not known.

Lessons from Fetal Pancreas Transplantation

It is clear from the above discussion that spontaneous diabetes in animal models is an immunologically mediated—possibly autoimmune—process. Such an immune response must involve the recognition of an islet-associated antigen. A question now arises: Is the disease process extrinsic to the pancreatic islet tissue or is there some factor intrinsic to the islet tissue of the NOD mouse that contributes to the development of disease? We have new data that bear on this question. Studies of pancreatic islet growth and differentiation following grafting suggest that there is a differentiation defect associated with the endocrine pancreas of the NOD mouse.

We first became aware of this phenomenon while investigating the effect of disease recurrence on the growth and development of NOD fetal pancreas grafted in spontaneously diabetic recipient mice. Table 9.6 shows the results of one such study. Here 17-day fetal pancreas from NOD donors was grafted in the kidney capsule of spontaneously diabetic NOD mice. The growth and differentiation of fetal tissue was followed by macroscopic and microscopic examination four weeks postgraft. Two sets of controls were included in this initial study. The first set was grafts of NOD fetal pancreas in nondiabetic NOD mice; the second was an isograft of BALB/c fetal pancreas in BALB/c recipients.

The BALB/c tissue showed normal growth and development with well-vascularized islets four weeks after grafting. No graft development was seen in diabetic NOD mice, which could reflect disease interfering with islet development. What

TABLE 9.6. Failure of NOD fetal pancreas to differentiate in NOD recipients.

			Fetal pancreas differentiation (N) / Total (N)[a]	
Group no.	Fetal donor	Recipient	Normal[b]	Abnormal[c]
1	NOD	Diabetic NOD	0/5	5/5[d]
2	NOD	Nondiabetic NOD	0/6	6/6[e]
3	BALB/c	BALB/c	6/6	0/6

[a] Fetal pancreas from 17-day-old donors was grafted in kidney capsules of recipient mice. Differentiation of fetal tissue was determined macroscopically 28 days postgrafting.
[b] Normal islet differentiation: Fifteen or more well-differentiated islets were observed macroscopically.
[c] Poor islet differentiation: Five or less poorly differentiated (tiny) islets were observed macroscopically.
[d] $P \leq .004$ when compared with group 3 by Fisher's exact test.
[e] $P \leq .002$ when compared with group 3 by Fisher's exact test.

was a little surprising, however, was the finding that islets failed to develop when NOD fetal pancreas was grafted to nondiabetic NOD mice. Could there be an intrinsic differentiation defect in the NOD fetal pancreas?

We investigated this question further by grafting NOD fetal pancreas under the kidney capsule of (NOD×CBA) F_1 mice. The (NOD×CBA) F_1 animals showed no signs of insulitis and did not spontaneously develop diabetes. Thus any failure of graft development could not be attributed to disease affecting the grafted tissue. Also the parental graft is accepted by the F_1 recipient because self-tolerance involves tolerance of both parental strains. The protocol for these studies is summarized in Figure 9.6. This protocol involved grafting 17-day NOD, CBA, and (NOD×CBA) F_1 fetal pancreas in (NOD×CBA) F_1 recipient animals. Graft development was studied four weeks postgraft. Results of this study are shown in Table 9.7. These data show a striking difference between the NOD fetal pancreas

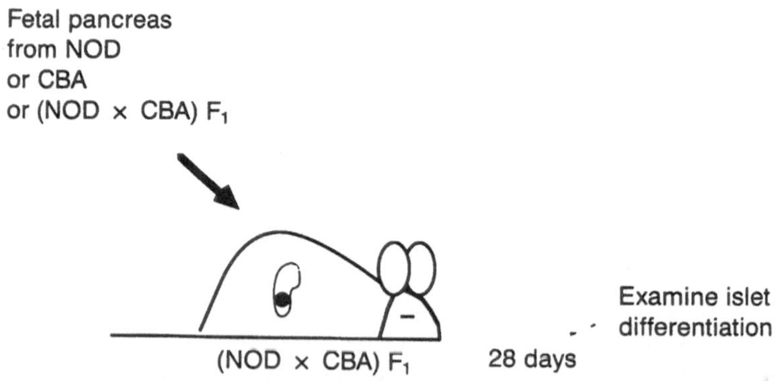

FIGURE 9.6. This protocol is used to determine the capacity of fetal pancreas from NOD, CBA, and (NOD × CBA) F_1 animals; differentiation was assessed by microscopically and macroscopically, 28 days postgraft.

TABLE 9.7. Abnormal differentiation is intrinsic to NOD fetal pancreas.

		Fetal pancreas differentiation (N) / Total (N)[a]	
Group no.	Fetal donor	Normal[b]	Abnormal[c]
1	(NOD × CBA) F_1	13/13	0/13[d]
2	CBA	15/15	0/15[e]
3	NOD	19/39	20/39

[a] Fetal pancreas from 17-day-old donors was grafted in kidney capsules of (NOD × CBA) F_1 recipient mice. Differentiation of fetal tissue was determined macroscopically, 28 days postgraft.
[b] Normal islet differentiation: Fifteen or more well-differentiated islets were observed macroscopically.
[c] Poor islet differentiation: Five or less poorly differentiated (tiny) islets were observed macroscopically.
[d] $P \le .0002$ when compared with group 3 by Fisher's exact test.
[e] $P \le .0005$ when compared with group 3 by Fisher's exact test.

and tissue from either CBA or F_1 donors. In the case of the NOD tissue, approximately half the grafts failed to develop. None or very few (less than five) very small islets were seen when the grafted kidney was examined. Upon histologic examination, a large cyst, lined with epithelial cells and scattered accumulations of fat cells, was observed. There was little or no evidence of islet budding from the epithelial component of the graft (Fig. 9.7). This abnormal differentiation of the NOD fetal pancreas suggests that some factor intrinsic to the NOD is responsible for this abnormal growth and differentiation. Abnormal NOD fetal pancreas development was not correlated with the sex of the tissue donor or the F_1 recipient (Table 9.8).

TABLE 9.8. Abnormal differentiation is not correlated with the sex of the donor or of the recipient.

			Fetal pancreas differentiation (N) / Total (N)[a]	
Group no.	Test	Sex	Normal[b]	Abnormal[c]
1a	Donor	Male	11/20	9/20
1b	Donor[d]	Female	9/20	11/20[e]
2a	Recipient	Male	15/29	14/29[f]
2b	Recipient[g]	Female	12/26	14/26

[a] Fetal pancreas from a 17-day-old NOD donor mouse was grafted in the kidney capsule of the (NOD × CBA) F_1 recipient mouse. Differentiation of fetal tissue was determined macroscopically, 28 days postgraft.
[b] Normal islet differentiation: Fifteen or more well-differentiated islets were observed macroscopically.
[c] Poor islet differentiation: Five or less poorly differentiated (tiny) islets were observed macroscopically.
[d] The sex of the fetus was determined before grafting.
[e] Groups 1a and 1b are not significantly different, $P \le .37$.
[f] Groups 2a and 2b are not significantly different, $P \le .44$.
[g] The sex of the (NOD × CBA) F_1 recipients was determined before grafting.

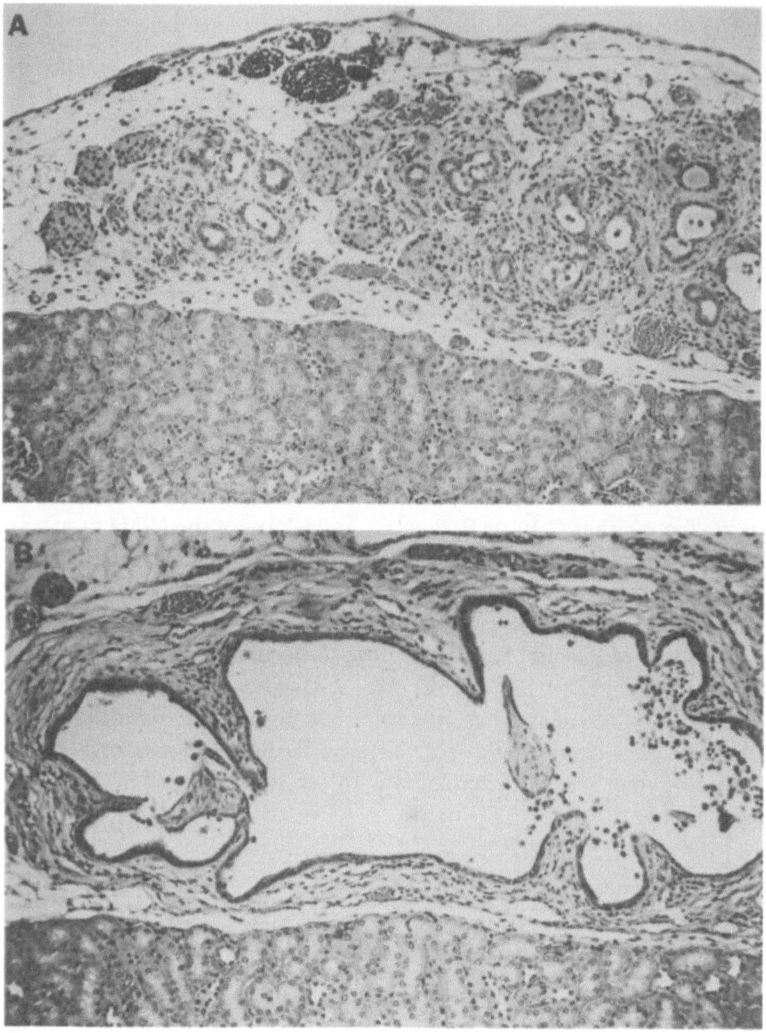

FIGURE 9.7. **A**: Normal islet development as seen following grafting of NOD fetal pancreas in (NOD × CBA) F_1 recipient animals. **B**: Abnormal development of NOD fetal pancreas to (NOD × CBA) F_1 recipients. Note the presence of an epithelial cell-lined cyst and the lack of islet development (H&E × 200).

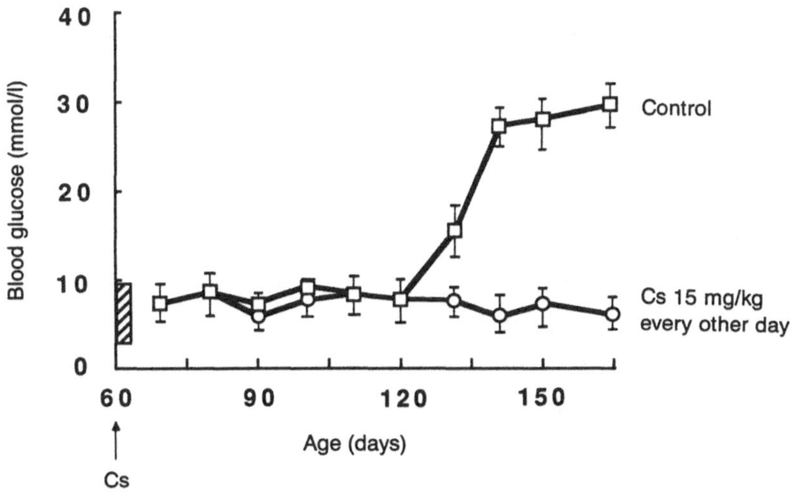

FIGURE 9.8. Low-dose cyclosporin prevents disease development in NOD mice. This figure is adapted from Mori and colleagues.[11] Fifteen nondiabetic NOD mice from 60 days of age were divided into two groups and treated orally with cyclosporin (Cs) (15 mg/kg every other day) or oil as a control until they were 160 days of age. Diabetes developed in five of eight oil-treated mice, with partial to complete islet destruction. None of the seven cyclosporin-treated NOD mice showed evidence of developing diabetes, as determined by blood glucose levels. Histologic examination showed no evidence of insulitis in these experimental animals.

Effect of Cyclosporin on Disease Recurrence

Our studies of disease recurrence in islet tissue grafted in spontaneously diabetic NOD mice led to the conclusion that the disease process involves CD4 T-cell–initiated inflammatory tissue damage. The exact cause of this inflammatory reaction is presently not understood, but the defect in NOD islet differentiation (see above) suggests that some aberration in the process of islet development could contribute to the disease process itself. Whatever the cause of this disease, it would seem likely that lymphokine production by the CD4 T-cell is the process that initiates the pathologic sequence. Lymphokine-dependent pathology is known to be sensitive to cyclosporin,[44] a potent immunosuppressive agent that functions by blocking message transcriptions for lymphokine synthesis in the activated T-cell (T′).[44–47] Cyclosporin does not block lymphokine synthesis from messages already formed in the cytoplasm of T′-cells.[46] That is the drug is relatively ineffective once T′-cells have been generated and express message for lymphokine production.[45,46] Mori and coworkers,[11] have reported that low-dose cyclosporin (15 mg/kg every other day) in NOD mice significantly delayed development of disease (Fig. 9.8). Similar studies were observed when BB rats were treated with cyclosporin prior to the onset of disease.[48,49] Because cyclo-

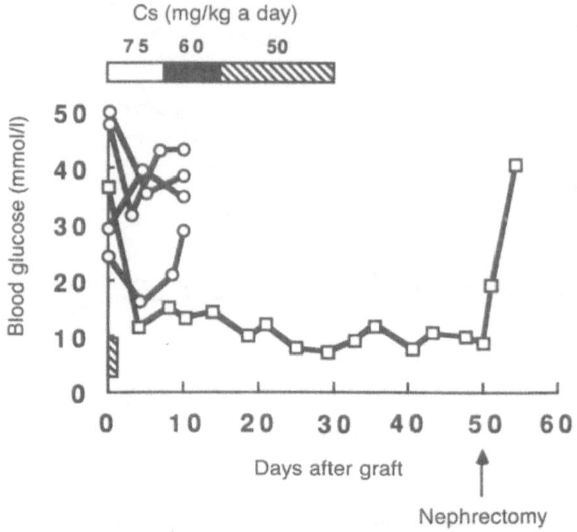

FIGURE 9.9. Use of cyclosporin to control disease recurrence in grafted islet tissue. Five diabetic NOD mice were grafted with cultured BALB/c islet tissue on day 0 and received cyclosporin (Cs) daily (75 mg/kg, then 60 mg/kg) for each of two consecutive weeks; the daily dose was lowered to 50 mg/kg for another two weeks. Graft function in the one normoglycemic animal was confirmed by the return of hyperglycemia upon removal of the grafted kidney.

sporin functions early in the sequence of T-cell activation by regulating lymphokine production and because cyclosporin does not inhibit lymphokine synthesis once message is already expressed in the T-cell,[45,46] we might expect cyclosporin to prevent the development of diabetes when it is administered before onset of disease (Fig. 9.8), since reactive T-cells have not been activated at this stage. However the active disease process might be relatively resistant to the effect of cyclosporin in the NOD mouse. When diabetic NOD mice were grafted with cultured BALB/c islets and treated with a high dose of cyclosporin (75 mg/kg each day for the first week, 60 mg/kg each day for the second week, and 50 mg/kg each day for another two weeks), only one of five animals showed any evidence of graft function (Fig. 9.9). Histologic examination showed mononuclear cell infiltration at the graft site, with extensive islet tissue destruction.[30]

This failure of cyclosporin to control disease recurrence in islet tissue grafted in acutely diabetic mice contrasts with the relatively potent effect of this agent in controlling the progression of the disease in man,[50] and its capacity to prevent disease if administered before the onset of pathology in the NOD mouse.[11] In the case of disease recurrence in grafted tissue, we may be dealing with a situation in which T′-cells have an active message for lymphokine production in their cytoplasm and are therefore resistant to the effect of this drug. Islet destruction in this model is acute and probably occurs before cyclosporin has had an affect on lymphokine production.

Acknowledgments. This work was supported in part by the Juvenile Diabetes Foundation, grant no. 187425 and the U.S. Public Health Service, grant no. p01 DK40144.

References

1. Bottazzo GF. Beta cell damage in diabetic insulitis: Are we approaching a solution? *Diabetologia* 1984;26:241.
2. Eisenbarth GS. Type I diabetes mellitus. A chronic autoimmune disease. *N Engl J Med* 1986;314:1360.
3. Rossini AA, Mordes JP, Like AA. Immunology of insulin-dependent diabetes mellitus. *Ann Rev Immunol* 1985;3:289–320.
4. Burnet FM. *Cellular Immunology.* Melbourne: Melbourne University Press; New York: Cambridge University Press, 1969.
5. Burnet FM. *The Clonal Selection Theory of Acquired Immunity.* London: Cambridge University Press, 1959.
6. Todd JA, Bell JI, McDevitt HO. A molecular basis for genetic susceptibility to insulin dependent diabetes mellitus (Review). *Trends Genet* 1988;4:129.
7. Miyazaki A, Hanafusa T, Yamada K, et al. Predominance of T lymphocytes in pancreatic islet and spleen of pre-diabetic non-obese diabetic (NOD) mice. A longitudinal study. *Clin Exp Immunol* 1985;60:622.
8. Mordes JP, Desemone- J, Rossini AA. The BB rat. *Diabetes Metab Rev* 1987;3: 725–750.
9. Logothetopoulos J, Valiquette N, Madura E, Cvet D. The onset and progression of pancreatic insulitis in the overt, spontaneously diabetic young adult BB rat studied by pancreatic biopsy. *Diabetes* 1984;33:33.
10. Harada M. Immunological manipulation of diabetes production in NOD mice, in Tarui S, Tochino Y, Nonaka K (eds): *Insulitis and Type I Diabetes: Lessons from the NOD Mouse.* Academic Press pp 143–153. Tokyo 1986.
11. Mori Y, Suko M, Okudaira H, et al. Preventive effects of cyclosporine on diabetes in NOD mice. *Diabetologia* 1986;29:244.
12. Like AA, Anthony M, Guberski DL, Rossini AA: Spontaneous diabetes mellitus in the BB/W rat: Effects of glucocorticoids, cyclosporin-A and antiserum to rat lymphocytes. *Diabetes* 1983;32:326.
13. Like AA, Kislauskis E, Williams RM, Rossini AA: Neonatal thymectomy prevents spontaneous diabetes mellitus in the BB/W rat. *Science* 1982;216:644.
14. Koevary S; Rossini A, Stoller W, Chick W, Williams RM. Passive transfer of diabetes in the BB/W rat. *Science* 1983;220:727–728.
15. Wicker LS, Miller BJ, Mullen Y. Transfer of autoimmune diabetes mellitus with splenocyte from nonobese diabetes (NOD) mice. *Diabetes* 1986;35:855.
16. Panitch HS, McFarlin DE: Experimental allegic encephalomyelitis: Enhancement of cell-mediated transfer by concanavalin A. *J Immunol* 1977;119:1134–1137.
17. Handler ES, Mordes JP, Seals J, Koevary S, Like AA, Nakano K, Rossini AA. Diabetes in the Bio-Breeding/Worcester Rat. Induction and acceleration by spleen cells-conditioned media. *J Clin Invest* 1985;76:1692–1694.
18. Mandrup-Poulsen T, Bendtzen K, Nerup J, Dinarello CA, Svenson M, Nielsen JH. Affinity-purified human interleukin I is cytotoxic to isolated islets of Langerhans. *Diabetologia* 1986;29:63–67.

19. Mandrup-Poulsen T, Bendtzen K, Dinarello CA, Nerup J. Human tumour necrosis factor potentiated human interleukin-1 mediated rat pancreatic β cell cytotoxicity. *J Immunol* 1987;139:4077.
20. Miller BJ, Appel MC, O'Neil JJ, Wicker LS. Both the lyt-2⁺ and L3T4 T cell subsets are required for the transfer of diabetes in nonobese diabetic mice. *J Immunol* 1988; 140:52–58.
21. Koike T, Itoch Y, Ishii T, Ito I, Takabayashi K, Maruyama N, Tomioka H, Yoshida S. Preventive effect of monoclonal anti-L3T4 antibody on development of diabetes in NOD mice. *Diabetes* 1987;36:539–541.
22. Like AA, Biron CA, Weringer EJ, Byman K, Sroczynski E, Guberski DL. Prevention of diabetes in Bio-Breeding/Worcester rats with monocolonal antibodies that recognize T lymphocytes or natural killer cells. *J Exp Med* 1986;164:1145–1159.
23. Lafferty KJ, Prowse SJ, Simeonovic CJ, Warren HS. Immunobiology of tissue transplantation: A return to the passenger leucocyte concept. *Ann Rev Immunol* 1983;1: 143–173.
24. Prowse SJ, Simeonovic CJ, Lafferty KJ, Bond BC, Magi CE, Mackie D. Allogenic islet transplantation without recipient immunosuppression. *Methods Diabetes Res* 1984;1:253–269.
25. Prowse SJ, Bellgrau D, Lafferty KJ. Islet allografts are destroyed by disease occurrence in the spontaneously diabetic BB rat. *Diabetes* 1986;35:110–114.
26. Prowse S, Nomikos IN, Pratt PF, Lafferty KJ. Islet transplantation in spontaneous diabetic animals, in Jaworski MA et al (eds): *The Immunology of Diabetes Mellitus*. Amsterdam: Elsevier Science Publishers, 1986, pp 175–180.
27. Weringer EJ, Like AA. Immune attack on pancreatic islet transplants in the spontaneously diabetic Bio-Breeding/Worcester (BB/W) rat is not MHC restricted. *J Immunol* 1985;134:2383–2386.
28. Woehrle M, Markmann JF, Silvers WK, Barker CF, Naji A. Transplantation of cultured pancreatic islets to BB rats. *Surgery* 1986;100:334–341.
29. Bendelac A, Carnaud C, Boitard C, Bach JF. Syngeneic transfer of autoimmune diabetes from diabetic NOD mice to healthy neonates. Requirement for both L3T4 and Lyt2 T cells. *J Exp Med* 1987;166:823–832.
30. Wang Y, McDuffie M, Nomikos IN, Hao L, Lafferty KJ. Effect of cyclosporine on immunologically mediated diabetes in NOD mice. *Transplantation* 1988 vol 46: 101S–106S, suppl.
31. Wang Y, Hao L, Gill RD, Lafferty KJ. Autoimmune Diabetes in NOD mouse is L3T4 T-lymphocyte dependent. *Diabetes* 1987;36:535.
32. Gill RG, Lafferty KJ. A novel form of T-T collaboration requiring the active participation of the antigen presenting cell. Proc. *NY Acad Sci* 1988;542:421–423.
33. Okamoto H. Regulation of proinsulin synthesis in pancreatic islets and a new aspect to insulin-dependent diabetes. *Mol Cell Biochem* 1981;37:43–61.
34. Okamoto H. The role of poly (ADP-ribose) synthetase in the development of insulin dependent diabetes and islet β cell regeneration. *Biomed Biochem Acta* 1985;44: 15–20.
35. Fantone JC, Ward PA. Role of oxygen-derived free radicals and metabolites in leucocyte dependent inflammatory reactions. *Am J Pathol* 1982;107:397–418.
36. Nathan CF. Neutrophil activation on biological surfaces. Massive secretion of hydrogen peroxide in response to products of macrophages and lymphocytes. *J Clin Invest* 1987;80:1550.

37. Wang Y, Goodman M, Lumerman J, Sussman KE, Dahl R, Lafferty KJ, Draznin B. In vivo administration of interleukin-1 inhibits glucose-stimulated insulin release. *Diabetes Research and Clinical Practice* 1989;7:205.
38. Yamada K, Nonaka K, Hanafusa T, Miyazaki A, Toyoshima H, Tarui S. Preventive and therapeutic effects of large-dose nicotinamide injections on diabetes associated with insulities. An observation in nonobese diabetic (NOD) mice. *Diabetes* 1982;31: 749–753.
39. Nomikos IN, Prowse SJ, Carotenuto P, Lafferty KJ. Combined treatment with nicotinamide and desferrioxamine prevents islet allograft destruction in NOD mice. *Diabetes* 1986;36:1302.
40. Nomikos IN, Wang Y, Lafferty KJ. Involvement of O2 radicals in "autoimmune" diabetes. Immunol. *Cell Biol* 1989;67:85–87.
41. Nakajima H, Yamada K, Hanafusa T, Fujino-Kurihara H, Miyagara J. Elevated antibody-mediated cytotoxicity and its inhibition by nicotinamide in the diabetic NOD mouse. *Immunol Lett* 1986;12:91.
42. Clark IA, Chaudhri G, Cowden WB. Interplay of reactive oxygen and tumor necrosis factor in tissue injury. Oxy-radicals in molecular biology and pathology. *UCLA Symposia on Molecular and Cellular Biology*, New Series, vol. 82. New York: Alan R. Liss, 1988.
43. Oschilewski U, Kiesel U, Kolb L. Administration of silica prevents diabetes in BB rats. *Diabetes* 1985;34:197–199.
44. Lafferty KJ, [illegible] LL. Cyclosporin A and the regulation of autoimmune disease. *J of Autoimmunity* 1988;1:519.
45. Granelli-Piperno A, Inaba K, Steinman RM. Stimulation of lymphokine release from T lymphoblasts, requirement for mRNA synthesis and inhibition by cyclosporine. A. *J Exp Med* 1984;160:1792.
46. Hodgkin PD, Hapel AJ, Johnson RM, Young IG, Lafferty KJ. Blocking of delivery of the antigen-mediated signal to the nucleus of T cells by cyclosporine. *Transplantation* 1987;43:685.
47. Andrus L, Lafferty KJ. Inhibition of T-cell activity by cyclosporine A. *Scand J Immunol* 1982;15:449.
48. Laupacis A, Stiller CR, Gardell C, et al. Cyclosporine prevents diabetes in BB/Wistar rats. *Lancet* 1983;i:10.
49. Brayman KL, Armstrong J, Shaw LM, et al. Prevention of diabetes in BB rats by intermittent administration of cyclosporine. *Surgery* 1987;102:235.
50. Bougneries PF, Carel JC, Castano L, Bottard C, Gardin JP, Landais P, Hors J, Mihatsch MJ, Paillard M, Chaussain JL, Bach JF. Factors associated with early remission of type I diabetes in children treated with cyclosporine. *N Engl J Med* 1988; 318:663–670.

10
Viral Pathogenesis of Insulin-Dependent Diabetes Mellitus

JI-WON YOON

Introduction

The familial occurrence of diabetes, or at least, a tendency to develop the disease, has been long recognized, but the true nature of the genetic mechanisms governing transmission of the diabetes trait still is not well understood. Concordance for insulin-dependent diabetes mellitus (IDDM) between identical twins approaches 50%, suggesting that nongenetic influences, such as environmental factors, contribute to the clinical expression of a genetic susceptibility. Triggering factors that might induce diabetes include diabetogenic viruses, chemicals, diet, and toxins. Such likely candidates act either as primary injurious agents to pancreatic beta-cells or as triggers of autoimmunity in man and animals (Fig. 10.1).[1,2]

The evidence that diabetes is virus induced comes largely from experiments in animals,[3-5] but several studies also implicate viruses as a trigger in human diabetes.[6-12] Numerous case reports describe a temporal relationship between the onset of certain viral infections (e.g., rubella, mumps, and Coxsackie B viruses) and the subsequent development of diabetes. Additional reports and epidemiologic studies have shown (1) viral specific antigens in the islets of Langerhans and destruction of beta-cells in the pancreas of a diabetic patient[6]; (2) the presence of viral antibodies with rising titers in paired sera from newly diagnosed IDDM patients[6,7]; (3) a high frequency of Coxsackie B virus-IgM–specific antibody in newly diagnosed diabetic children[8,9]; (4) beta-cell damage in children who died of well-documented, overwhelming viral infections[12]; (5) the isolation of viruses from patients with acute-onset diabetes and the demonstration that these viruses induce diabetes in mice[6,7]; and (6) the association of autoantibody production with certain viral infections [e.g., congenital rubella and persistent cytomegalovirus (CMV) infection].[10,11]

The best experimental evidence supporting an etiologic role of viruses in the pathogenesis of IDDM derives from studies of mice infected with encephalomyocarditis (EMC) virus.[13,14] In genetically susceptible mice, the M-variant of EMC virus can directly infect and destroy pancreatic beta-cells. This produces a diabetes-like syndrome characterized by hypoinsulinemia, hyperglycemia, glycosuria, polydipsia, and polyphagia. The severity of the diabetes correlates

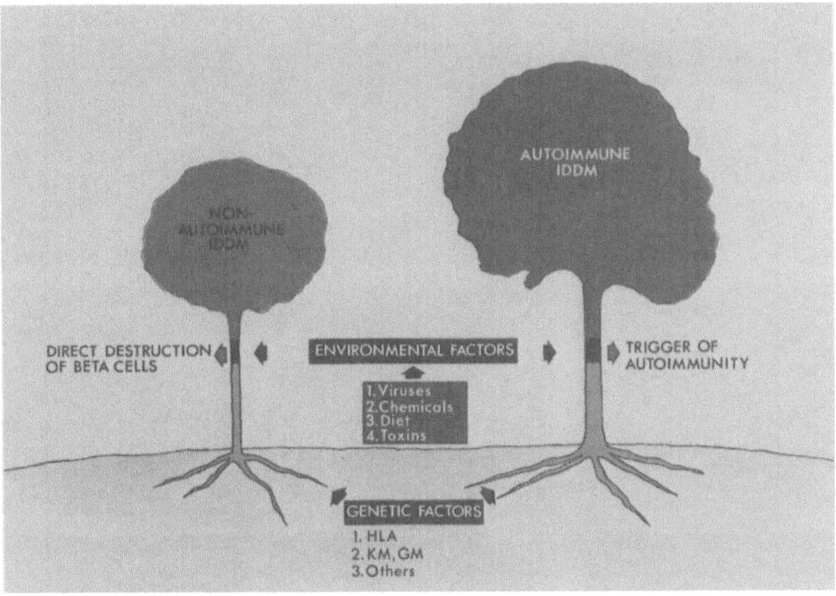

FIGURE 10.1. Possible environmental factors influencing the clinical expression of genetic susceptibility. Diabetogenic viruses, beta-cell toxic chemicals, diet, and toxins are likely candidates either as agents for the direct destruction of beta-cells or as triggers of autoimmunity.

closely with the degree of virus-induced, beta-cell damage. Also murine retrovirus is associated with autoimmune IDDM in NOD mice.[15] This review will focus on two possible pathogenic mechanisms for virus-induced diabetes: a direct cytolytic viral infection of beta-cells and a virus-induced, autoimmune-mediated destruction of beta-cells.

Direct Cytolytic Viral Infection of Beta-Cells

Murine Models

A group of viruses, including EMC virus, Mengo 2T virus and Coxsackie B virus, can directly infect murine pancreatic beta-cells and replicate in these cells, thus destroying the cells (Fig. 10.2).[14,16,17] The infected animals subsequently develop hypoinsulinemia and hyperglycemia.

Encephalomyocarditis Virus

Encephalomyocarditis virus, a member of the picornaviridae with a positive-sense, single-stranded RNA genome, has been most extensively studied in an

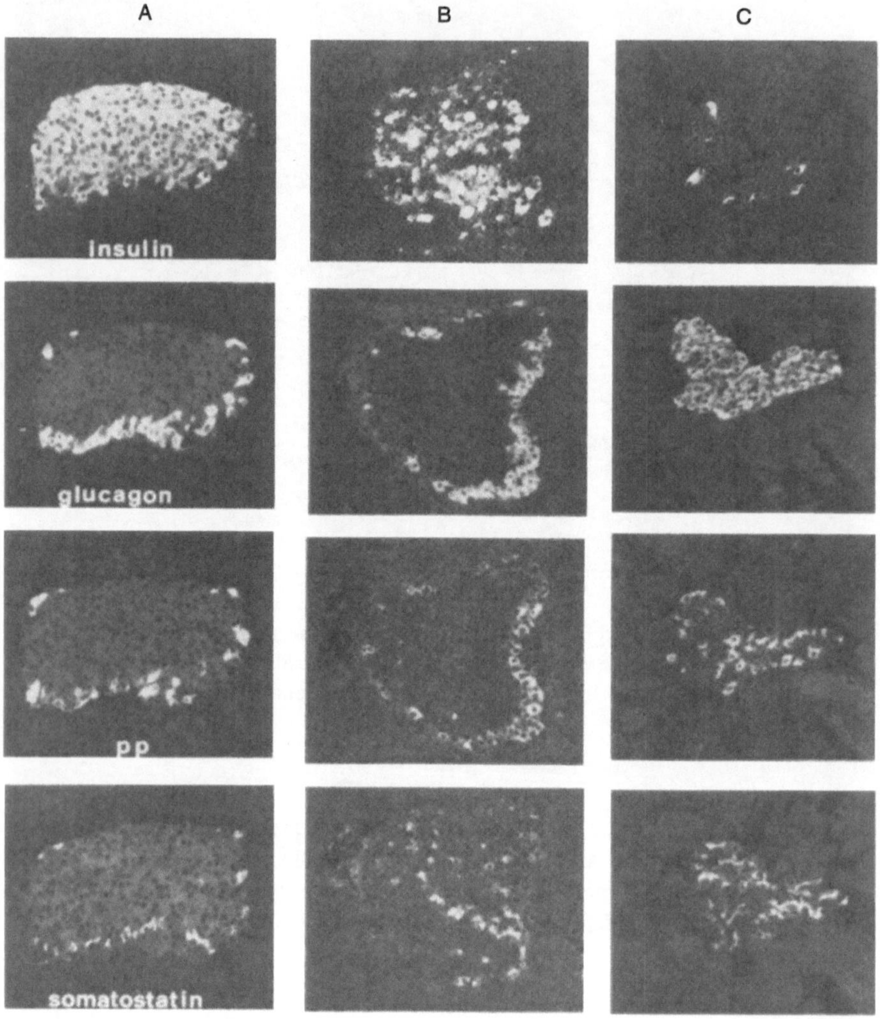

FIGURE 10.2. Immunofluorescent reactions in serial section of pancreas of (**A**) A control mouse (left panel). Note the peripheral localization of the non-B–cells. (**B**) A hypoglycemic mouse three days after infection (middle panel). Glucagon and PP-immunofluorescent cells are still located in the islet periphery, and several somatostatin immunofluorescent cells can be seen in the center of the islet. (**C**) A hyperglycemic mouse 21 days after infection (right panel). Note the reduction of islet size, scarcity of insulin-immunofluorescent cells, and abundance of glucagon-immunofluorescent cells. PP- and somatostatin-immunofluorescent-cells are scattered throughout the islet[99] (× 165). (Reprinted from ref. 99 with permission of Springer-Verlag Heidelberg.)

TABLE 10.1. Induction of Diabetes in Athymic Nude Mice.[a]

Strains	Virus infected	Glucose index	Diabetes (%)
CD-1 nu/nu	+	339 ± 109	86
CD-1 nu/nu	+	135 ± 14	0
CD-1 +/nu	+	249 ± 90	71
CD-1 +/nu	−	142 ± 17	0
NIH Swiss nu/nu	+	211 ± 90	45
NIH Swiss nu/nu	−	132 ± 10	0
NIH Swiss +/nu	+	200 ± 72	33
NIH Swiss +/nu	−	125 ± 8	0

[a]An animal was scored as diabetic when its calculated glucose index was 5 SD above the mean of uninfected controls.

animal model for virus-induced diabetes. The destruction of beta-cells in EMC virus-infected mice does not depend on autoimmune responses[18] but rather on the genetic background of the host[19-22] and the genetic makeup of the virus.[3,14]

Pathogenic Mechanism

Since several investigators have suggested that immune mechanisms may be involved in EMC (M-variant) virus-induced diabetes,[23-26] we recently investigated the role of immune function in the pathogenesis of EMC (D-variant) virus-induced diabetes by studying the susceptibility of athymic nude mice, thymectomized mice, and immunosuppressed mice and the segregation of H-2 haplotypes in crosses of resistant and susceptible mice.[18] Athymic nude mice infected with EMC virus showed a nearly identical diabetogenic response compared with their heterozygous littermates (Table 10.1). Immunosuppression by antilymphocyte serum did not prevent induction of EMC-D virus-induced diabetes. Passive transfer of lymphocytes from the spleen or peripheral blood of highly diabetic SJL/J mice into normal SJL/J mice, however, failed to produce diabetes. Treatment of EMC virus-infected mice with anti-L_3T_4 antibody did not prevent diabetes. In crosses of susceptible and resistant mice, the inheritance of susceptibility did not show an association with H-2 haplotype (Table 10.2). These results were similar to those of Vialettes and colleagues who failed to find any evidence for a thymus-dependent, cell-mediated immune response in EMC (M-variant) virus-induced diabetes.[27]

Additional evidence argues against an immune component in the pathogenesis of EMC virus-induced diabetes mellitus. Like other picornaviruses, EMC virus rapidly infects and lyses host cells. In the case of EMC virus infection of pancreatic beta-cells, evidence of this lysis can be seen within 24 to 48 hours. The virus destroys large numbers of beta-cells and many of the surviving cells have viral antigens in their cytoplasm, as demonstrated by immunofluorescence (Fig. 10.3). In this early phase of the infection, there is little or no evidence of an inflammatory response. There is also no indication that EMC virus modifies the

TABLE 10.2. Relationship of Diabetes and H-2 Haplotypes in Crosses of SJL/J and C57BL/6J Mice.[a]

Strain	H-2 Genotype[b]	Animals tested (*N*)	Diabetic animals (*N*)[c]	Diabetes (%)
SJL/J	s/s	77	75	97
C57BL/6J	b/b	50	0	0
F_1	b/s	45	5	11
F_1 × SJL/J	s/s	65	33	51
F_1 × SJL/J	b/s	55	29	53
F_1 × C57BL/6J	b/b	52	2	4
F_1 × C57BL/6J	b/s	37	2	5

[a] Male mice were inoculated (IP) with 1 × 10^5 PFU of EMC-D virus. Control mice were injected with an equal volume of Eagle's minimal essential medium with 5% fetal bovine serum (the vehicle used for the virus).
[b] H-2 genotypes were determined by hemagglutination.
[c] An infected animal was scored as diabetic when its glucose index was 3 SD above the mean of uninfected controls.
(Reproduced from ref. 18 with permission of the American Diabetes Association, Inc.)

host cell surface by inserting viral antigens or that it triggers an autoimmune response (e.g., there is no evidence of an islet-cell cytoplasmic antibody or islet-cell surface antibody).[18]

The severity of EMC-D virus-induced diabetes correlated closely with the degree of beta-cell damage, and the degree of beta-cell damage correlated closely with the extent of virus multiplication in beta-cells. Furthermore EMC virus-induced diabetes in mice can be prevented by a live attenuated vaccine.[28] It thus appears that EMC-D virus can infect, multiply within, and directly destroy pancreatic beta-cells in genetically susceptible mice (e.g., SJL/J) in the absence of an autoimmune response. However the possibility that *T*-lymphocytes might participate in the control of EMC-M virus infection and the development of diabetes in certain strains of mice (e.g., Balb/c ByJ) cannot be excluded.[29,30]

Genetic Control of EMC Virus-Induced Diabetes

When mice were infected with the M-variant of EMC virus, only certain inbred strains, such as SJL/J, SWR/J, DBA/1J, and DBA/2J, developed diabetes, whereas other strains, such as C57BL/6J, CBA/J, and AKR/J, did not.[19,20,22,31] When diabetes-prone SWR/J mice were crossed with diabetes-resistant C57BL/6J mice, the F_1 generation were resistant to diabetes upon infection with the virus.[19,22,31] More than 20% of the F_2 generation, however, developed diabetes upon exposure to the virus, indicating that susceptibility was inherited as an autosomal recessive trait. When resistant F_1 generations were back-crossed to the resistant C57BL/6J parents, this F_1 generation was also resistant to the development of EMC virus-induced diabetes. In contrast when the resistant F_1 generation were back-crossed to the susceptible SWR/J parents, approximately 50% of the F_1 generation developed diabetes (Table 10.3). Although the situation may be more complex, the data are consistent with the idea that EMC virus-

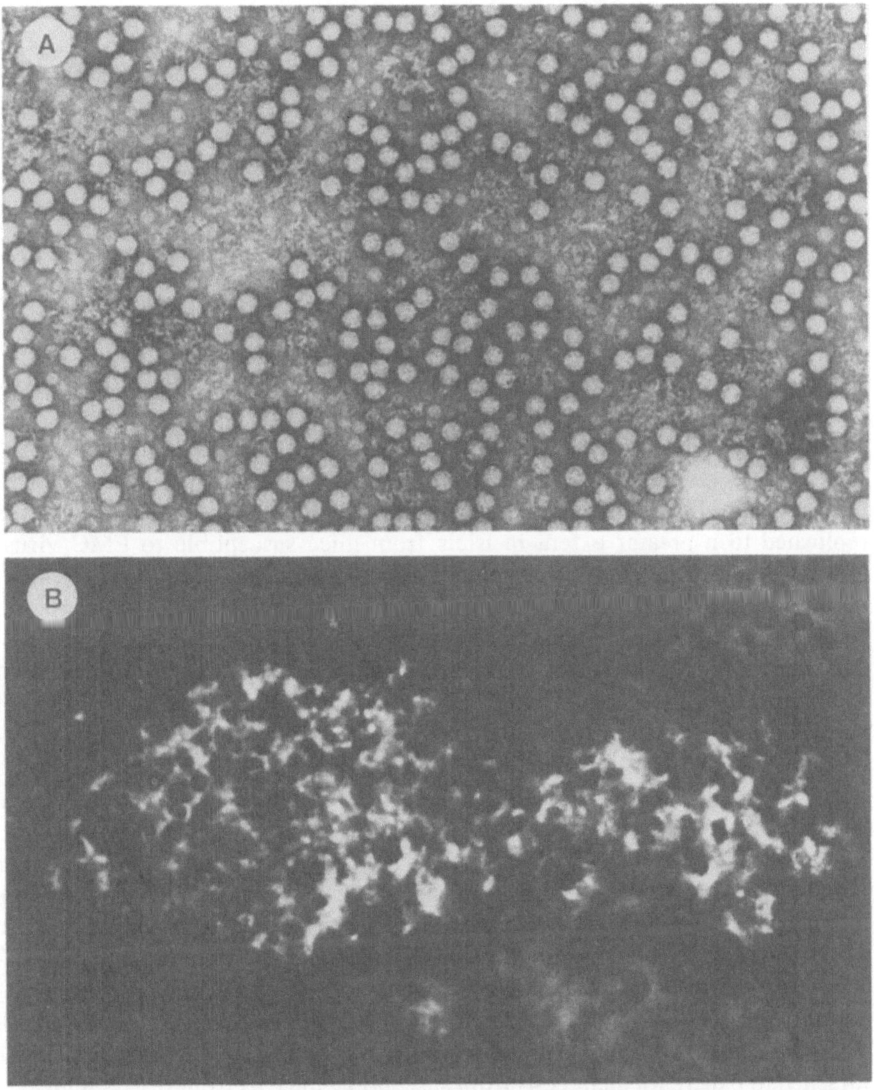

FIGURE 10.3. (**A**) Electron micrograph of purified D-variant of EMC virus particles prepared by CsCl gradient centrifugation. The diameter of the virion is approximately 25 nm. (**B**) Section of mouse pancreas obtained three days after infection with the D-variant of EMC virus and stained with FITC-labeled, anti-EMC-D virus antibody. The majority of the cells in the islets contain viral antigens. The surrounding acinar cells show little fluorescence (× 700).

TABLE 10.3. Genetic Control of EMC Virus-Induced Diabetes.[a]

Mouse strain	Mice (*N*)	Diabetes (%)
SWR/J	33	88
C57BL/6J	33	0
(SWR/J × C57BL/6J)F1	45	0
F_1 × C57BL/6J	70	5
F_1 × SWR/J	93	46

[a] An animal was scored as diabetic when its glucose index was 3 SD above the mean of 346 uninfected animals.

induced diabetes follows Mendelian inheritance and that susceptibility is primarily controlled by a single locus.[19]

What is the nature of the genetic factors that control the development of virus-induced diabetes in mice? One possibility is that mouse hosts vary in viral attachment to beta cells or replication in the cells. Viral titers revealed that EMC virus replicated to a greater extent in islets from mice susceptible to EMC virus-induced diabetes than in islets from resistant mouse strains.[20] Moreover the number of beta-cells infected with EMC virus in a susceptible strain was about ten times higher than the number of cells in a resistant strain.[20,21] Additional evidence that the severity of virus-induced diabetes is secondary to the degree of viral replication in islet cells comes from studies on the F_1 and F_2 generation of susceptible and resistant strains of mice.[22] These studies showed that blood glucose and plasma insulin levels in EMC virus-infected F_1 mice were within the normal range and that viral titers in F_1 mice islets were similar to those found in resistant mice islets. However, F_2 mice segregated into two groups, one with plasma insulin and blood glucose levels within the normal range and the other with low plasma insulin and high blood glucose levels. One group had viral titers approaching those of susceptible mice; the other group had titers similar to those of resistant mice. Thus viral replication and the subsequent development of diabetes are positively correlated. Why are beta-cells from SWR/J mice more susceptible to infection than beta-cells from C57BL/6J mice? Experiments in which viral attachment was determined showed that at least twice as much virus attached to pancreatic beta-cells from susceptible mice as from resistant mice.[21] These data suggest that there may be more receptors for virus on the surface of beta-cells from susceptible mice.

Isolated Diabetogenic and Nondiabetogenic EMC Virus Variants

In one experiment, development of diabetes after infection with the M-variant of EMC virus was not consistent, and statistically significant differences were found upon repetition of the experiment, as well as between cages of mice within experimental groups.[31] When the M-variant of EMC virus was grown in mouse embryo fibroblast cultures, the diabetogenic activity of the virus markedly diminished compared with that of virus passaged in mice (Table 10.4).[32] These findings suggested that the stock pool of EMC virus was made up of at least two

TABLE 10.4. Induction of Diabetes in Swiss NIH Miss with Different Doses and Preparation of Virus.

Source of virus	Dose (PFU/mouse)	Glucose index	Diabetes (%)[a]
Mouse passaged	10^2	245 ± 97	40
Mouse passaged	10^3	213 ± 90	31
Mouse passaged	10^4	330 ± 117	63
Mouse passaged	10^6	172 ± 32	10
Tissue culture passaged	10^2	139 ± 46	5
Tissue culture passaged	10^3	159 ± 71	3
Tissue culture passaged	10^4	165 ± 27	0
Tissue culture passaged	10^6	260 ± 80	45

[a] An animal was scored as diabetic when its glucose index was 3 SD above the mean of the uninfected animals.

populations: one diabetogenic and tropic for insulin-producing beta-cells, the other nondiabetogenic and not tropic for insulin-producing beta-cells. In a further experiment, individual plaques selected from the stock pool were cloned several times and inoculated into mice.[14] As shown in Table 10.5, two clones (No. 82 and No. 108) produced diabetes, two clones (No. 5 and No. 16) produced only minimal changes in blood glucose level, and two clones (No. 125 and No. 162) gave intermediate results. Clones 108 and 16 were plaque-purified two more times. When clone 108-D-II (the D-variant) was injected into SJL/J male mice, diabetes developed in more than 90% of the animals. In contrast none of the mice inoculated with clone 16-B-I (the B-variant) developed diabetes (Fig. 10.4).

Physiologic Differences and Similarities Between D- and B-Variants of EMC Virus

Light microscopy revealed that the D-variant, but not the B-variant, severely damaged beta-cells (Fig. 10.5). Fluorescent microscopy using fluorescein-labeled

TABLE 10.5. Isolation of Diabetogenic and Nondiabetogenic Variants.

Virus	Glucose index[a] (Mean ± SD)	Diabetes (%)
EMC-M	252 ± 136	47
Clones[b]		
5	191 ± 44	15
16	174 ± 45	10
82	447 ± 75	95
108	441 ± 49	100
125	213 ± 89	45
162	209 ± 43	30
Uninfected	145 ± 19	0

[a] The mean glucose index of 110 uninfected SJL/J was 145 ± 19 mg/dl. An animal was scored as diabetic when its glucose index was 5 SD above the mean (>240 mg/dl).
[b] Each clone was used to inoculate ~20 mice.

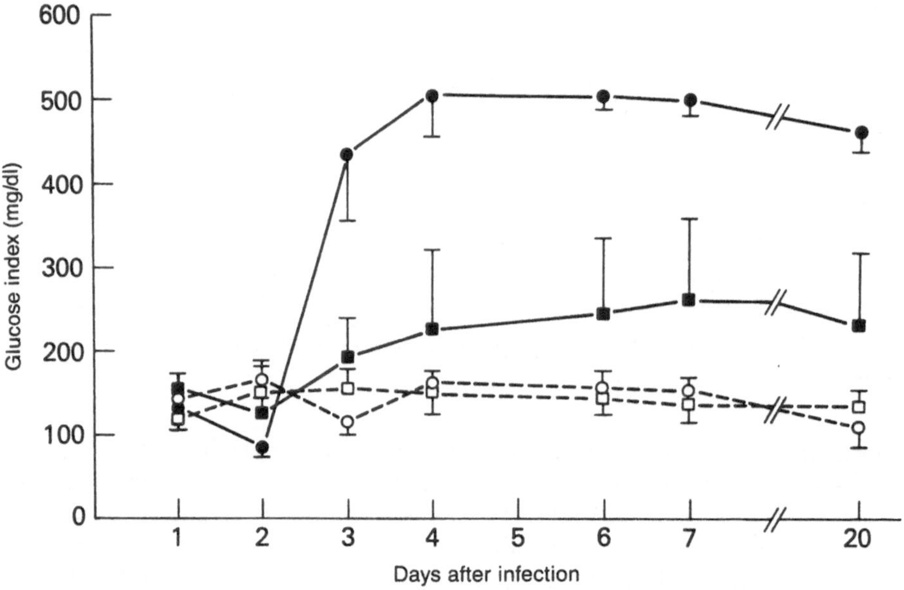

FIGURE 10.4. Blood glucose levels of mice infected with the D- or B-variant of EMC virus. Each mouse received 10^5 PFU of virus. At the times indicated, the mice were bled, and nonfasting glucose (NFG) levels were determined. Each point represents the mean of 10 animals, the vertical bars show the standard deviation. Key: D-variant, -●-; B-variant, -○-; M-variant, -■-; uninfected control, —□—.

anti-EMC antibody showed that approximately ten times more beta-cells became infected in mice inoculated with the D-variant than in mice inoculated with the B-variant. Moreover upon measuring infectious virus levels, 10 to 100 times more virus was recovered from the islets of mice infected with the D-variant than from those infected with the B-variant. Tissue culture experiments showed that the D-variant induced little if any interferon, whereas substantial amounts of interferon were induced by the B-variant.[14,34]

Despite these differences, D- and B-variants could not be distinguished antigenically by a sensitive plaque neutralization assay.[14] Antibody made against the D-variant neutralized both the D- and B-variants. Conversely antibody made against the B-variant also neutralized both the B- and D-variants. These studies illustrate how difficult it is to identify naturally occurring diabetogenic viruses. Based on standard serologic tests, it would be difficult to distinguish the diabetogenic D-variant (EMC-D) from the nondiabetogenic B-variant (EMC-B). Competitive radioimmunoassays also failed to reveal any major differences between the two variants.

Biochemical Similarities and Differences between EMC-D and -B Virus

Despite the differences noted above, these variants could not be distinguished by size, morphology, or capsid polypeptides.[3,33] Molecular hybridization studies with

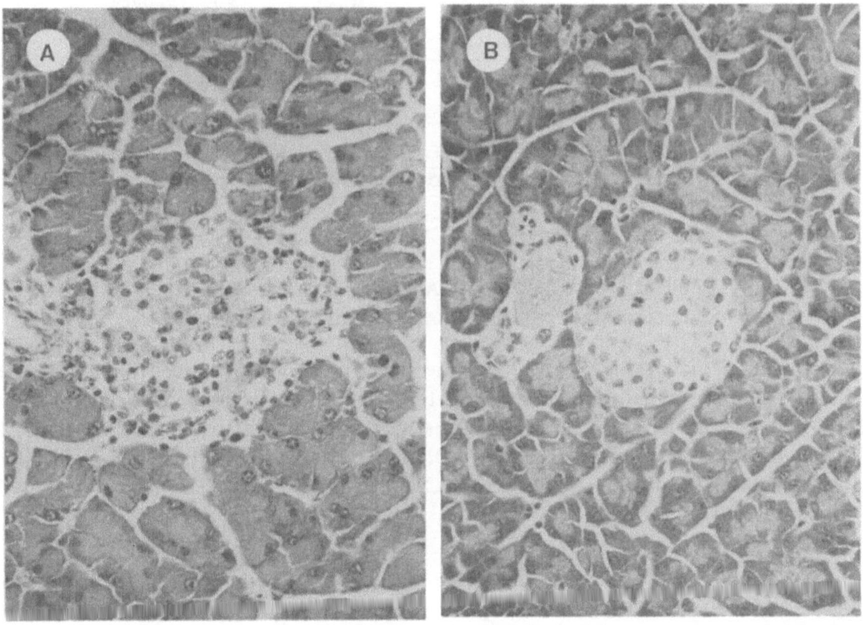

FIGURE 10.5. Sections of pancreas from mice infected with the B- or D-variant of EMC virus. (**A**) Section taken seven days after infection with the D-variant shows an inflammatory infiltrate in the islets of Langerhans and beta-cell necrosis (H&E, × 550. (**B**) Section taken seven days after infection with the B-variant shows little or no infiltration of inflammatory cells or beta-cell necrosis (H&E, × 550). (Reproduced from *The Journal of Experimental Medicine*,[14] by copyright permission of The Rockefeller University Press.)

radiolabeled DNA complementary to EMC-D and EMC-B viral RNAs failed to distinguish between EMC-D and EMC-B.[35] In addition many similarities between EMC-D and EMC-B have been reported.[36,37] A 103 nucleotide sequence from the 5′ non-coding region was shown to be identical for EMC-D and EMC-B,[37] and a comparison of restriction maps of the entire open reading frame of both variants, using 12 different enzymes, did not reveal any differences.[37]

However oligonucleotide fingerprinting after T1-digestion of the RNAs from these two variants revealed a difference in one spot.[35] An oligonucleotide that was missing in the RNA of the B-variant was present in the RNA of the D-variant. The oligoribonucleotide specific to EMC-D was isolated from a two-dimensional polyacrylamide gel and sequenced as 5′ACAAUCUCACUUUUCCAACAACAG -3′.[36] Molecular hybridizations of EMC-D and EMC-B genomic RNAs with a DNA primer complementary to the EMC-D–specific oligoribonucleotide revealed that the absence of a corresponding spot in EMC-B was the result of a point mutation, not a deletion. By sequencing a cloned cDNA of EMC-B corresponding to the EMC-D–specific oligoribonucleotide, the point mutation was identified as a "G" for EMC-B and an "A" for EMC-D transversion at base 9 of the

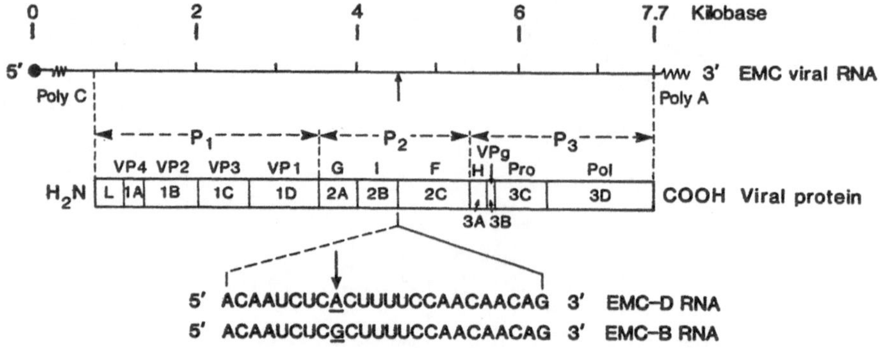

FIGURE 10.6. Structure of the EMC genome. The viral RNA and encoded peptides are schematically illustrated. The position of the oligonucleotide unique to diabetogenic EMC-D was identified near the 3′ end of the I(2B) protein. The nucleotide sequence of EMC-B RNA was converted from the cDNA corresponding to the EMC-D-specific oligonucleotide. The different bases in EMC-D and EMC-B are underlined. (Reprinted with permission from ref. 36.)

oligonucleotide (Fig. 10.6). A comparative sequence analysis of eight randomly picked RNA segments around the EMC-D–specific oligoribonucleotide revealed that there were no base changes between EMC-D and EMC-B. From the results of this study, it can be concluded that the diabetogenic EMC-D viral genome differs from the nondiabetogenic EMC-B viral genome by at least one point mutation.[36] The role of this point mutation is not known. It is unlikely to affect the attachment of the virus to cells, since the site of mutation is not located in any of the outer capsid proteins.

The EMC-D– and EMC-B–specific monoclonal antibodies, which were directed against the *VP1* genes of the virus, were used to demonstrate the antigenic differences between EMC-D and EMC-B in a recent study.[38] Thus we were particularly interested in cloning and sequencing the *VP1* genes of both the EMC-D and EMC-B viruses to see whether there were any differences between them. The genes for the major capsid protein VP1(ID) of both diabetogenic EMC-D and nondiabetogenic EMC-B viruses were cloned by using two synthetic primers that are common to both variants. The cloned cDNAs were mapped for major restriction enzyme sites, including Accl, BamHI, EcoRI, Hincll, KpnI, PvuII, SstI, TaqI, and XbaI. Among those nine restriction enzyme sites, only the TaqI site distinguished the EMC-D genome from its counterpart on the EMC-B genome. The complete nucleotide sequences (831 bases) of the *VP1* genes revealed five amino acid differences between the two variants (Fig. 10.7).[39] Three of the changes, at positions 41, 58, and 152, were Thr (EMC-B) to Ala (EMC-D). Two additional changes occurred at position 63 [Gln (EMC-B) to Glu (EMC-D)] and position 181 [Thr(EMC-B) to Ser (EMC-D)]. All these amino acid changes were due to point mutations at the first base of each codon. Some of these point mutations identified in the *VP1* gene may affect either the attachment of the virus

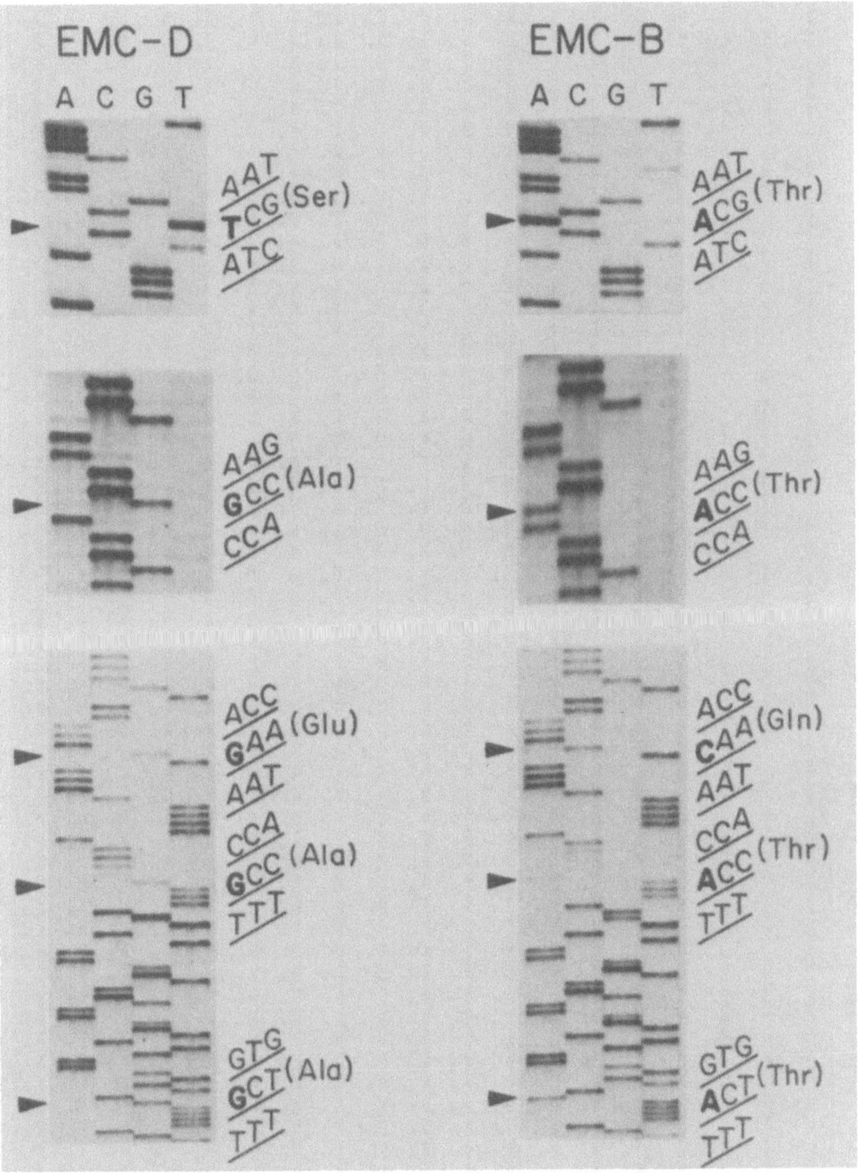

FIGURE 10.7. Specific-sequence autoradiograms of the area of point mutations in the VP1 gene of the EMC-D and EMC-B viruses. The side arrows indicate the nucleotide position that is different for EMC-D and EMC-B. The resultant amino acid substitution is given beside the triplet codon.

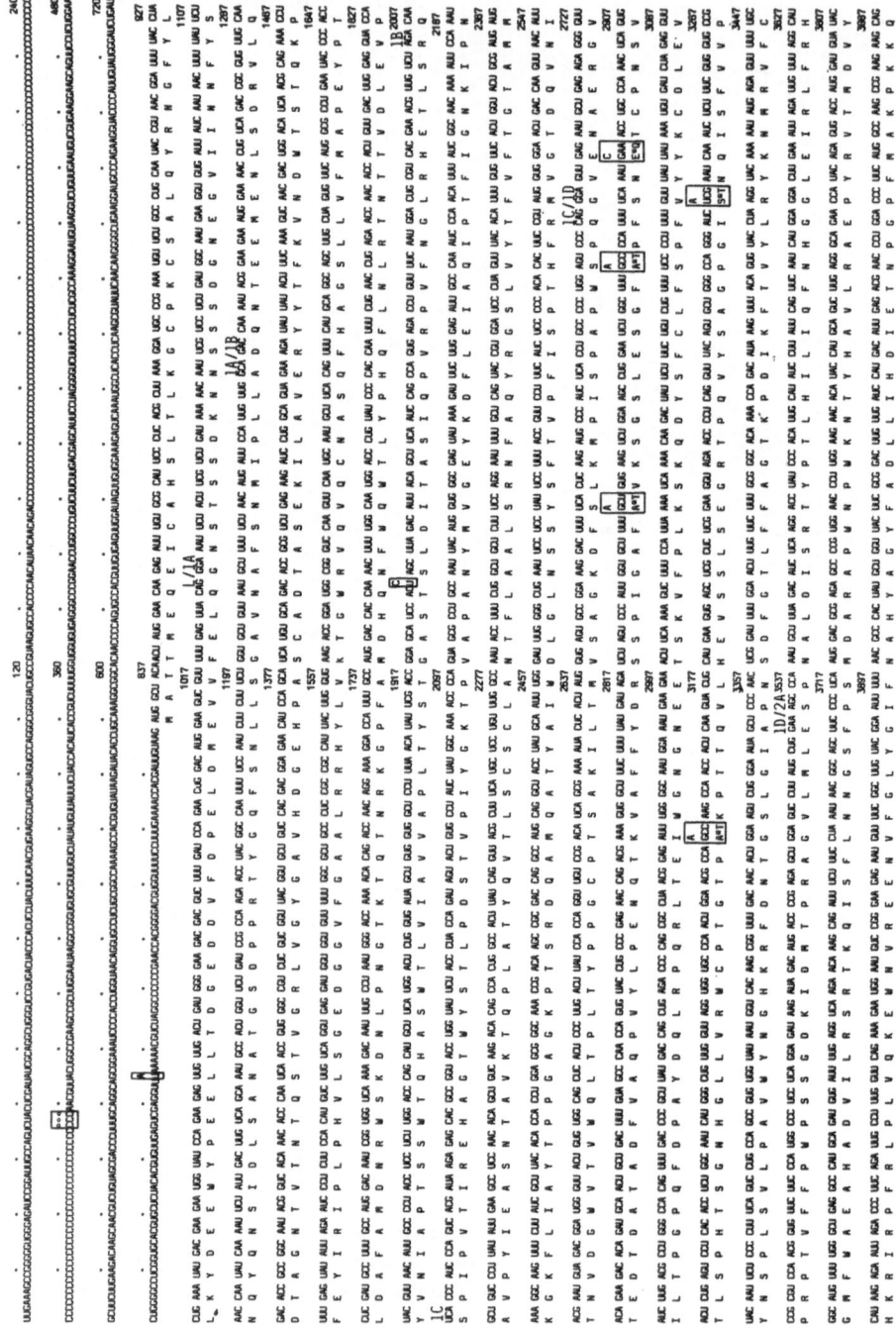

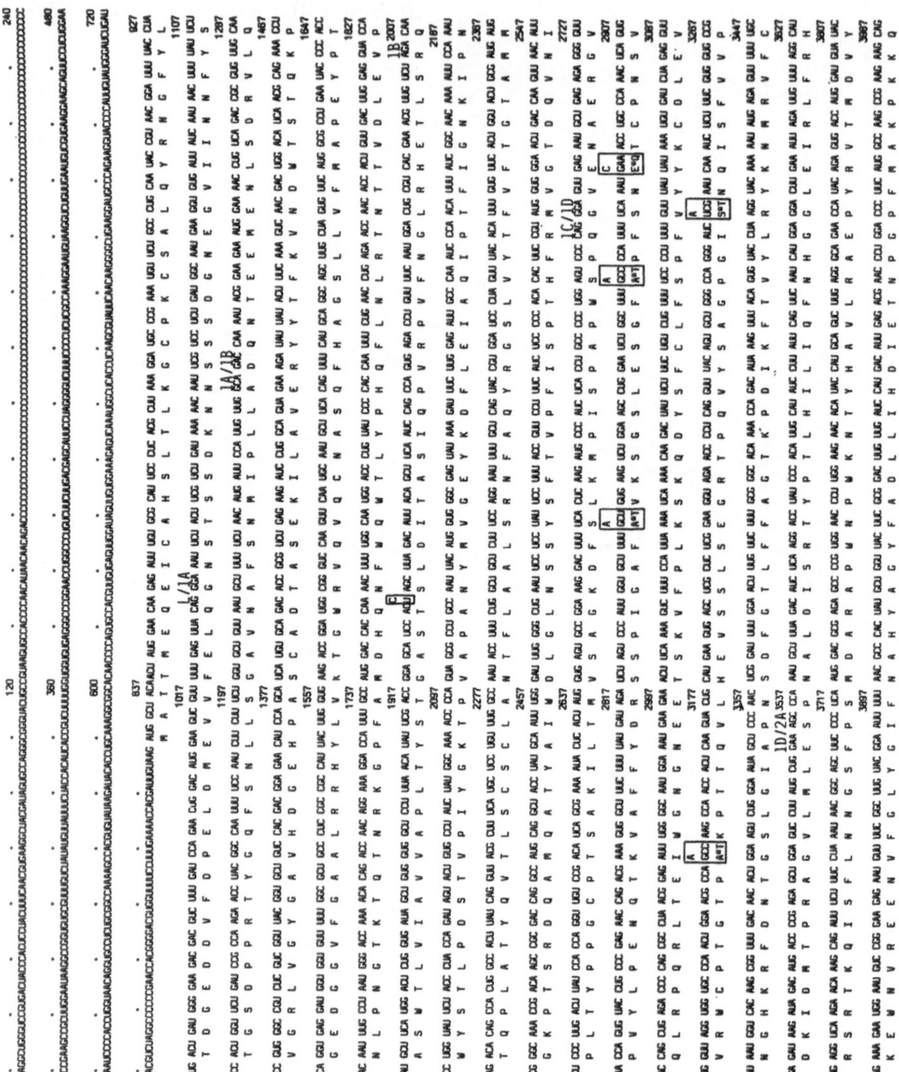

FIGURE 10.8. Complete nucleotide and deduced amino acid sequences of EMC-D and EMC-B viral genomes. The backbone is the EMC-D sequence (7829 bases), and the sequence of EMC-B (7825 bases) is identical to that of EMC-D except for the boxed area. Upper nucleotides and the amino acids on the right side of the asterisk in the box are

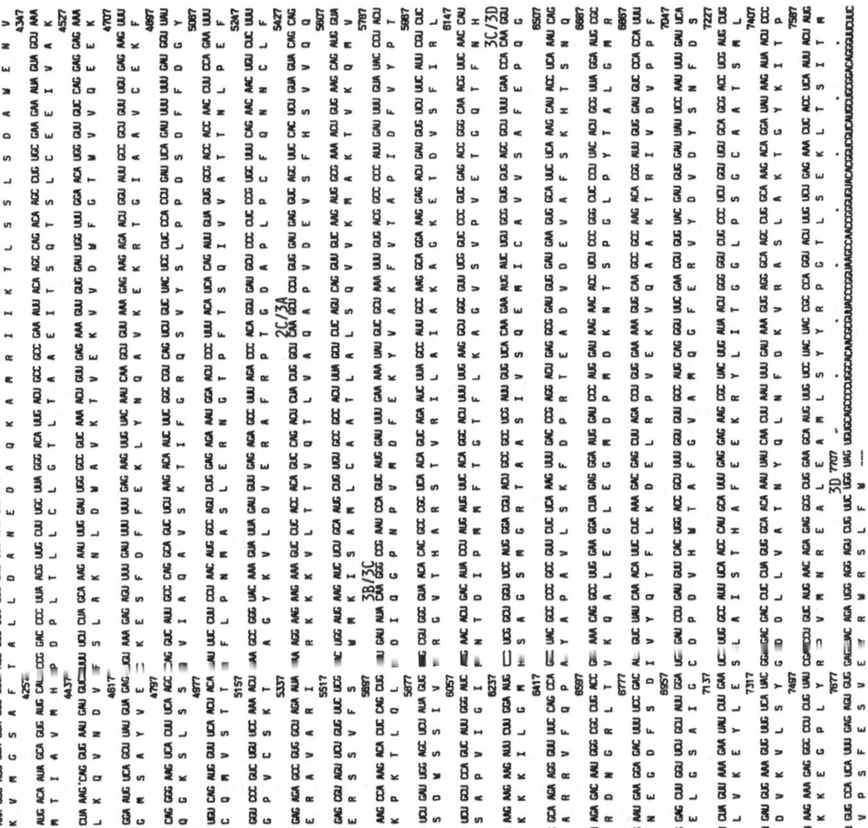

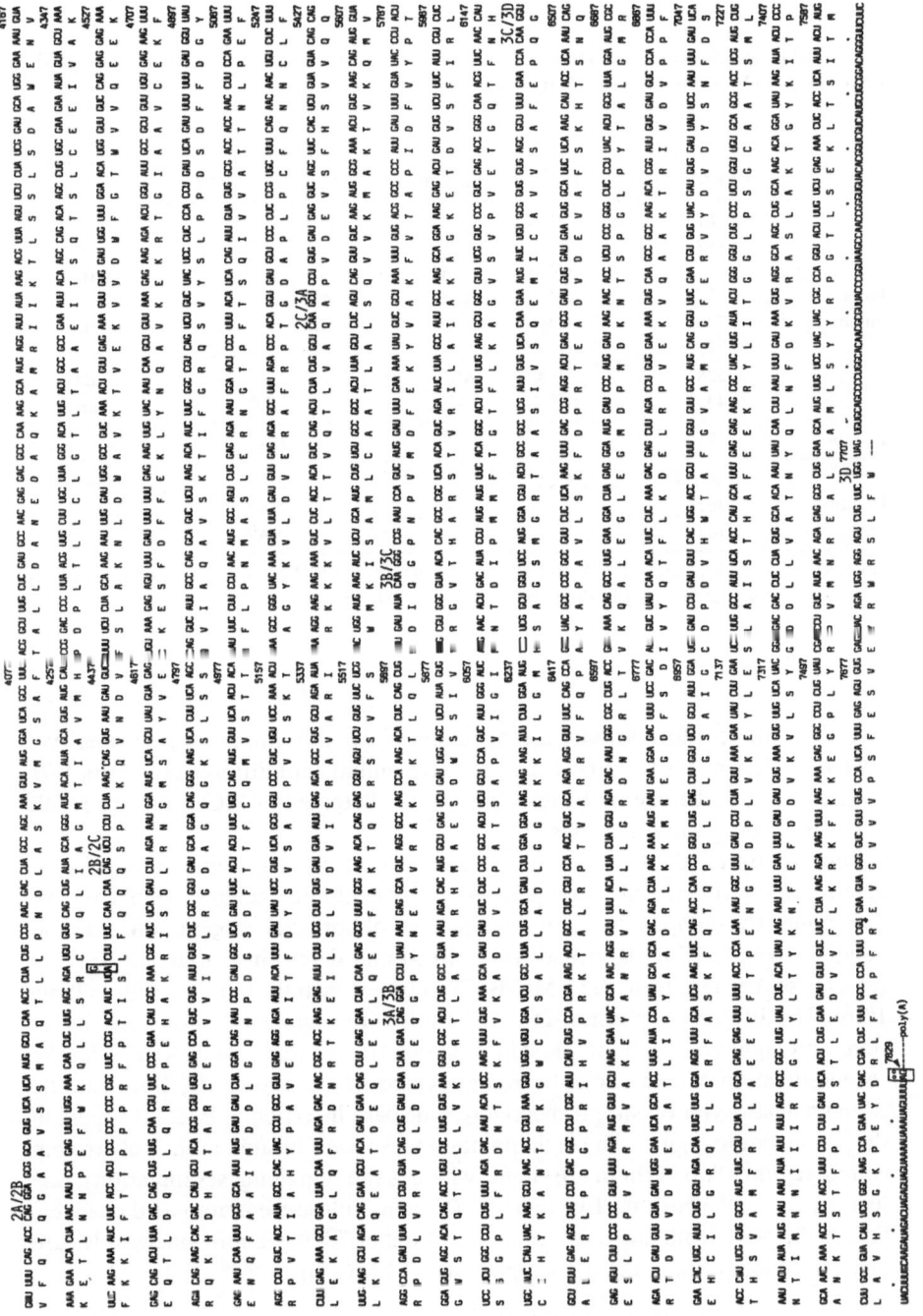

specific to EMC-B. The deletion loci in EMC-B are marked by upper open circles, and the corresponding locus in EMC-D for the insertion at 756 (U) of EMC-B is denoted by a hyphen. The polyprotein processing scheme of L434 is shown by letters on the backbone sequence at starting and stop sites, on the basis of other picornavirus processing.

TABLE 10.6. Nucleotide and Deduced Amino Acid Differences Between Diabetogenic EMC-D and Nondiabetogenic EMC-B Viruses.

	Sequence position based on the EMC-D	Nucleotide difference		Amino acid difference		Genomic position
		EMC-D	EMC-B	EMC-D	EMC-B	
Point mutation	874	UUU	CUU	Phe	Leu	L
	1929	ACU	ACC	ND[a]		1B
	2839	GCU	ACU	Ala	Thr	1D
	2872	GCC	ACC	Ala	Thr	1D
	2887	GAA	CAA	Glu	Gln	1D
	3154	GCC	ACC	Ala	Thr	1D
	3241	UCG	ACG	Ser	Thr	1D
	4395	UCA	UCG	ND[a]		2B
Insertion	765	-	-	NA[b]		5′-Noncoding
Deletion	149–278	CCC	U	NA[b]		5′-Noncoding
	7828	A	-	NA[b]		3′-Noncoding
	7829	G	-	NA[b]		3′-Noncoding

Note: Deletion and insertion are denoted for the sequence of EMC-B on the basis of the EMC-D sequence. Three C are deleted in the poly (C) track of EMC-B.
[a] ND; no difference.
[b] NA, not applicable.

to cells or the conformational arrangement of possible immunogenic sites. We have thus shown, for the first time, that five amino acid differences in the *VP1* gene between diabetogenic EMC-D and nondiabetogenic EMC-B viruses result from first-base point mutations.[40]

Information on the complete sequences of both viral genomes is needed to determine the functional consequences of the genomic changes. Thus we have determined the complete nucleotide sequences of both diabetogenic EMC-D and nondiabetogenic EMC-B viruses (Fig. 10.8) and determined that the EMC-D (7829 bases) differs from the EMC-B (7825 bases) by only 14 nucleotides (Table 10.6). The differences consist of two deletions of five nucleotides, one base insertion, and eight point mutations. The first deletion of three nucleotides and the second deletion of two nucleotides are located in the 5′ poly (C) tract and the 3′-end polyadenylation site, respectively. One base insertion in EMC-B occurs in the 5′ noncoding region. The eight point mutations are located in the polyprotein coding region. Two of them are silent, whereas the remaining six mutations, one located on the *L* gene and five on the *VP1* gene, introduce amino acid changes. Our findings indicate that a maximum of 14 out of 7829 genomic nucleotides are critical in determining the diabetogenicity of EMC virus.[40]

Long-Term Complications of EMC Virus-Induced Diabetes

The separation of the M-variant of the EMC virus into D- and B-variants has made it possible to study some of the long-term complications of diabetes. The D-variant, in the absence of the B-variant, produces far more severe and prolonged diabetes than the original M-variant of the EMC virus. The kidneys of mice that had been diabetic for six months showed both diffuse and nodular types of glomerulosclerosis (Fig. 10.9A), and electron microscopy revealed a two- to fourfold increase in the thickness of the glomerular basement membrane (Fig. 10.9B).[41,42] These findings are typical of those seen in humans with the Kimmelstiel–Wilson type of diabetic glomerulosclerosis. The diabetic animals also showed changes in the cornea and retinal vessels (e.g., a decrease in the number of pericytes) similar to those found in patients with diabetes mellitus (Fig. 10.10).[41] Furthermore, there was a four- to sixfold increase in mortality of the diabetic animals compared to the control animals. A decrease in bone formation and mineralization also were seen in diabetic mice 30 to 180 days after infection with EMC-D virus.[43] These endochondral bone changes are not the result of virus-induced tissue damage but rather of chronic metabolic changes. Thus the animal model is valid in the sense that virus can produce both early metabolic changes and at least some of the long-term complications of diabetes.

Inhibition of Virus-Induced Diabetes

The nondiabetogenic B-variant of EMC virus completely inhibited induction of diabetes by the D-variant of the EMC virus. 10^4 PFU of the D-variant produced diabetes in 95% of the infected mice, but 10^4 PFU of the B-variant failed to produce diabetes in any mice. When the B-variant was injected along with the D-variant at a 1:1 ratio, only 60% of the mice developed diabetes; when the B- and D-variants were mixed at a ratio of 9:1 and then injected, only about 11% of the mice developed diabetes; and when the B- and D-variants were mixed at a ratio of 99:1 none of the injected mice developed diabetes.[14] When the mice were first inoculated with the B-variant and then inoculated with the D-variant at different times thereafter, none of the animals developed diabetes.[14]

To see whether inhibition of D-variant–induced diabetes by the B-variant was the result of differences in a capacity to induce interferon, mice were infected with the D- or B-variant and, then, at various times, sera were drawn and assayed for interferon. Substantial amounts of interferon were induced by the B-variant within 12 hours. In contrast peak interferon titers were not reached until 30 hours after infection with the D-variant; the maximum titer then was approximately 30% of that with the B-variant.[14] These results suggest that induction of interferon by the B-variant may inhibit D-variant replication at an early stage of infection. Because interferon disappeared from the circulation within four days of inoculation of the B-variant, protection observed more than four days after infection is most likely produced by a cross-reacting neutralizing antibody. Interferon (early) and antibody (late) thus act in combination and appear to contribute to the B-variant–induced inhibition of diabetes in mice infected with the D-variant of the EMC virus.

Recent studies have shown that virus-induced diabetes can be prevented by a live attenuated vaccine.[28] Mice were immunized with the nondiabetogenic B-variant of EMC virus and challenged 30, 43, or 90 days later with the diabetogenic D-variant

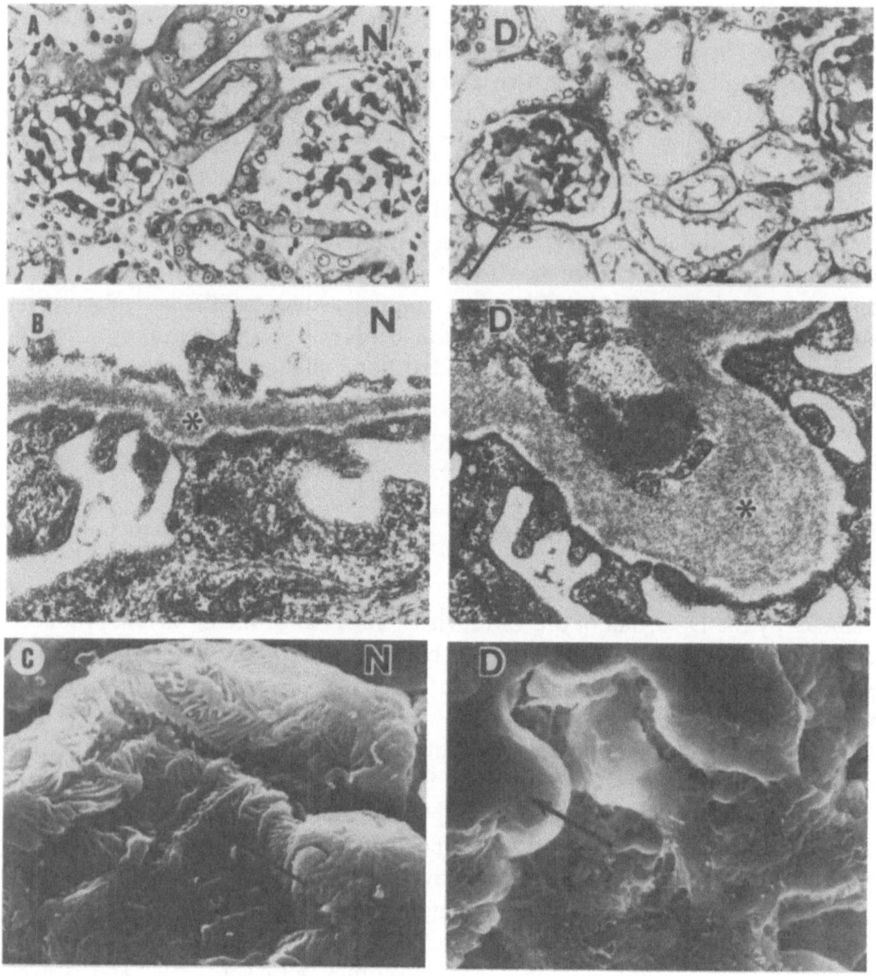

FIGURE 10.9. Kidney sections from uninfected controls and EMC-infected mice that were diabetic for six months. Light microscopy of (**A**) control mouse (N) shows normal glomeruli, tubules, and Bowman's capsule; diabetic mouse (D) (six months duration) showing prominent nodular glomerulosclerosis (long arrow) and thickening of Bowman's capsule (periodic acid-Schiff, × 145). Transmission electron micrographs (**B**) of kidney from a diabetic mouse (D) shows marked thickening of the peripheral glomerular capillary basement membrane, compared with an uninfected control (N) (× 15,750). Scanning electron micrograph (**C**) of kidney from diabetic mouse shows focal effacement of the normal glomerular epithelial cells with loss of foot processes (**D**) compared with the usual surface pattern of epithelial cells having abundant arborizing foot processes in normal kidney (N) (× 3,700). Sections were prepared by standard methods and examined using a Philips 400 electron microscope or a JEOLSM35 scanning electron microscope. (Reprinted by permission from *Nature*.[41] Copyright © 1982 Macmillan Magazines Ltd.)

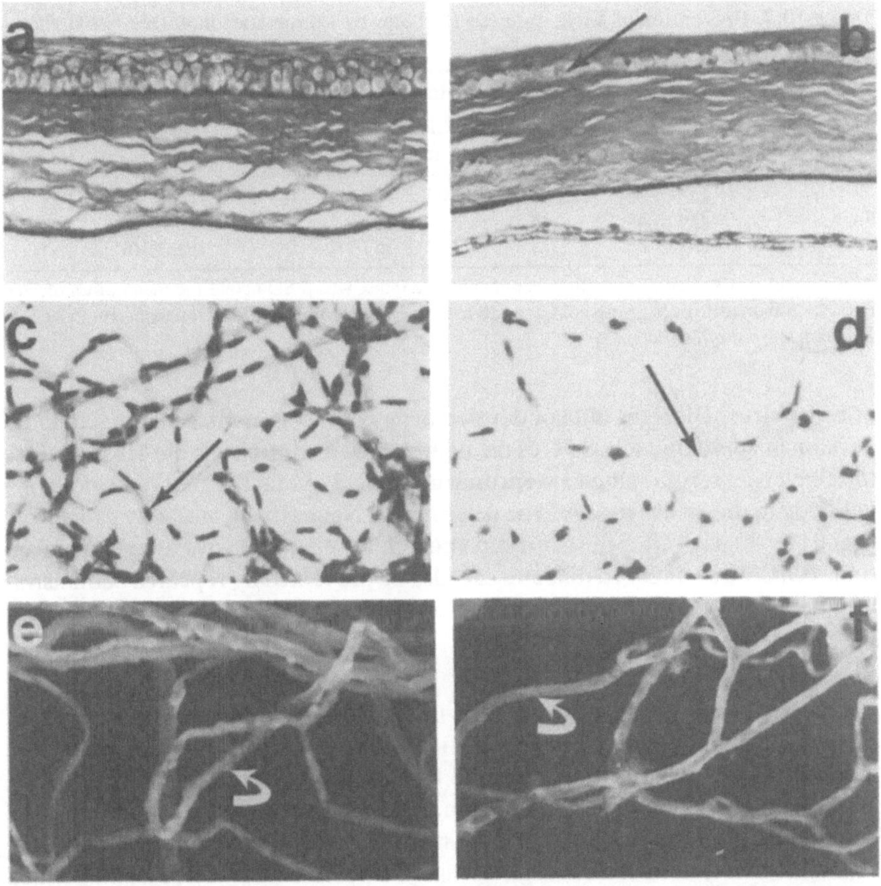

FIGURE 10.10. The corneal epithelium of uninfected mice (A) shows the usual cross-stratifications and glycogen content (periodic acid-Schiff; × 236). Cornea from mice that were diabetic for six months (B) show irregular stratification and moderate edema of the epithelium, particularly of the basal layer (long arrow) (periodic acid-Schiff, × 230). The stroma and Descemet's membrane (short arrow) are normal. (C) Micrograph of retinal capillaries, obtained by trypsin digestion from uninfected mice, shows the normal pericyte distribution (arrow). (D) A significant decrease in the number of pericytes is observed in retinal capillaries from the diabetic mice. (E) Scanning electron micrograph of retinal capillaries obtained by trypsin digestion from uninfected mice, shows normal pericyte distribution (arrows). (F) A moderate decrease in the number of pericytes is observed in the trypsin digest of retinal capillaries from mice that were diabetic for six months (× 308).

TABLE 10.7. Prevention of Virus-Induced Diabetes by Immunization with a Nondiabetogenic Variant.[a]

Days after immunization	Infected, unimmuized		Infected, immunized		Uninfected	
	Glucose index	Diabetes (%)	Glucose index	Diabetes (%)	Glucose index	Diabetes (%)
30	339 ± 78	87	162 ± 12	0	155 ± 14	0
43	378 ± 84	78	172 ± 13	0	158 ± 13	0
90	307 ± 89	75	168 ± 15	0	161 ± 16	0

[a] An aniimal was scored as diabetic when its glucose index was 5 SD above the mean of uninfected mice (>240 mg/dl). Each group contained 10 to 20 mice. (Reprinted with permission from *The New England Journal of Medicine*.[28])

of EMC virus. Diabetes did not develop in any of the immunized mice, but it did develop in approximately 80% of the unimmunized controls (Table 10.7). Before the challenge, serum samples were drawn and held so that the titer of neutralizing antibody to the D-variant could be determined. Neutralizing antibody titers were over 1000 30 days after immunization and close to 500 90 days after immunization. EMC-virus–induced diabetes can also be prevented by repeated administration of interferon or an interferon inducer (e.g., poly I:C).[34]

MENGO 2T VIRUS

Mengo virus, a member of the cardiovirus genus of the picornaviridae, causes fatal encephalitis in mice. Plaque-purified Mengo 2T virus infects and destroys pancreatic beta-cells, as demonstrated by immunofluorescent techniques (Fig. 10.11) and histopathologic examination.[16] EMC virus, the other member of the cardiovirus genus, induces diabetes in only certain strains of mice, but is not neuropathologically lethal. Although Mengo 2T virus could not be distinguished antigenically from EMC-D virus, its spectrum of host susceptibility is strikingly different. Mengo 2T virus causes diabetes even in EMC-D–resistant strains of mice (C57BL/6J, CBA/J, C3H/J, CE/J, and AKR/J).[16] The precise mechanism by which Mengo virus infects pancreatic beta-cells in strains of mice resistant to EMC virus is not known. One of the many possibilities is that Mengo virus and EMC virus are distinct viruses that bind to different receptors on the beta-cell surface.[44] This difference also might be the result of differing degrees of virus replication in the beta-cells.

Mengo virus can infect, replicate within, and destroy pancreatic beta-cells in the absence of an autoimmune response. We have not found any evidence of changes in T-cell subpopulations or of the production of autoantibodies against pancreatic beta-cells in Mengo virus-induced diabetes in animals. We cannot, however, exclude the possibility that the stress induced by encephalitis may affect glucose homeostasis.

COXSACKIE B VIRUSES

Coxsackie B4 virus was the first of the six serotypes of Coxsackie B viruses shown to be capable of producing diabetes in animal models.[17,45] In early studies,

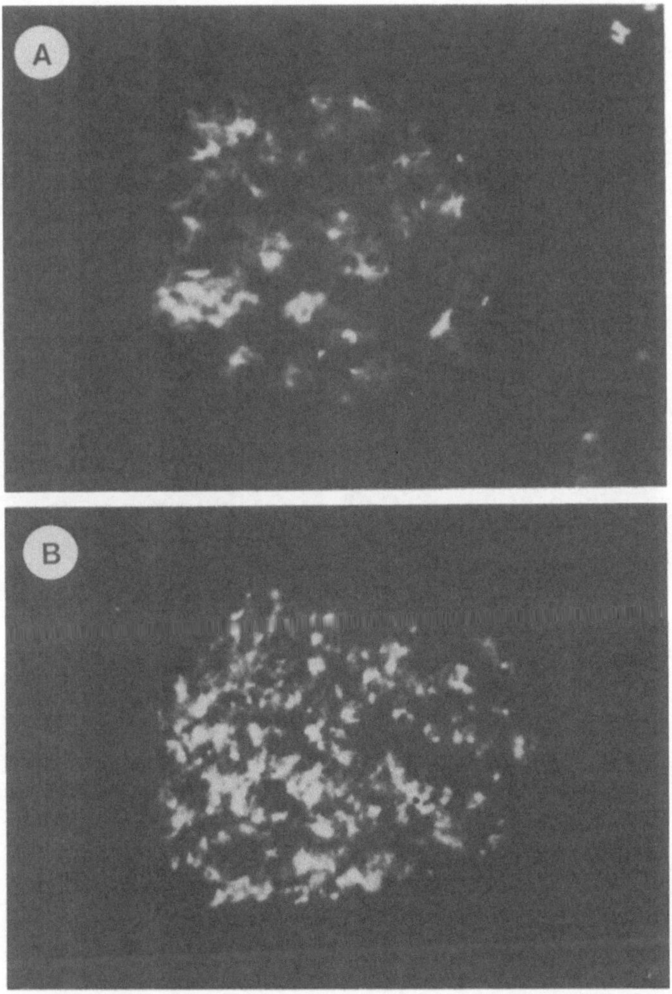

FIGURE 10.11. Section of mouse pancreas obtained three days after infection with Mengo 2T virus and stained with FITC-labeled, anti-Mengo virus antibody. Virus-specific antigens were found in the islets of Langerhans. **A**: A few cells contained viral antigens. **B**: Most islet cells contained viral antigens.

Coxsackie B4 virus did not produce diabetes when it was injected into mice.[46] Naturally acquired Coxsackie virus infection of the pancreas, however, produces a predominantly acinar-cell pancreatitis. Passage of Coxsackie virus B4 through a murine-enriched, pancreatic beta-cell culture enhances the tropic properties of the beta-cells (Fig. 10.12) and the diabetogenic capacity of the Coxsackie virus. Thus enhanced, the virus destroys the pancreatic beta-cells to produce a decrease in the insulin content of the pancreas, which causes hypoinsulinemia and, subsequently, hyperglycemia.[17,98] In the majority of animals infected with Coxsackie

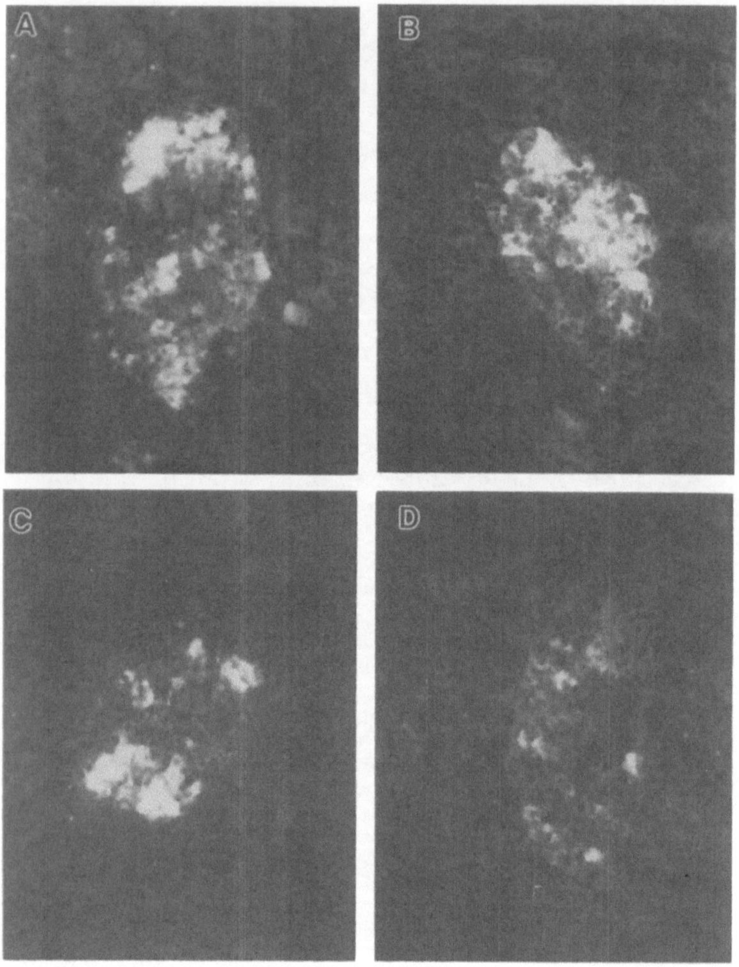

FIGURE 10.12. Section of mouse pancreas obtained three days after infection with Coxsackie B4 virus and stained with FITC-labeled anti-Coxsackie B4 virus antibody. **A** and **B**: The majority of the cells in the islets contain viral antigens. **C**: One portion of the islet contained viral antigens. **D**: Only a few cells contained viral antigens. The surrounding acinar cells show little fluorescence (× 500).

B4 virus, the hyperglycemia is transient.[17] This may be due to the fact that the remaining beta-cells compensate by proliferation and/or hypertrophy. In some studies, pretreatment of susceptible mice with a subdiabetogenic streptozotocin dose before Coxsackie B virus inoculation increases the severity of the diabetes.[47,48]

As in the case of EMC virus, only certain inbred strains of mice developed diabetes when exposed to Coxsackie B4 virus, and the diabetes was more severe in male than in female mice. The susceptibility of mice strains to the diabetogenicity of Coxsackie B virus is similar to that to EMC virus, with the exception of DBA/1J and DBA/2J, which are susceptible to EMC but not to Coxsackie B4 virus. Patas monkeys also are susceptible to Coxsackie B4, but not EMC virus-induced diabetes.[49]

Nonhuman Primates

In mice EMC virus, Mengo virus, and the Coxsackie B virus group can infect beta-cells and cause diabetes. In humans there is evidence that Coxsackie B viruses may either cause or serve as the final insult in triggering some cases of IDDM.[4] The number of well-documented cases is small, however, and a nonhuman primate model would be of value. Recently, glucose and insulin levels were determined in several species of monkeys, including Cynomolgus, Rhesus, Cebus, and Patas after infection with EMC virus. Although the glucose and insulin levels of the infected monkeys were generally within the normal range,[49] a slight glucose elevation was noted in the Cebus monkey at 7 days but not at 14 or 21 days after inoculation.

Similarly Cynomolgus, Rhesus, and Cebus monkeys had normal glucose tolerance and insulin secretion curves after infection with Coxsackie B4 virus (Fig. 10.13).[49] In the Patas monkey, however, the glucose tolerance curves were clearly elevated (Fig. 10.13), and the insulin secretion curves were markedly depressed (Fig. 10.14).

In an attempt to increase the severity of the diabetes, Patas monkeys were treated with a subdiabetogenic dose of streptozotocin (15 mg/kg body weight). The monkeys were then infected with Coxsackie B4 virus and several weeks later with Coxsackie B3 virus. Streptozotocin alone produced a barely detectable elevation in glucose levels (CTTT). When streptozotocin-treated monkeys were then infected with Coxsackie B4 virus, however, blood glucose levels (GTT) were significantly elevated. An even greater increase was observed if the monkeys were then infected with Coxsackie B3 virus. Immunoreactive insulin levels also were markedly depressed. The glucose tolerance curves returned to normal within four months of infection with Coxsackie B3 virus. Only in Patas monkeys infected with Coxsackie B4 virus or pretreated with streptozotocin and then infected with Coxsackie B4 and B3 viruses did glucose appear in the urine.

Serum samples from all the infected monkeys were analyzed at 7, 14, 21, or 38 days after infection for the presence of islet-cell cytoplasmic antibodies and islet-cell surface antibodies. No antibody activity was found.[49]

Abnormal glucose and insulin levels seemed to occur more often in the Patas monkeys that received multiple insults, but clearly Coxsackie B4 virus alone can produce such abnormal levels. Although it is tempting to argue that the appearance of virus-induced diabetes in Patas monkeys, as in mice, relates to the number of viable beta-cells, it was not possible to kill the monkeys and look for lesions and viral antigens in the pancreas during the acute phase of the infection. Because some strains of Coxsackie virus can infect cells of the central nervous system, it is possible that abnormal secretion of neurologically regulated

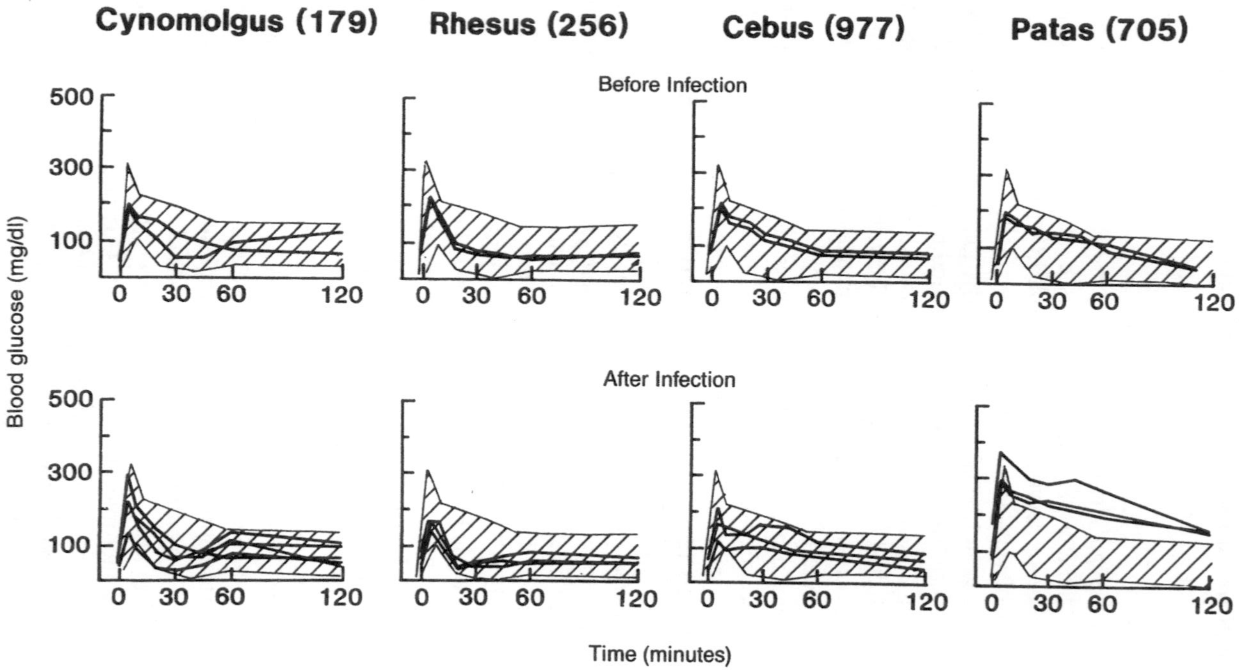

FIGURE 10.13. Blood glucose levels after intravenous glucose tolerance tests in monkeys before infection (top) and after infection (bottom) with Coxsackie B4 virus. Hatched areas represent mean ± 3 SD of 11 monkeys before infection. Samples were obtained two times (−7 and −14 days) before infection and three to five times (+7, +14, +21, +31, and +38 days) after infection. At 31 and 38 days after infection, a GTT for Cynomolgus was done.

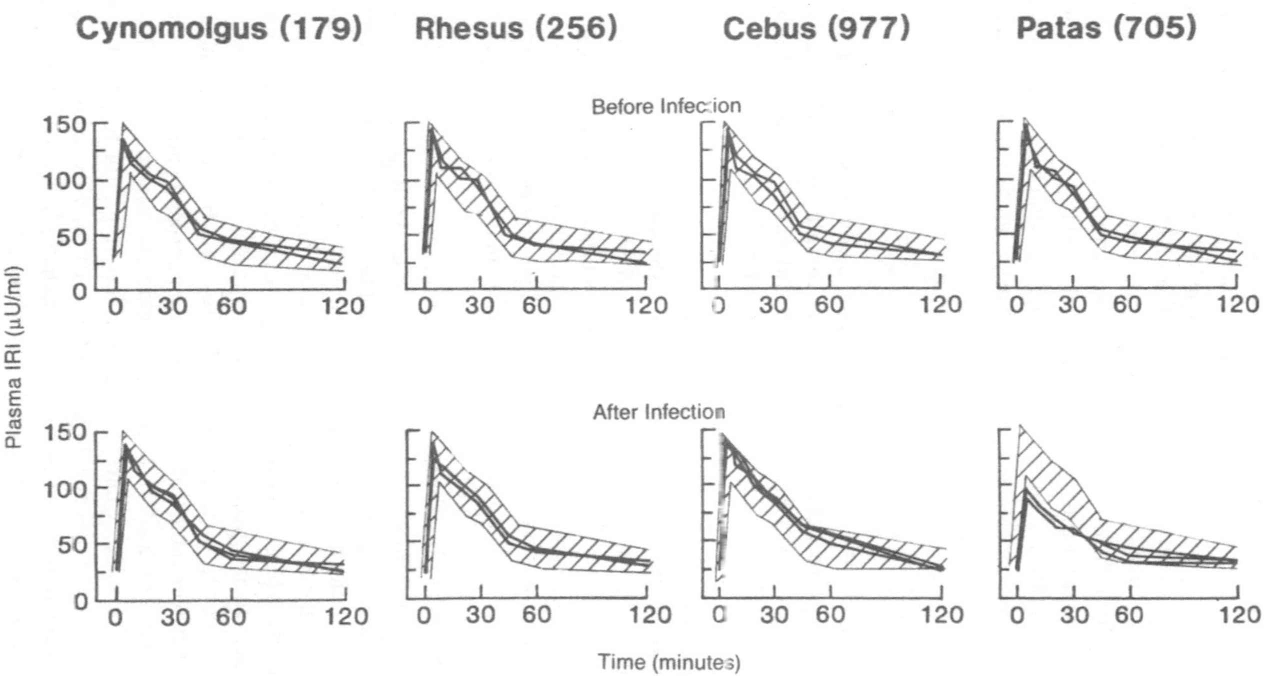

FIGURE 10.14. Insulin levels after intravenous glucose tolerance tests in monkeys before infection (top) and after infection (bottom) with Coxsackie B4 virus. Hatched areas represent mean ± 3 SD of 11 monkeys before infection. Samples were obtained two times (−7 and −14 days) before infection and three times (+7, +14, and +4 days) after infection. IRI: Immunoreactive insulin.

hormones (e.g., β-endorphin, catecholamines) may contribute to the abnormal glucose and insulin levels.

Whatever the mechanism, evidence from studies in mice, humans, and nonhuman primates indicates that Coxsackie viruses can affect glucose homeostasis. Recent studies on Coxsackie B4 virus have demonstrated that antigenic changes at the epitope level occur at a frequency greater than 10^{-2}.[50,51] This suggests that even within the same virus pool there may be many antigenic variants and that these variants may have different tissue tropism and different physiologic properties. This could account for the wide spectrum of clinical disease produced by the Coxsackie viruses. A rare variant may be diabetogenic, which might explain why only an occasional case of IDDM appears to be associated with a Coxsackie B virus infection.

Human Diabetes

Epidemiologic Studies. An early epidemiologic study reported that IDDM patients with disease of recent onset (within three months) had higher neutralizing antibody titers to Coxsackie B4 virus than did either normal subjects or patients who had had IDDM longer than three months.[52] More recent epidemiologic studies have not confirmed this finding.[53,54] The possible relationship between virus infections and subsequent development of IDDM has been a subject of some controversy, but recent studies have shown a positive correlation between Coxsackie B virus infection and IDDM.[8,9,55] King and colleagues detected Coxsackie B virus-specific IgM responses in 39% of children aged 3 to 14 years with newly diagnosed IDDM.[8] In contrast only 6% of an age-matched control group had Coxsackie B virus-specific IgM antibody. More recently Banatvala and colleagues confirmed this observation in different European countries, including Austria and in Australia.[9]

Several hypotheses can be advanced to account for lack of uniformity in some of the foregoing epidemiologic survey results. For example people could have different genetic backgrounds in different geographic areas, techniques used by different scientists could vary, or sampling techniques could be biased. There is also a fundamental heterogeneity to the concept of "virus as etiologic agent." Each virus designation (e.g., Coxsackie B4 virus) includes many variants that are sufficiently similar to be grouped under the same name. Prabhaker and colleagues identified over 13 variants of Coxsackie B4 virus by specific monoclonal antibodies, and they were able to demonstrate major antigenic differences among naturally occurring isolates.[50,51] Some Coxsackie B4 variants (e.g., B4-BP) produced diabetes in mice, whereas other variants (e.g., B4-NP, B4-FP, and B4-MK) did not produce diabetes in mice (Table 10.8). Coxsackie B4 and B5 viruses isolated from diabetic patients produced diabetes in certain inbred strains of mice,[6,7] but many other Coxsackie B4 variants isolated from nondiabetic patients did not produce diabetes in these mice. Furthermore Mengo 2T virus induced diabetes in mice, but its prototype did not. The most convincing animal model of diabetogenic and nondiabetogenic variants of a virus comes from

TABLE 10.8. Differences in Diabetogenic Activity Among Variants of EMC and Coxsackie B4 Virus.

Virus	Variants	Glucose index (mean ± SD)	Diabetes (%)[a]
EMC	EMC-M	252 ± 136	47
	EMC-B	164 ± 12	0
	EMC-D	430 ± 51	100
Coxsackie B4[b]	Coxsackie B4-NP	165 ± 27	0
	Coxsackie B4-FP	156 ± 19	0
	Coxsackie B4-MK	149 ± 21	0
	Coxsackie B4-BP	245 ± 103	60

[a] The mean glucose index of 110 uninfected SJL/J mice was 145 ± 19 mg/dl. An animal was scored as diabetic when its glucose index was 5 SD above the mean (>240 mg/dl). Each viral variant (5×10^5 PFU/mouse) was injected into 20 five-week-old mice.

[b] Coxsackie B4-NP, -FP, -MK, and -BP variants were obtained by plaque-purification of unpassaged prototype (NP), murine fibroblast cell-passaged (FP), monkey kidney cell-passaged (MK), and pancreatic beta-cell–passaged (BP) Coxsackie B4 virus, respectively.

studies on the EMC virus, discussed earlier. The D-variant of EMC virus produced diabetes in up to 100% of infected susceptible animals, but none of the mice infected with the B-variant of EMC virus developed diabetes (Table 10.8).[14] If an animal is infected with the B-variant first and then challenged with the D-variant, it will not become diabetic,[14,28] although the D-variant and the B-variant cannot be distinguished by hyperimmune antisera.

Some of the discrepancies in reports regarding the relationship of viruses and IDDM can be explained by the outbreak of different virus variants. It is quite possible that antigenic drift occurs over time and from area to area. If children are first exposed to a nondiabetogenic virus variant, they may be immune to subsequent exposure to potentially diabetogenic variants of the same virus. Given this situation, it is impossible to demonstrate correlation or lack of correlation between infection with a virus and the development of IDDM. Evidence from animal models suggests that positive and negative correlations between virus infection and IDDM development may well depend on the spread of diabetogenic and nondiabetogenic variants in a given geographic area and the genetic background of the population concerned.

Viral Infections of Human Pancreatic Beta-Cells and Cell Destruction. The possibility that a virus might cause some cases of IDDM by infecting and destroying pancreatic beta-cells has received considerable attention, but it is difficult to demonstrate in vivo that viruses replicate in human beta-cells and produce diabetes in man. As a practical model, a new in vitro system has been used to determine if viruses can infect and destroy human beta-cells in culture. Because the cells in this system were not pure beta-cells, a double-label immunofluorescent antibody technique was used to show unequivocally that the beta-cells, not the contaminating nonbeta-cells, became infected. Antibody to virus was labeled with fluorescein isothiocyanate (FITC) and antibody to insulin was labeled with

TABLE 10.9. Susceptibility of Cultured Human Beta-Cells to Viruses.

Virus	Infectivity (%)	Condition for infection	References
Coxsackie B3	30	1 MOI* for 24 hr	57
Coxsackie B4 virus	37	0.1 MOI for 60 hr	6
Mumps (ABC strain)	60	1 MOI for 72 hr	56
Reovirus (type 3)			
unpassaged virus	6	1 MOI for 48 hr	58
Reovirus (type 3)			
human beta-cell passaged virus	27	1 MOI for 48 hr	58

*Multiplicity of infection.

tetramethyl rhodamine isothiocyanate (TRITC). This system clearly showed that several common human viruses, including mumps,[56] Coxsackie B3 virus,[57] Coxsackie B4 virus,[6] and reovirus type 3,[58] infect human beta-cells (Fig. 10.15; see color insert) (Table 10.9). It was also shown by radioimmunoassay that the infection markedly decreased the insulin content of beta-cells. Thus, at least under in vitro conditions, human beta-cells are not inherently resistant to viral infection. There is no way to prove that these viruses actually infect and destroy human pancreatic beta-cells and cause diabetes in man, however, since some viruses will grow in cultured cells derived from animals that are resistant to infection.[59]

Isolated Viruses from Patients with Acute-Onset Diabetes can Cause Diabetes in Animals. As discussed above, in vitro susceptibility to beta-cells may not truly reflect in vivo susceptibility. A more cogent approach entails the isolation of virus from the pancreas of patients dying from acute-onset IDDM. Several years ago, we obtained material from a previously healthy, 10-year-old boy,[6] who was admitted to the hospital in diabetic ketoacidosis within three days after onset of a flu-like illness. Despite intensive therapy, his condition deteriorated and he died seven days later. At autopsy lymphocytic infiltration of the islets of Langerhans and necrosis of beta-cells were observed (Fig. 10.16). The picture was similar to that seen in the islets of Langerhans of mice that developed diabetes after infection with EMC or Coxsackie B4 virus. A small piece of the child's pancreas was removed at autopsy and homogenized and then incubated in a number of cultures with different cell lines. Several days later, cytopathology was observed in several of the cultures and a variant of Coxsackie B4 virus was isolated. The patient's serum had shown a rise in the titer of neutralizing antibody to this isolate from less than 4 on the second hospital day, to 32 on the day of death. In addition sections of the child's brain stained with fluorescein-labeled anti-Coxsackie B4 virus antibody revealed a few cells that contained viral antigens. These findings indicated that the virus had actually come from the child's tissues and was not a laboratory contaminant.

The possibility still remained that the Coxsackie B4 virus infection did not cause the child's diabetes. This is where our animal models proved invaluable.

When several inbred strains of mice were injected with the human isolate, only the SJL/J male mice developed diabetes; the CBA/J, C57BL/6J, and BALB/c

Insert I

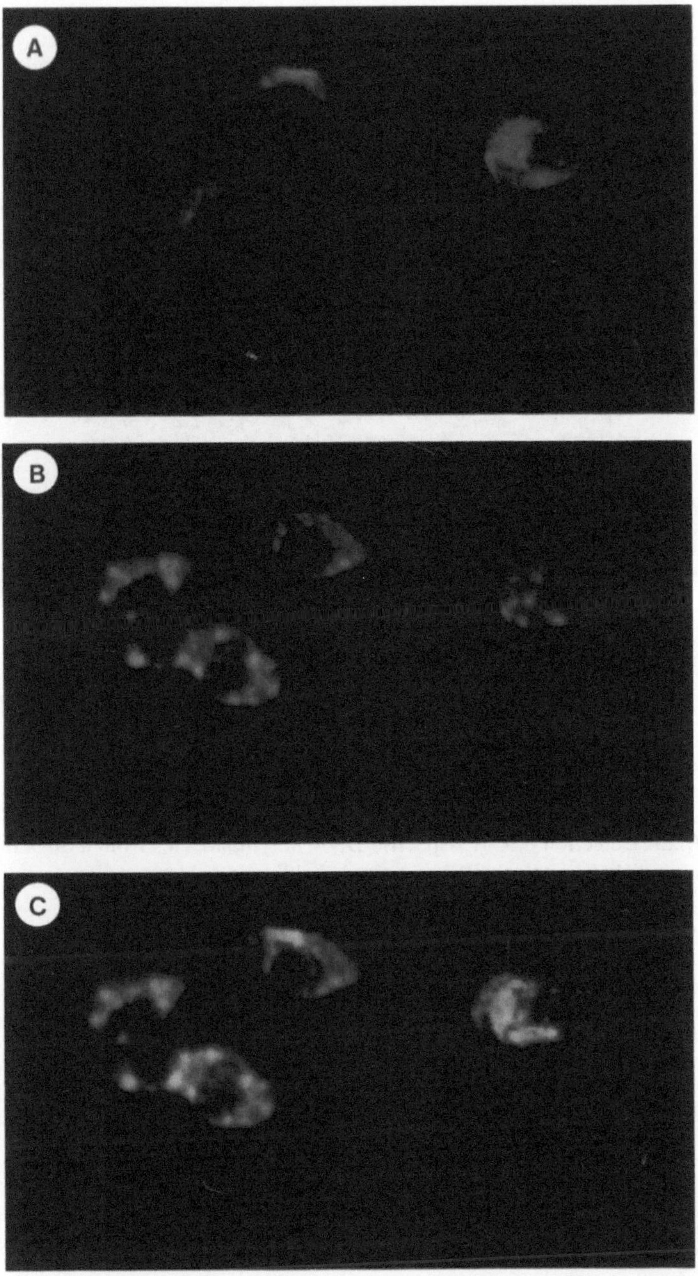

FIGURE 10.15. Human pancreatic beta cell cultures infected with reovirus type 3 at a virus-to-cell ratio of 1:1. At the end of 30 hours, the cultures were stained with TRITC-labeled anti-insulin and FITC-labeled anti-reovirus antibody. **A**: Photograph taken with rhodamine filters. **B**: Photograph of same area but taken with fluorescein filters. **C**: Photograph of same area but taken first with rhodamine filters and then with fluorescein filters (× 720).

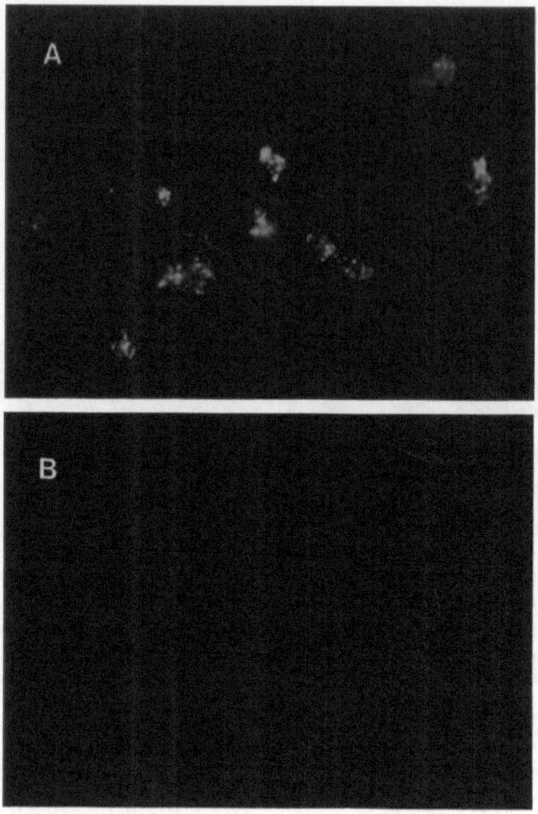

FIGURE 10.25. Detection of CMV-specific DNA in lymphocytes from a diabetic patient (**A**) and a nondiabetic control (**B**) by in situ hybridization.

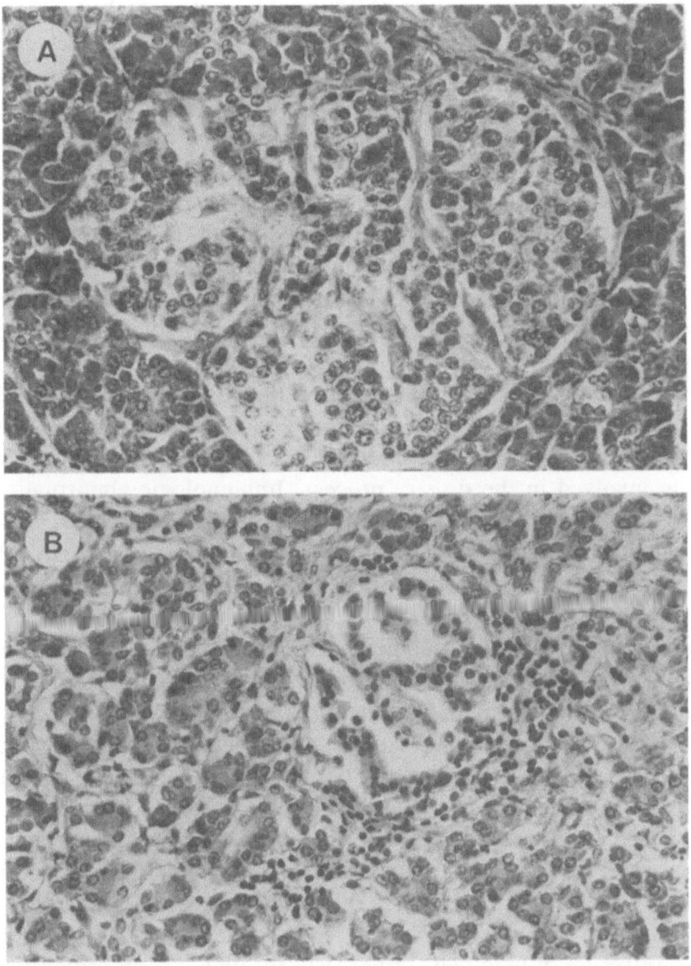

FIGURE 10.16. Sections of normal human pancreas at autopsy show a single islet of Langerhans surrounded by acinar cells (**A**) (H&E × 320). A section of pancreas from the IDDM patient shows extensive inflammatory infiltrate, loss of islet architecture, and islet-cell degeneration (**B**) (H&E × 320) (**A** and **B** reprinted by permission of *The New England Journal of Medicine*.[6])

mice did not. The pancreatic tissue from the diabetic mice showed extensive inflammatory cell infiltration and viral antigens in the islets of Langerhans. Many beta-cells were also destroyed. Based on the patient's clinical picture, the virus isolation studies, and the induction of diabetes in mice, it was concluded that the patient's clinical diabetes was virus induced.

The idea that viruses can trigger some cases of diabetes in man has been strengthened by two recent reports. The first case was a 16-month-old child with a Coxsackie B5 virus infection who, a few days later, developed diabetes.[7] This virus, isolated from the feces of the child, produced abnormal glucose tolerance tests following injection into mice. The second case was a five-year-old girl who had myocarditis and diabetes two to three weeks after open heart surgery.[60] At necropsy her islets showed a lymphocytic infiltrate and beta-cell necrosis. By immunofluorescence, Coxsackie B4 virus antigens were found in the islets, and high levels of antibody to Coxsackie B4 virus were found in the serum. In other studies, pancreatic sections from four of seven neonates who died of Coxsackie B virus infections showed insulitis and beta-cell damage.[12] The insulitis and beta-cell damage observed at autopsy do not prove that these children would have developed diabetes if they had survived, although these studies supported the idea that, under certain circumstances, some viruses can infect beta-cells in vivo.

Diabetes is not a common consequence of Coxsackie B virus infections, although from 5 to 20% of IDDM appears to be associated with a Coxsackie B virus.[8,9,55] It is still not known whether Coxsackie B viruses are primarily or secondarily involved in the disease process. Immunologic abnormalities (e.g., ICA, ICSA, and Ia-positive T-cells) often precede the onset of IDDM, but they are absent in about 5 to 20% of IDDM patients at the time of diagnosis.[61] Therefore one could speculate that some patients who fail to show immunologic abnormalities may have a cytolytic, Coxsackie B viral infection of their pancreatic beta-cells. The induction of antibodies to Coxsackie B viruses has not been correlated with the production of autoantibodies in patients with IDDM. Our current understanding of the pathogenesis of Coxsackie B virus-induced diabetes also argues against an autoimmune component. As do other picornaviruses, Coxsackie B viruses may rapidly infect and lyse pancreatic beta-cells before autoimmunity develops. Also there is no indication that Coxsackie B viruses modify the host cell surface by inserting viral antigens or that the virus triggers an autoimmune response (i.e., there is no evidence that islet cytoplasmic antibody or islet-cell surface antibody are present). Therefore if Coxsackie B viruses are involved in the pathogenesis of human IDDM, these viruses could induce diabetes through a cytolytic infection of beta-cells rather than through an autoimmune response. The possibility that chronic Coxsackie B viral infection of beta-cells could effect the synthesis and release of alpha-interferon, which in turn induce class I MHC hyperexpression on adjacent endocrine cells, cannot be excluded.[62]

Association of Virus Infection with Autoimmune IDDM

The possible association of an acute cytolytic infection of beta cells and the development of IDDM has been discussed earlier. However it is believed that a chronic process usually precedes the clinical onset of IDDM. Therefore, it is important to consider the possibility that certain virus infections trigger islet-cell autoimmunity. In this section we would like to discuss several candidate viruses that might trigger autoimmune IDDM in animals and man.

Animal Models

In contrast to encephalomyocarditis virus, Mengo 2T virus and Coxsackie B4 virus, rubella virus, retrovirus and reovirus seem to be associated with autoimmunity in the pathogenesis of diabetes in neonatal golden Syrian hamsters,[63] nonobese diabetic (NOD) mice,[15] and suckling SJL mice,[64,65] respectively.

Rubella Virus

The congenital rubella (CRS) syndrome in humans demonstrates that a viral infection is associated with the subsequent development of autoimmune type I diabetes. Recently we developed an animal model in which neonatal golden Syrian hamsters are infected with rubella virus passaged in beta-cells; this model closely parallels the diabetes observed with CRS.[63] Seven- to ten-day-old Syrian hamsters developed hyperglycemia, hypoinsulinemia, a mononuclear cell infiltrate of the islets, and positive immunofluorescence for rubella virus antigen in beta-cells following inoculation with a passaged variant of rubella virus. Very weak cytoplasmic islet cell antibodies were also present in 8 of the 20 infected animals. Our preliminary data indicate that when the passaged variant of rubella virus is injected in fetal hamsters during the first trimester of gestation, congenital anomalies similar to CRS in humans (absent hindlimbs) develop.

The mechanism of rubella virus-induced diabetes is not known. In contrast to EMC virus, rubella virus is not highly lytic. Our preliminary finding of insulitis in 34.5% of hamster islets and measurable circulating islet-cell antibodies is compatible with an autoimmune response. Rubella virus belongs to the togavirus family. This enveloped virus has a lipoprotein coat that is the result of the budding of the maturing virus through the host cell membrane.[66] Thus rubella virus might insert, expose, or alter antigens in the plasma membrane of the host cell during intracellular infection.[67] Alternatively the virus might induce an autoimmune syndrome by affecting T-cell subpopulations (helper or suppressor T-cells) that regulate the host's immune response.[68,69]

Retrovirus

Nonobese diabetic mice spontaneously develop a diabetic syndrome that, in many respects, resembles human type 1 (insulin-dependent) diabetes.[70] Diabetes in the NOD mice is characterized by an insulitis that precedes beta-cell destruction and that culminates in hypoinsulinemia and hyperglycemia. The precise events that trigger the onset of the disease are largely unknown. Various effector systems, including macrophages, T-lymphocytes, and/or humoral mediators, have been implicated as the possible effectors of an immune response.

Recent studies from our laboratory and other laboratories have shown that upon cyclophosphamide administration, NOD mice (about 70%) rapidly progress to overt diabetes with severe insulitis within two to three weeks (Fig. 10.17).[71,72] None of the control rats became diabetic. Cyclophosphamide significantly

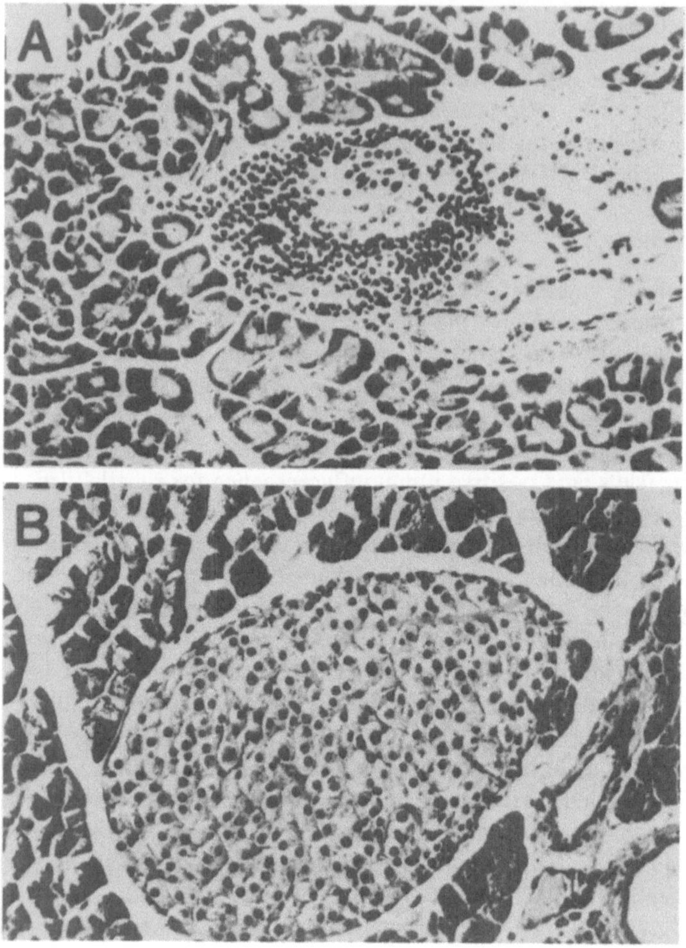

FIGURE 10.17. Pancreatic islet of a nonobese diabetic mouse treated with cyclophosphamide (A) and with silica and cyclophosphamide (B) (H&E × 300). (Reproduced from ref. 71 with permission of The American Diabetes Association, Inc.)

increases the incidence of diabetes in NOD mice by inhibiting suppressor T-cells or by activating cytotoxic T-cells.[73] Depletion of macrophages by silica treatment prevents insulitis and diabetes in cyclophosphamide-treated NOD mice, however.[71] These findings suggested that (1) macrophages play a major role in the initiation of organ-specific autoimmunities in NOD mice, (2) the presentation of autoantigen(s) to specific target cells (beta-cells) by macrophages could be the initial step in the development of insulitis, and (3) the elimination of macrophages inhibits the immune process. We recently initiated investigations to determine whether there are any specific changes in the beta-cells that might attract

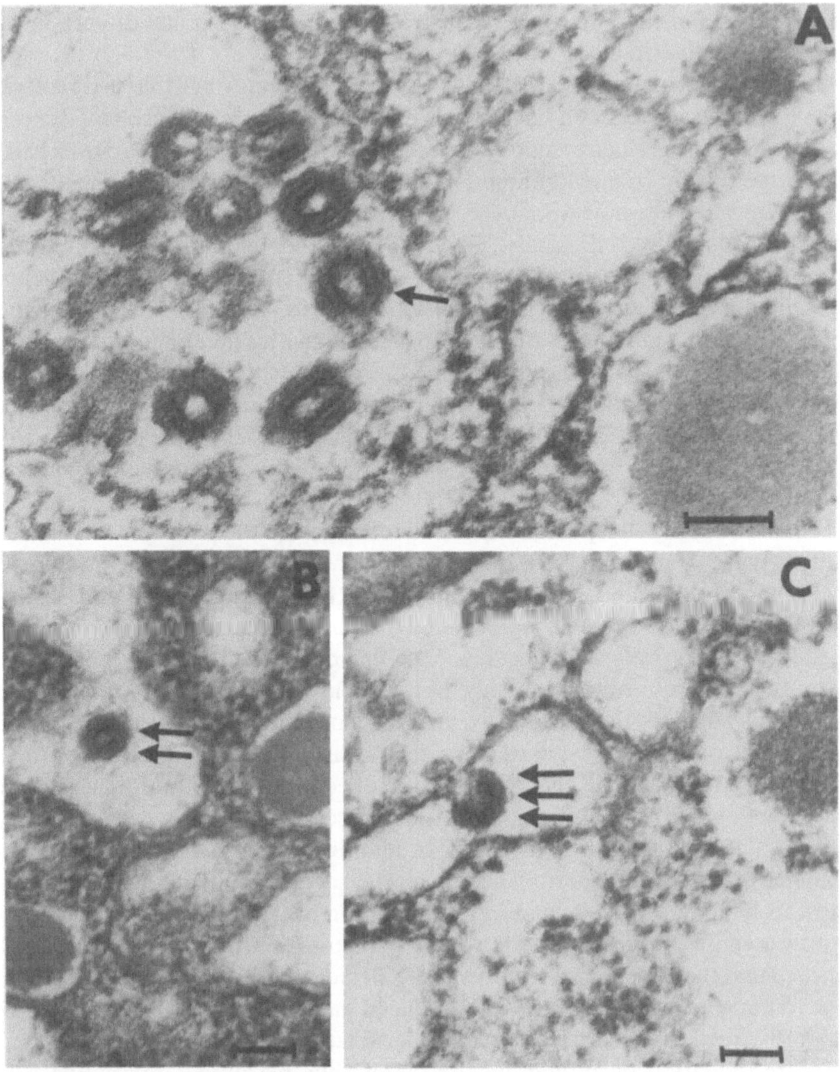

FIGURE 10.18. Endogenous type A retrovirus particles found in beta-cells from silica-treated NOD male mice after cyclophosphamide administration (**A** and **C**), and without cyclophosphamide (**B**). A cluster of virus particles (one arrow) is seen in the cytoplasm of the beta-cells (**A**), whereas a single particle (two arrows) is seen in the beta-cell (**B**). A particle that is budding (three arrows) from the inner membrane surface of the rough endoplasmic reticulum is also seen (**C**). Horizontal bar, 0.1 μm. (Reproduced from ref. 105 with permission of The American Diabetes Association, Inc.)

macrophages and thereby trigger beta-cell–specific autoimmune disease in the cyclophosphamide-treated, NOD male mice.

When thin sections of pancreatic islets from NOD male mice that had received silica to preserve islets and had subsequently received cyclophosphamide were examined under the electron microscope, clusters of endogenous retrovirus particles (A type) were frequently found in beta-cells (Fig. 10.18A and Fig. 10.19). Retrovirus particles, however, were rarely found in beta-cells from NOD male mice that had received silica only (Fig. 10.18B). Other endocrine cells, including alpha, delta, and pancreatic polypeptide-producing (pp) cells and exocrine acinar cells, did not contain these virus particles. Also virus particles were not found in spleen, liver, or kidney of either cyclophosphamide-treated or untreated NOD male mice. The presence of retrovirus particles in beta-cells was correlated with the presence of insulitis lesions in the cyclophosphamide-treated mice (Fig. 10.20). On the basis of these observations, we concluded that specific expression of endogenous retrovirus by beta-cells is associated with the development of insulitis and diabetes in NOD mice.[105]

The role of beta-cell–specific expressed retrovirus in the pathogenesis of autoimmune type 1 diabetes in NOD mice is not clear. These endogenous retroviruses might induce either an aberrant expression of class II major histocompatibility complex (MHC) antigen on the beta-cells, or beta-cell–specific autoantigen(s) (Fig. 10.21). Our preliminary data indicate that beta-cells from these NOD mice are not stained by anti-Ia–antibody (K. Amano and J.W., Yoon, unpublished data). This suggests that the retrovirus does not induce class II antigen under these conditions. Work in several laboratories has indicated that retrovirus (e.g., MuLV) gp 70 may be present on the surface of cells but may not produce detectable infectious virus.[74–76] Using immunofluorescent techniques, Lerner and his coworkers found that a protein similar or identical to the retrovirus (MuLV) gp 70 major envelope glycoprotein occurs on the surface of mouse (127/J) thymocytes, in lymphoid tissues, and in murine epithelial lining cells.[77] Since retroviruses (xenotropic viruses) were regularly recovered from mouse tissues, it is believed that these immunofluorescent tests detected viral antigen on the cell surface. Whether viral antigen or virus-induced autoantigen will be expressed on beta-cells from these animals is under investigation.

The presentation of antigen on the target cells by antigen-presenting cells (APC) is critical in distinguishing elicitation of the immune response from elicitation of immunologic tolerance.[78] The immune response to a specific antigen on the target tissue involves the activation of a T-helper cell subset of lymphocytes. The T-helper cell is activated only when it interacts with antigen presented on the surface of a macrophage or other APC (Fig. 10.22). Direct exposure of lymphocytes to an antigen in the absence of APCs is required to induce immunologic tolerance to the specific antigen.[79] Since elimination of APCs prevented insulitis in NOD mice, we can postulate that the presentation of autoantigens on the target cells by macrophages is the initial step in the development of beta-cell–specific autoimmune disease in NOD mice (Fig. 10.22). Amplification of immune

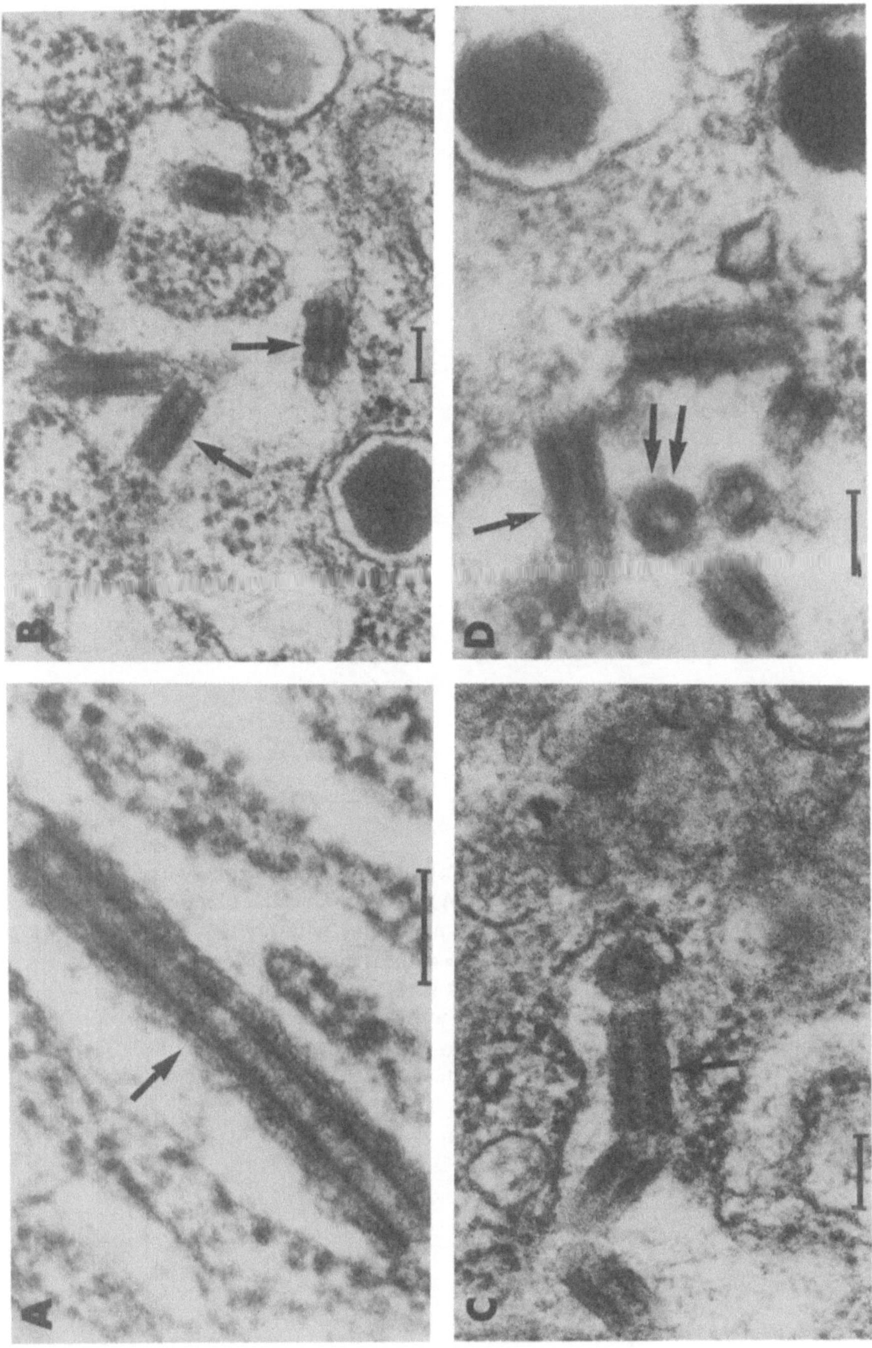

FIGURE 10.19. The heterogeneous shapes of endogenous retrovirus particles include a long tubular structure (**A**) (one arrow), several short tubular structures (**B**) (one arrow), a broken form of tubular structure (**C**) (one arrow), and a mixture of circular (two arrows) and tubular (one arrow) forms (**D**). All these particles are seen in the cytoplasm of beta-cells from cyclophosphamide-treated NOD male mice. Horizontal bar, 0.1 μm. (Reproduced from ref. 105 with permission of The American Diabetes Association, Inc.)

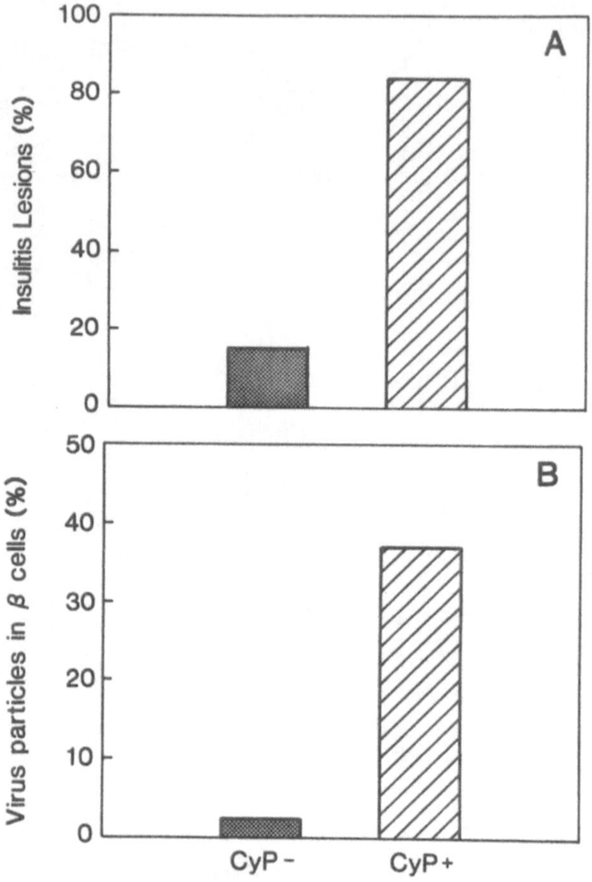

FIGURE 10.20. Correlation between incidence of insulitis in NOD male mice with (cyp+) and without (cyp−) cyclophosphamide treatment (**A**), and the frequency of retrovirus particles in the beta-cells from silica-treated NOD male mice with (cyp+) and without (cyp−) cyclophosphamide treatment (**B**). Over 80% of the islets from the cyp+ mice show advanced insulitis lesions, whereas only 15% of the islets from the cyp− mice show similar lesions. The percentage of retrovirus particles was calculated with the following formula: Number of virus particles observed divided by Number of beta-cells examined times 100 equals percent. Approximately 150 to 200 beta-cells per group were examined.

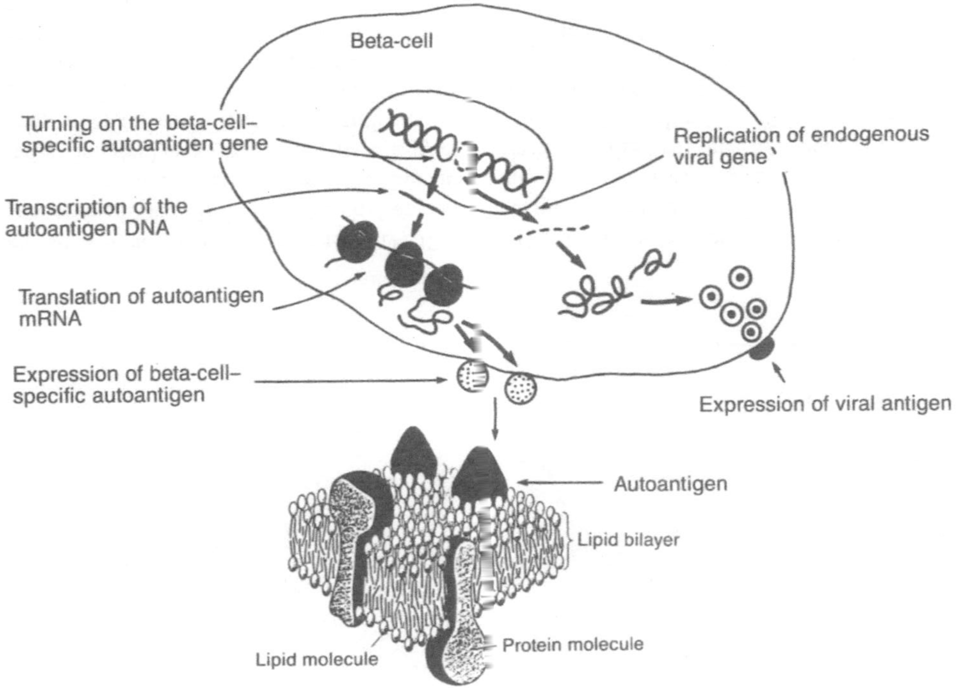

FIGURE 10.21. Schematic presentation of a hypothesis of beta-cell–specific expression of autoantigen gene. Endogenous retrovirus can be transmitted vertically if the virus integrates its genome (provirus) into the host DNA. Under certain circumstances, the viral genome can replicate and express the viral antigen on the surface of the beta-cell membrane. Replication of endogenous retroviral genome may initiate expression of a beta-cell–specific autoantigen gene, if any.

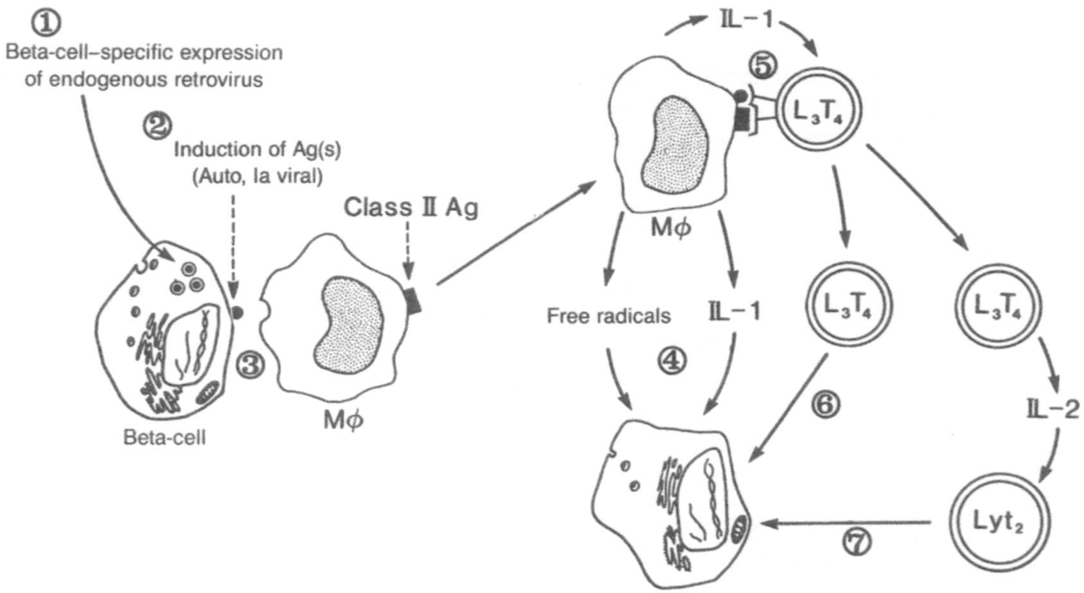

FIGURE 10.22. Possible pathways for beta-cell destruction in NOD mice. (1) The endogenous retrovirus may induce a beta-cell–specific autoantigen, class II antigen, or viral antigen (or two of these three or all three). (2) The expressed autoantigen(s) may attract macrophages. (3) The macrophages secrete free radicals (superoxide, hydroxyl radicals) and/or interleukin-1, which may have a cytotoxic effect on the beta-cells. (4) During this process, the macrophages may present the processed antigen to L3T4+ helper T-lymphocytes. (5) The L3T4+ cells may activate cytotoxic T-lymphocytes and/or stimulate B-cell response to produce antibodies against islet cells. (6) A fraction of L3T4+ T-cells may exert a direct cytotoxic effect on beta-cells. (7) Lyt2 cells developed during the further immune process may have additional cytotoxic effects on residual beta-cells, resulting in a beta-cell–specific autoimmune disease.

responses by T-lymphocytes (e.g., L_3T_4 and LyT_2), however, is required for the clinical expression of overt diabetes in NOD mice (Fig. 10.22).

Like and Rossini earlier reported induction of an aberrant retrovirus (C type) budding into the rough endoplasmic reticulum of beta-cells two to three days before insulitis was observed in male CD-1 mice treated with multidose streptozotocin.[100] In addition to type C retrovirus, intracisternal type A particles represent the most enigmatic retrovirus associated with murine diabetes.[2] Intracisternal type A particles are normally expressed during oogenesis and early embryogenesis. The 73-kD core protein (p73) may be a differentiation antigen.[2] If p73 is a differentiation antigen normally expressed before an immune surveillance system is constituted, then its expression by beta-cells in adult hyperglycemic mice might lead to its perception as a neoantigen. Such a neoantigen might induce an autoimmune reaction against beta-cells.

These retrovirus particles were first observed in necrotic beta-cells of a C57BL/KSJ–db/db mouse by Like and Chick.[101] Biochemical and immunocytochemical studies established that these particles in cells from C57BL/KS diabetes (db/db) mice were intracisternal type A particles.[102,103] In a study analyzing the response of different inbred strains of mice to the db mutation, only those strains that constitutively expressed beta-cell, intracisternal type A particles developed severe diabetes.[103] Transcription and translation of the viral genome in beta-cells of diabetes-susceptible strains, but not of diabetes-resistant strains, were inducible by elevated glucose levels.[104] In addition to C57BL/KS–db/db mice, it was recently found that beta-cell–specific expression of endogenous retrovirus in NOD mice was associated with insulitis and diabetes.[105]

Reovirus Type I & 3

Since reoviruses produce a variety of lesions in newborn mice, we passaged reovirus type 3 in cultured pancreatic beta-cells to see whether the virus could be adapted to beta-cells. When the beta-cell–passaged virus was injected into suckling SJL/J male mice, some of the infected animals showed abnormal glucose tolerance curves 10 days after infection.[80] Immunofluorescence demonstrated specific viral antigens in some beta-cells, as well as in acinar cells. Viral particles could be seen in the cytoplasm of some beta-cells on electron micrographs (Fig. 10.23). The surviving animals were mildly hyperglycemic for about three weeks and then returned to normal.

Onodera and colleagues later showed that mice infected with reovirus type 1, which had been passaged in pancreatic beta-cell cultures, developed transient diabetes and a runting syndrome.[64] The runting syndrome consisted of retarded growth, oily hair, alopecia, and steatorrhea. Inflammatory cells and viral antigens, as well as virus particles, were found in the islets of Langerhans (alpha-, beta-, and delta-cells), as well as in the growth hormone-producing cells of the anterior pituitary. Sera from the infected mice contained autoantibodies that reacted with cytoplasmic antigens in the islets of Langerhans, anterior pituitary,

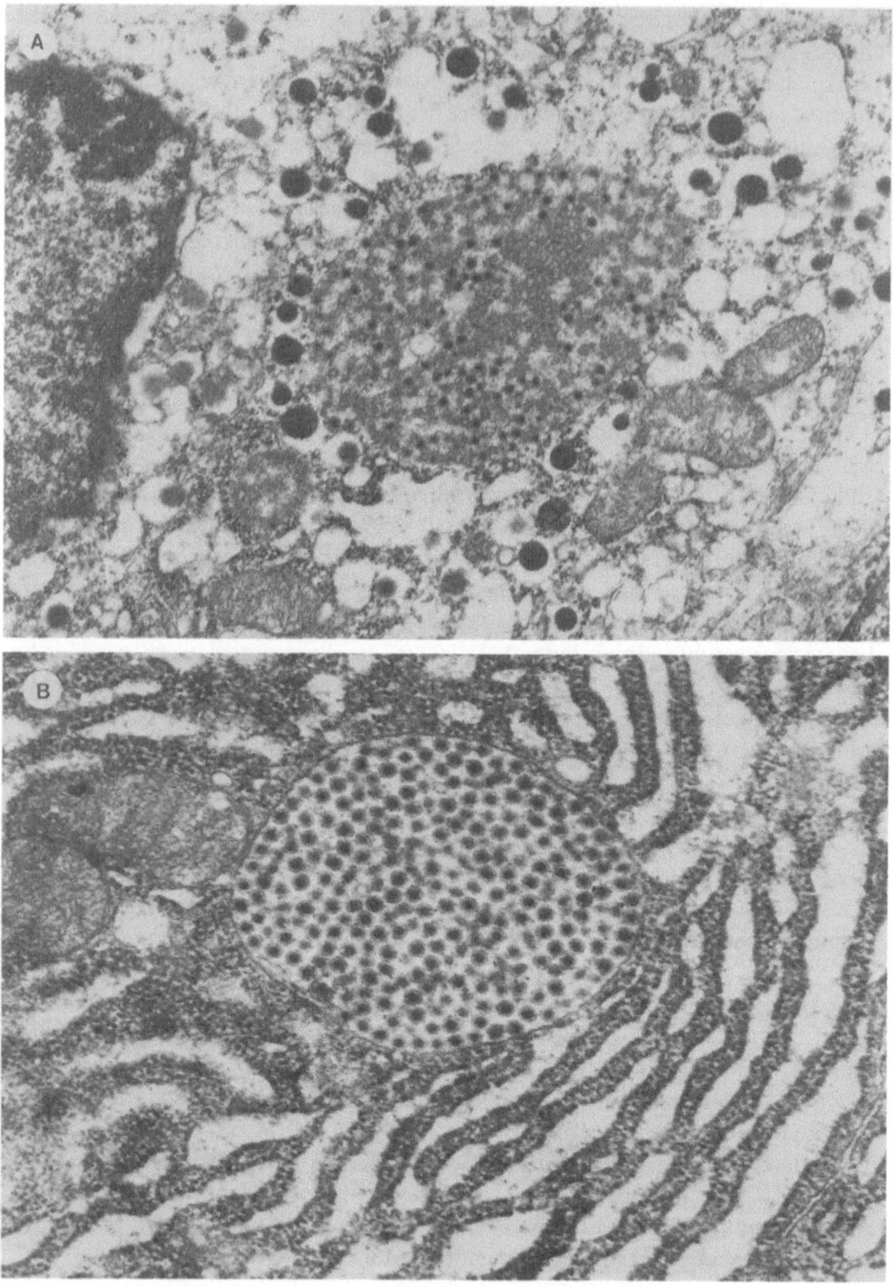

FIGURE 10.23. Electron micrograph of reovirus particles in pancreatic beta-cells (**A**) and exocrine acinar cells (**B**) of SJL/J mice five days after infection with reovirus. The diameter of the virus is approximately 75 nm.

and gastric mucosa of uninfected mice. Administration of antilymphocyte serum, antithymocyte serum, or cyclophosphamide reduced the severity or prevented the development of reovirus-induced diabetes.[65] Also, virus-infected immunosuppressed mice gained weight at almost the same rate as uninfected controls, and mortality was greatly decreased. These investigators concluded that autoimmunity plays a role in the pathogenesis of reovirus-induced diabetes.[64]

Precisely how reovirus infection triggers the development of autoantibodies is still unclear, but viruses have often been suspected to cause autoimmune disease. In contrast to reovirus type I, reovirus type 3 does not induce autoantibodies in mice and does not infect the pituitary. The critical difference between reovirus type 1 and reovirus type 3 seems to be at the level of the sigma-1 polypeptide responsible for virus tropism. Therefore it has been postulated that a single viral molecule appears to control pituitary infection and autoantibody production. In view of the histopathologic changes seen in the lymphoid organs of reovirus-infected mice, it is possible that specific subsets of lymphocytes have receptors for such a virus. Thus an infection of hormone-producing cells and an interaction of the virus with immune system cells may be required to initiate autoantibody production.[64] The relationship of infection of reovirus type 1, induction of autoantibodies, and development of the diabetic syndrome is not clear.

Rubella Virus

The CRS has emerged as an important human model for type I diabetes. Congenital rubella syndrome diabetes takes 5 to 20 years to develop.[81] Although the diabetes is the most common of the delayed manifestations of CRS, autoimmune thyroid disease,[82] Addison's disease,[82] and growth hormone deficiency[83,84] have also been reported. In patients with CRS and diabetes the frequency of HLA DR3 is significantly increased and that of HLA DR2 is significantly decreased.

Epidemiologic studies have shown that the prevalence of type I diabetes among CRS patients in both the United States and Australia is high (about 20%),[10,81] whereas studies in the United Kingdom have failed to confirm this association.[85] Rubella virus has been isolated from the pancreas in a few cases of congenital infection,[86,87] however, and insulitis with severe beta-cell depletion was also reported in the case of an infant with CRS dying of acute-onset diabetes.[81]

An autoimmune basis for the endocrine abnormalities in CRS is suggested by several findings: an increase in islet cell surface antibody (21% of CRS patients, 50–80% of patients with abnormal carbohydrate metabolism),[88,89] an increase in antithyroid microsomal and/or antithyroglobulin antibody (26% of CRS patients)[88,89] and an increase in antiinsulin autoantibody (13% of CRS patients as opposed to less than 1% of control).[90] The most convincing evidence that a persistent viral infection may cause IDDM and produce autoantibody comes from studies of patients with the CRS, but it is not known how rubella virus triggers autoimmunity and IDDM. Animal models of rubella could be useful in elucidating the underlying pathogenetic mechanism.

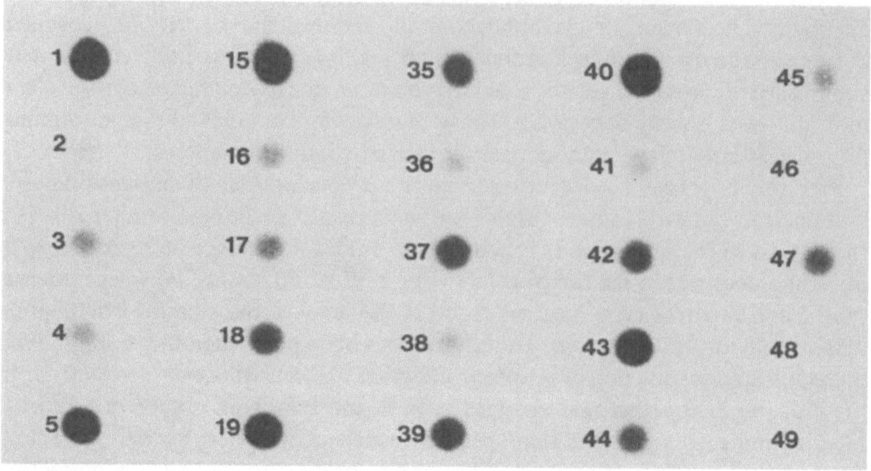

FIGURE 10.24. Dot blot analysis of CMV genome in lymphocyte DNA from about one-half of the IDDM patients tested. Patients 1, 5, 15, 18, 19, 35, 37, 39, 40, 42, 43, 44, and 47 were counted as positive. (Reprinted from ref. 11 with permission.)

CYTOMEGALOVIRUS

One case report describes a child with congenital cytomegalovirus (CMV) infection who developed diabetes mellitus at the age of 13 months.[91] He presented with a two-week history of polydipsia, vomiting, and weight loss and was found to be severely dehydrated and ketoacidotic, with a blood glucose of 50 mmol/l (900 mg/dl). In other reports, characteristic inclusion bodies have been found in the beta-cells of infants and children who died with disseminated CMV infections.[12] Insulitis and CMV-like particles have been observed in the pancreas of a rodent (*Octodon degu*) that manifests spontaneous diabetes.[92]

Human CMV infection is ubiquitous and largely subclinical. In many persistent viral infections, the initial viral infection occurs before birth or very early in life, although the disease may not appear until later. The infection can be passed through the sperm or ovum if a virus has integrated its genome into the host DNA. Viral infections can also be transmitted transplacentally, perinatally, or postnatally through close contact or breast milk. The immaturity of the immune systems of infants also favors establishment of persistent viral infections.

Nucleic acid hybridization is an especially useful technique for studies of persistant viral infections, since it can detect even inactive or incomplete viral genomes in a cell. Since persistent viral infections are often local, in situ hybridization can be used to locate the cells that contain the viral genome. Our finding, with both dot (Fig. 10.24) and in situ (Fig. 10.25, see color insert) hybridization techniques, that about 15% of newly diagnosed type I diabetes patients had a human, CMV-specific viral genome in their lymphocytes and islet-cell autoantibodies in their sera suggests that autoimmune type I diabetes is sometimes associated with persistent CMV infection.[11]

What might be the link between persistent CMV infection and autoimmune type I diabetes? Autoimmune disease could result from an immune response to viral antigens in the host cells or to host-cell–specific antigens that are exposed as a result of infection. If the CMV infection persists in the beta-cells and either expresses viral antigens or induces an aberrant expression of class II MHC antigen on the cells, under certain circumstances, such as a particular genetic background or environment (e.g., drug, diet, toxins, infections), it might trigger a beta-cell–specific autoimmune disease such as type I diabetes (Fig. 10.26).

Mumps Virus

It has been postulated that an earlier infection with mumps virus may also trigger at least some cases of IDDM. Recent reports suggest that some children with mumps may develop islet-cell antibody.[93] Gamble showed that mumps infection apparently preceded the development of diabetes in some newly diagnosed diabetic children.[94] The results of other epidemiologic studies, however, do not support the hypothesis that mumps infection is associated with IDDM onset.[95]

Persistent Infection of Pancreatic Beta-Cells and Mild Hyperglycemia

Lymphocytic Choriomeningitis Virus

Oldstone and colleagues reported that lymphocytic choriomeningitis (LCM) virus persistently infects murine pancreatic islet cells.[96] In their studies, viral nucleoprotein was found to predominate in pancreatic beta-cells, by the use of the double-labeled immunofluorescent antibody technique. Electron microscopy confirmed these findings by showing virions budding from the beta-cells. Persistent infection was associated with such chemical evidence of diabetes as hyperglycemia and abnormal glucose tolerance. However the virus-infected islet cells were normal, anatomically and cytomorphologically. Neither beta-cell destruction nor lymphocytic infiltration was routinely observed. Thus the chemical and morphologic picture is similar to that observed in the early stages of adult-onset diabetes mellitus.

One possible mechanism by which this virus might cause diabetes is through the establishment of an infection that turns off the "luxury functions" of insulin-producing beta-cells.[96,97] Another possible mechanism is that persistent infection might gradually reduce the number of functioning beta-cells, since the regenerative capacity of beta-cells is thought to be poor.[97] The precise mechanism by

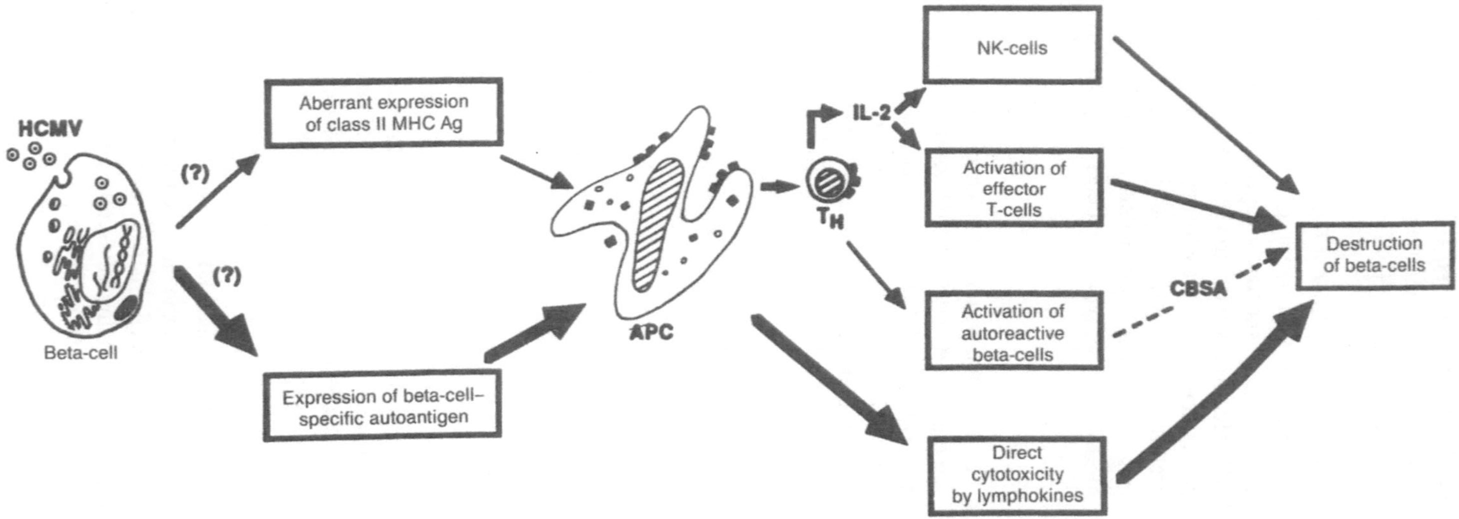

FIGURE 10.26. Possible pathways for the initiation of beta-cell destruction in CMV-associated IDDM patients. (**A**) Persistent CMV infection in beta-cells may result in the expression of viral antigens, autoantigens, and/or Class II MHC antigen (aberrant) on the cells under certain conditions, such as a particular genetic background, environment (e.g., drug, diet, toxins, infections), and physiologic changes (e.g., puberty, stress). (Reprinted from ref. 11 with permission.) (**B**) The beta-cell–specific autoantigen (?), Class II MHC antigen (?), and/or viral antigen may attract macrophages and activate them. (**C**) Macrophages secrete free radicals and/or interleukin-1, which may have a cytotoxic effect on beta-cells as an initial event. During this process, some macrophages may penetrate the islets. (**D**) Simultaneously the macrophages may present the processed antigen to helper T-cells. The helper T-cells may activate cytotoxic T-cells and/or stimulate B-cell response to produce antibodies against islet cells. (**E**) A fraction of the helper T-cells and cytotoxic T-cells may have an additional cytotoxic effect on the residual beta-cells, for the clinical expression of a beta-cell–specific autoimmune disease.

which LCM virus induces diabetes is not known. Nevertheless this new finding is very interesting and potentially important for studies on diabetes in humans and animals.

Conclusion

This chapter discusses two possible pathogenic mechanisms for virus-induced diabetes. In animal models the D-variant of EMC virus and Mengo 2T virus can induce diabetes by infecting and destroying pancreatic beta-cells in the absence of an autoimmune response, whereas retrovirus and rubella virus appear to be associated with autoimmune IDDM.

There appears to be no correlation in man between the presence of anti-Coxsackie B viral antibodies and islet-cell autoantibodies in newly diagnosed IDDM patients, whereas rubella and CMV infections appear to correlate with the presence of autoantibodies and IDDM. The precise mechanisms by which these viruses contribute to the pathogenesis of IDDM remain to be defined.

Acknowledgment. This work was supported by the Medical Research Council of Canada (MA9584), the Canadian Diabetes Association, and the Alberta Children's Hospital Foundation. The author is a Heritage Medical Scientist Award recipient. The secretarial help of Daphne Kwan is gratefully acknowledged.

References

1. Yoon JW, Kim CJ, Pak CY, McArthur RG. Environmental factors and IDDM. *Clin Invest Med* 1987;1457–469.
2. Leiter EH, Wilson GL. Viral interactions with pancreatic beta cells, in Lefebvre PJ, Pipeleers DG (eds): *The Pathology of the Endocrine Pancreas in Diabetes*. New York: Springer-Verlag, 1988, pp 85–105.
3. Yoon JW, Notkins AL. Virus-induced diabetes mellitus in mice. *Metabolism* 1983; 32(Suppl):37–40.
4. Notkins AL, Yoon JW, Onodera T, Toniolo T, Janson AB. Virus-induced diabetes mellitus. *Perspect Virol* 1981;11:141–62.
5. Craighead JE. The role of viruses in the pathogenesis of pancreatic diseases and diabetes mellitus. *Prog Med Virol* 1975;19:161–214.
6. Yoon JW, Austin M, Onodera T, Notkins AL. Virus-induced diabetes mellitus. Isolation of a virus from the pancreas of a child with diabetic ketoacidosis. *N Engl J Med* 1979;300:1173–1179.
7. Champsaur H, Bottazzo G, Bertrams J, Assan R, Bach C. Virologic, immunologic and genetic factors in insulin-dependent diabetes mellitus. *J Pediatr* 1982;100: 15–20.
8. King ML, Shaikh A, Bidwell B, Voller A, Banatvala JE. Coxsackie-B-virus specific IgM responses in children with insulin-dependent diabetes mellitus. *Lancet* 1983;i: 1397–1399.

9. Banatvala JE, Schernthaner G, Schober E, et al. Coxsackie B, mumps, rubella and cytomegalovirus specific IgM responses in patients with juvenile-onset insulin-dependent diabetes mellitus in Britain, Austria and Australia. *Lancet* 1985;i:1409–1412.
10. Rubinstein P, Walker ME, Fedun B, Witt ME, Cooper LZ, Ginsberg-Fellner F. The HLA system in congenital rubella patients with and without diabetes. *Diabetes* 1982; 31:1088–1091.
11. Pak CY, Eun HM, McArthur RG, Yoon JW. Association of cytomegalovirus infection with autoimmune type I diabetes. *Lancet* 1988;ii:1–4.
12. Jenson AB, Rosenberg HS, Notkins AL. Pancreatic islet cell damage in children with fatal viral infections. *Lancet* 1980;2:354–358.
13. Craighead JE, McLane MF. Diabetes mellitus: Induction in mice by encephalomyocarditis virus. *Science* 1968;162:913–915.
14. Yoon JW, McClintock PR, Onodera T, Notkins AL. Virus-induced diabetes mellitus. Inhibition by a non-diabetogenic variant of encephalomyocarditis virus. *J Exp Med* 1980;152:878–892.
15. Suenaga K, Yoon JW. Association of beta cell specific expression of endogenous retrovirus with the development of insulitis and diabetes in NOD mice. *Diabetes* 1988;37:1722–1726.
16. Yoon JW, Morishima T, McClintock PR, Austin M, Notkins AL. Virus-induced diabetes mellitus: Mengovirus infects pancreatic beta cells in strains of mice resistant to the diabetogenic effect of encephalomyocarditis mice. *J Virol* 1984;50:684–690.
17. Yoon JW, Onodera T, Notkins AL. Virus-induced diabetes mellitus: Beta cell damage and insulin-dependent hyperglycemia in mice infected with Coxsackie virus B4. *J Exp Med* 1978;148:1068–1080.
18. Yoon JW, McClintock PR, Bachurski LJ, Longstretch JD, Notkins AL. Virus-induced diabetes. No evidence for immune mechanisms in the destruction of beta cells by encephalomyocarditis virus. *Diabetes* 1985;34:922–925.
19. Onodera T, Yoon JW, Brown KS, Notkins AL. Evidence for a single locus controlling susceptibility to virus-induced diabetes mellitus. *Nature* 1978;274:693–695.
20. Yoon JW, Notkins AL. Virus-induced diabetes mellitus VI. Genetically determined host differences in the replication of encephalomyocarditis virus in pancreatic beta cells. *J Exp Med* 1976;143:170–185.
21. Chairez R, Yoon JW, Notkins AL. Virus-induced diabetes mellitus. X. Attachment of encephalomyocarditis virus and permissiveness of cultured pancreatic beta cells to infection. *Virology* 1978;85:606–611.
22. Yoon JW, Lesniak MA, Fussganger R, Notkins AL. Genetic differences in the susceptibility of pancreatic B cells to virus-induced diabetes mellitus. *Nature* 1976; 265:178–180.
23. Jansen FK, Müterfering H, Schmidt WAK. Virus-induced diabetes and the immune system: Suggestion that appearance of diabetes depends on immune reactions. *Diabetologia* 1977;13:545–549.
24. Jansen FK, Turnoyssen O, Münterfering H. Virus-induced diabetes and the immune system. I. Evidence for an immunopathogenesis of the acute phase of diabetes. *Biomedicine* 1979;31:1–2.
25. Buschard K, Rygaard J, Lund E. The inability of a diabetogenic virus to induce diabetes mellitus in athymic (nude) mice. *Acta Pathol Microbiol Scand* 1976;C84: 299–303.

26. Buschard K, Hastrup N, Rygaard J. Virus-induced diabetes mellitus in mice and the thymus-dependent immune system. *Diabetologia* 1983;24:42–46.
27. Vialettes B, Baume D, Charpin C, De Maeyer-Guignard J, Vague P. Assessment of viral and immune factors in EMC virus-induced diabetes: Effects of cyclosporin A and interferon. *J Clin Lab Immunol* 1983;10:35–40.
28. Notkins AL, Yoon JW. Virus-induced diabetes in mice prevented by a live attenuated vaccine. *N Engl J Med* 1982;306:486.
29. Babu PG, Huber SA, Craighead JE. Contrasting features of T lymphocyte mediated diabetes in encephalomyocarditis virus infected Balb/c By and Balb/c Cum mice. *Am J Pathol* 1986;124:193–198.
30. Haynes MK, Huber SA, Craighead JE. Helper-induced T-lymphocyte mediated dia betes in EMC-infected Balb/c ByJ mice. *Diabetes* 1987;36:877–881.
31. Ross ME, Onodera T, Brown KS, Notkins AL. Virus-induced diabetes mellitus. IV. Genetic and environmental factors influencing the development of diabetes after infection with the M variant of encephalomyocarditis virus. *Diabetes* 1976;25: 190–197.
32. Yoon JW, Onodera T, Notkins AL. Virus induced diabetes mellitus. Passage of encephalomyocarditis virus and severity of diabetes in susceptible and resistant strains of mice. *J Gen Virol* 1977;37:225–232.
33. Yoon JW. Viruses and the pathogenesis of insulin-dependent diabetes mellitus, in *Genetic Environmental Interaction in Diabetes Mellitus* (ed. John S. Melish, J. Hanna and S. Baba) Excerpta Medica, Amsterdam-Oxford-Princeton International Congress Series 1982;549:227–34.
34. Yoon JW, Cha CY, Jordan G. Role of interferon in virus-induced diabetes in mice. *J Infect Dis* 1983;147:155–159.
35. Ray U, Aulakh G, McClintock PR, Yoon JW, Notkins AL. Virus-induced diabetes mellitus: Difference in the RNA fingerprints of diabetogenic and nondiabetogenic variants of encephalomyocarditis virus. *J Gen Virol* 1983;64:947–950.
36. Yoon JW, Wong AKC, Bae YS, Eun HM. An apparent deletion of an oligonucleotide detected by RNA fingerprint in the nondiabetogenic B variant of encephalomyocarditis virus is due to a point mutation. *J Virol* 1988;62:637–640.
37. Jordan GW, Cohen SH, Dandekar S, Vande-Brink KM. The genomic RNA of diabetogenic encephalomyocarditis virus: Characterization and molecular cloning. *Virology* 1987;159:120–125.
38. Yoon JW, Ko W, Bae YS, Pak CY, Amano K, Eun HM, Kim MK. Identification of antigenic differences between the diabetogenic and nondiabetogenic variants of encephalomyocarditis virus by monoclonal antibodies. *J Gen Virol* 1988;69:1085–1090.
39. Eun HM, Bae YS, Yoon JW. Amino acid differences in capsid protein, VP1, between diabetogenic and nondiabetogenic variants of encephalomyocarditis virus. *Virology* 1988;163:369–373.
40. Bae YS, Eun HM, Yoon JW. Molecular identification of diabetogenic viral gene. *Diabetes* 1989;38:316–320.
41. Yoon JW, Rodrigues MM, Currier C, Notkins AL. Long-term complications of virus-induced diabetes in mice. *Nature* 1982;296:567–569.
42. Rodrigues M, Currier C, Yoon JW. Electron microscopy of renal and ocular changes in virus-induced diabetes mellitus in mice. *Diabetologia* 1983;24:293–299.
43. Yoon JW, Reddi AH. Decreased bone formation and mineralization in virus-induced diabetes mellitus. *Am J Physiol* 1984;246:C177–C179.

44. Morishima T, McClintock PR, Billups LC, Notkins AL. Expression and modulation of virus receptors on lymphoid and myeloid cells: Relationship to infectivity. *Virology* 1982;116:605–618.
45. Toniolo A, Onodera T, Jordan G, Yoon JW, Notkins AL. Virus-induced diabetes mellitus: Glucose abnormalities produced in mice by the six members of the Coxsackie B virus group. *Diabetes* 1982;31:496–499.
46. Ross ME, Hayashi K, Notkins AL. Virus-induced pancreatic disease: Alterations in concentration of glucose and amylase in blood. *J Infect Dis* 1974;129:669–676.
47. Toniolo A, Onodera T, Yoon JW, Notkins AL. Induction of diabetes by cumulative environmental insults from viruses and chemicals. *Nature* 1980;288:383–385.
48. Wegner U, Kewitsch A, Madauss M, Dohner L, Zuhlke H. Hyperglycemia in BALB/c mice after pretreatment with one subdiabetogenic dose of streptozotocin and subsequent infection with Coxsackie B4 strain. *Biomed Biochem Acta* 1985;44: 21–27.
49. Yoon JW, London WT, Curfman BL, Brown RL, Notkins AL. Coxsackie virus B4 produces transient diabetes in nonhuman primates. *Diabetes* 1986;35:712–716.
50. Prabhakar BS, Haspel MV, McClintock PR, Notkins AL. High frequency of antigenic variants among naturally occurring human Coxsackie B4 virus isolates identified by monoclonal antibodies. *Nature* 1982;300:374–376.
51. Prabhakar BS, Menegus MA, Notkins AL. Detection of conserved and nonconserved epitopes on Coxsackie virus B4: Frequency of antigenic change. *Virology* 1985;146:302–306.
52. Gamble DR, Kinsley ML, Fitzgerald MG, Bolton R, Taylor KW. Viral antibodies in diabetes mellitus. *Br Med J* 1969;3:627–630.
53. Hierholzer JC, Farris WA. Follow-up of children infected in a Coxsackie B3 and B4 outbreak: No evidence of diabetes. *J Infect Dis* 1974;129:741–746.
54. Dippe SE, Bennet PH, Miller M. Coxsackie B virus and diabetes. *Br Med J* 1974; 2:443–444.
55. Friman G, Fohlman J, Frisk G, Diderholm H, Ewald U, Kobbah M, Tuvemo T. An incidence peak of juvenile diabetes. Relation to Coxsackie B virus immune response. *Acta Peadiatr Scand* 1985;320(Suppl):14–19.
56. Prince G, Jenson AB, Billups L, Notkins AL. Infection of human pancreatic beta cell cultures with mumps virus. *Nature* 1978;271:158–161.
57. Yoon JW, Onodera T, Notkins AL. Virus-induced diabetes mellitus. XI. Replication of Coxsackie virus B3 in human pancreatic beta cell cultures. *Diabetes* 1978;27: 778–782.
58. Yoon JW, Selvaggio S, Onodera T, Wheeler J, Jenson AB. Infection of cultured human pancreatic B cell with reovirus type 3. *Diabetologia* 1981;20:462–468.
59. McLaren LC, Holland JJ, Syverton JT. The mammalian cell–virus relationship. *J Exp Med* 1959;109:475–483.
60. Gladisch R, Hofmann W, Waldherr R. Myocarditis and insulitis in Coxsackie virus infection. *Z Kardiol* 1976;65:873–881.
61. Eisenbarth GS. Type I diabetes mellitus. A chronic autoimmune disease. *N Engl J Med* 1986;314:1360–1368.
62. Foulis AK, Farquharsaon MA, Meager A. Immunoreactive α-interferon in insulin-secreting B cells in type I diabetes mellitus. *Lancet* 1987;ii:1423–1427.
63. Rayfield EJ, Kelly KJ, Yoon JW. Rubella virus-induced diabetes in hamsters. *Diabetes* 1986;35:1278–1281.

64. Onodera T, Toniolo A, Ray UR, Jenson AB, Knazek RA, Notkins AL. Virus-induced diabetes mellitus. *J Exp Med* 1981;153:1457–1465.
65. Onodera T, Ray UR, Melez KA, Suzuki H, Toniolo A, Notkins AL. Virus-induced diabetes mellitus. Autoimmunity and polyendocrine disease prevented by immunosuppression. *Nature* 1982;297:66–69.
66. Schlesinger MJ, Kaarianen L. Translation and processing of alpha virus proteins, in, Schlesinger RW (ed): *Togavirus*. New York: Academic Press, 1980, pp 371–389.
67. Allison AC. Mechanisms by which autoimmunity can be produced, in Mendel NE, Cheers C, Hoskins CS, McKenzie IFC, Nossal H (eds): *Progress in Immunology, vol. III*. Amsterdam: Elsevier, 1977, p 512.
68. Fauci AC. Immunoregulation of autoimmunity. *J Allergy Clin Immunol* 1980;66: 5–17.
69. Reinherz EL, Schlossman SF. Regulation of the immune response inducer and suppressor T-lymphocytes subsets in human beings. *N Engl J Med* 1980;303:370–373.
70. Kataoka S, Satoh J, Fujiya H, Toyota T, Suzuki R, Itoh K, Kumagai K. Immunologic aspects of the nonobese diabetes (NOD) mouse: Abnormalities of cellular immunity. *Diabetes* 1983;32:247–253.
71. Lee KU, Amano K, Yoon JW. Evidence for initial involvement of macrophage in development of insulitis in NOD mice. *Diabetes* 1988;37:989–991.
72. Harada M, Makino S. Promotion of spontaneous diabetes in nonobese diabetes-prone mice by cyclophosphamide. *Diabetologia* 1984;27:604–606.
73. Harada M, Sueishi T, Misaki R, Makino S. Possible role of Lyt 2+ T cells in cyclophosphamide-enhanced diabetes in NOD mice. Presented at the International Research Symposium "The Immunology of Diabetes," Oct 27–30, Woods Hole, MA, 1987.
74. Ikeda H, Pincus T, Yoshiki T, Strand M, August JT, Boyse EA, Mellors RC. Biological expression of antigenic determinants of murine leukemia virus proteins gp 69/71 and p 30. *J Virol* 1974;14:1274–1280.
75. Del Villano BC, Nave B, Croker BP. The oncornavirus glycoprotein gp 69/71: A constituent of the surface of normal and malignant thymocytes. *J Exp Med* 1975;141: 172–187.
76. Kennel SJ, Feldman JD. Distribution of viral glycoprotein gp 69/71 on cell surfaces of producer and nonproducer cells. *Cancer Res* 1976;36:200–208.
77. Lerner RA, Wilson CB, Del Villano BC, McCohahey PJ, Dixon FJ. Endogenous oncornaviral gene expression in adult and fetal mice: Quantitative, histologic, and physiologic studies of the major viral glycoprotein, gp 70. *J Exp Med* 1976;143: 151–166.
78. Unanue ER, Allen PM. The basis for the immunoregulatory role of macrophages and other accessory cells. *Science* 1987;236:551–557.
79. Toews GB, Bergstresser PR, Streilein JW. Epidermal Langerhans cell density determines whether contact hypersensitivity or unresponsiveness follows skin painting with DNFB. *J Immunol* 1980;124:445–453.
80. Onodera T, Jenson AB, Yoon JW, Notkins AL. Virus-induced diabetes mellitus. Reovirus infection of pancreatic beta cells in mice. 1978;301:529–531.
81. Menser MA, Forrest JM, Bransby RD. Rubella infection and diabetes mellitus. *Lancet* 1978;i:57–60.
82. Schopfer K, Matter L, Flueler U, Werder E, Diabetes mellitus, endocrine autoantibodies and prenatal rubella infection. *Lancet* 1982;ii:159.

83. Underwood LE, Van Wyk JJ. Hormones in normal and aberrant growth, in Williams RH (ed): *Textbook of Endocrinology* (6th ed). Philadelphia: Saunders, 1981, pp 1149–1184.
84. Preece MA, Kearney PJ, Marshall WC. Growth-hormone deficiency in congenital rubella. *Lancet* 1977;ii:842–844.
85. Smithsells RW, Sheppard S, Marshall WC, Peckham C. Congenital rubella and diabetes mellitus. *Lancet* 1978;i:439.
86. DePrins F, VanAssche FA, Desmyter J. Congenital rubella and diabetes mellitus. *Lancet* 1978;i:439–440.
87. Patterson K, Chandra RS, Jenson AB. Congenital rubella, insulitis and diabetes mellitus in an infant. *Lancet* 1981;i:1048–1049.
88. Ginsberg-Fellner F, Witt ME, Yagihashi S, et al. Congenital rubella syndrome as a model for type I (insulin-dependent) diabetes mellitus: Increased prevalence of islet cell surface antibodies. *Diabetologia* 1984;27:87–89.
89. Ginsberg-Fellner F, Witt ME, Fedun B, et al. Diabetes mellitus and autoimmunity in patients with congenital rubella syndrome. *Rev Infect Dis* 1985;7(Suppl 1):S170–S175.
90. Ginsberg-Fellner F, Fedun B, Cooper Z, et al. Interrelationships of congenital rubella and type I insulin dependent diabetes mellitus, in Jaworski JA, Molnar GD, Rajotte RV, Singh B (eds): *The Immunology of Diabetes Mellitus*. Amsterdam: Elsevier, 1987, pp 279–286.
91. Ward KP, Galloway WH, Auchterlonie IA. Congenital cytomegalovirus infection and diabetes. *Lancet* 1979;i:497.
92. Fox GJ, Murphy JC. Cytomegalic virus-associated insulitis in diabetic *Octodon degus*. *Vet Pathol* 1979;16:625–628.
93. Helmke K, Otten A, Willems W. Islet cell antibodies in children with mumps infection. *Lancet* 1980;ii:211–212.
94. Gamble DR. Relation of antecedent illness to development of diabetes in children. *Br Med J* 1980;2:99–101.
95. Ratzman KP, Strese J, Witt S, Berling H, Keilacker H, Michaelis D. Mumps infection and insulin-dependent diabetes mellitus (IDDM). *Diabetes Care* 1984;7:170–173.
96. Oldstone MBA, Southern P, Rodriguez M, Lampert P. Virus persists in B cells of islets of Langerhans and is associated with chemical manifestations of diabetes. *Science* 1984;224:1440–1443.
97. Notkins AL, Yoon JW. Virus-induced diabetes, in Notkins AL, Oldstone MBA (eds): *Concepts in Viral Pathogenesis*. New York: Springer-Verlag, 1984; pp 241–247.
98. Gerling I, Nejman C, Chatterjee NK. Effect of Coxsackie B4 virus infection in mice on expression of 64,000-Mr autoantigen and glucose sensitivity of islets before development of hyperglycemia. *Diabetes* 1988;37:1419–1425.
99. Stefan Y, Malaisse-Lague F, Yoon JW, Notkins AL, Orci L. Virus-induced diabetes in mice: A quantitative evaluation of islet cell population by immunofluorescence technique. *Diabetologia* 1978;15:395–401.
100. Like AA, Rossini AA. Streptozotocin-induced pancreatic insulitis: New model of diabetes mellitus. *Science* 1976;193:415–417.
101. Like AA, Chick WL. Studies in the diabetic mutant mouse. II. Electron microscopy of pancreatic islets. *Diabetologia* 1970;6:216–242.
102. Leiter EH, Bedigian HG. Intracisternal type A particles in genetically diabetic mice: Identification in pancreas and induction in cultured B-cells. *Diabetologia* 1979;17:175–185.

103. Leiter EH, Kuff EL. Intracisternal type A particles in murine pancreatic B cells: Immunocytochemical demonstration of increased antigen (p73) in genetically diabetic mice. *Am J Pathol* 1984;114:46–55.
104. Leiter EH, Fewell JW, Kuff EL. Glucose induces intracisternal type A retroviral gene transcription and translation in pancreatic B-cells. *J Exp Med* 1986;163: 87–100.
105. Suenaga K, Yoon JW. Association of beta cell specific expression of endogenous retrovirus with the development of insulitis and diabetes NOD mouse. *Diabetes* 1988;37:1722–1726.

11
Immune Intervention in Diabetes: State of the Art and Future Directions

J. DUPRÉ AND C.R. STILLER

Introduction

The suggestion that insulin-dependent diabetes mellitus (IDDM) in humans results from the destruction of pancreatic beta-cells by an autoimmune process depends on three main lines of evidence. First the association of IDDM with markers of the major histocompatibility complex (MHC) indicates not only that genetic factors are important in determining susceptibility to the disease but also that this may be related to immune system functions.[1,2] Second the disease is associated with evidence of antibody-mediated and cell-mediated immunologic reactions during progression to insulin dependence.[3-5] Third the inflammatory process in the islets of Langerhans in IDDM is consistent with the autoimmune hypothesis and perhaps is best exemplified by pancreas transplantation in discordant identical twins, during which the histology of the pancreatitis that occurs in the recipient in the absence of immunosuppression has been studied.[6] Investigations concerning each of these lines of evidence continue, as discussed elsewhere in this volume.

The autoimmune hypothesis was greatly strengthened when it was recognized that IDDM can occur in animals as the apparent result of an autoimmune attack on the pancreatic beta-cells, in which the features closely paralleled aspects of the human disease described above.[7] Furthermore this process in animals can be modified as a result of immunomodulatory intervention.[8-10] In view of this, and of the continuing use of immunomosuppressive drugs to control allograft rejection, several studies of the effects of immunotherapy early in the course of clinically overt IDDM in man have been undertaken. The timing of such interventions relative to the course of the disease has been vigorously debated and must be reviewed in light of the results. In this account, the findings in cyclosporin trials, which to date outnumber studies with other immunosuppressive agents, are emphasized. The results of experimental immunotherapy in IDDM will be summarized before reviewing the conclusions that may be drawn and considering future directions.

Experimental Immunotherapy in IDDM of Recent Onset

Preliminary Open Studies: Rationale and Background

A total of more than 1500 patients with IDDM have been given a variety of immunomodulatory agents and have undergone a variety of procedures. The rationale for such interventions was that suppression of the postulated autoimmune attack on beta-cells might increase insulin secretion to produce clinical remission of the disease. The potential for such beneficial effects is supported by the results of studies of endogenous secretion of insulin in IDDM, which show that residual beta-cell function, assessed by immunoreactive blood C-peptide (CP) levels, commonly increases from very low values at the time of diagnosis to values in the low-normal range, with a mean optimum at three or four months after diagnosis.[11,12] During this period, exogenous insulin requirements are at their lowest, as are the levels of glycated hemoglobin in the blood, in groups of patients who are undergoing conventional insulin therapy.[13] Clinical studies over a period of time suggest that 3 to 10% of conventionally treated patients experience a clinical remission in which metabolic control can be maintained without insulin therapy one year after diagnosis.[13] These considerations led us to undertake several open (uncontrolled) trials in the expectation that markedly favorable or unfavorable effects on the course of the disease would be observed.

Definition of Remission in IDDM

The term *remission* has long been used in the context of IDDM to describe the phase of apparent amelioration of the disease during the first few months after diagnosis. Remission in the strict sense of the word, in which all the features of a disease by which it can be recognized are resolved, is virtually nonexistent in IDDM. The working definitions of clinical remission in studies of immunomodulatory intervention in IDDM in humans have been based variously on insulin dose and/or insulin secretory status in terms of blood CP levels in response to various stimuli, qualified or not qualified by control of hyperglycemia. The conditions under which reduction of insulin dose or withdrawal of insulin therapy is used to identify clinical remission must be defined. As discussed below, strict normoglycemia usually has not been required of noninsulin-receiving (NIR) patients in these studies. Investigators have taken the view that, in the absence of insulin therapy, maintenance of blood glucose levels below those diagnostic of diabetes mellitus represents a reasonable operating definition of a NIR remission. Use of the terms *noninsulin-requiring* or *complete* to describe such remissions probably should be avoided.

Open Studies of Immunomodulatory Therapies in IDDM

Results of Open Studies with Agents Other Than Cyclosporin

The first attempt to induce clinical remission made use of a short-term treatment that combined the administration of azathioprine, prednisone, and antilymphocytic globulin with plasmaphoresis in patients with early clinical diabetes.[14] These studies were not pursued because this treatment did not consistently yield clinical remissions of the NIR type. An earlier study in diabetic children employed low-dose prednisone through one year after diagnosis. No clinical benefit was apparent, but urinary CP excretion was enhanced for one or two years after entry into the study, when compared with the clinical records. This work was extended to a study in which patients were randomly allocated conventional insulin therapy with prednisone or conventional insulin therapy alone; an apparent effect on urinary CP output was maintained for three years of study, though only in children over the age of 10 years.[15] A similar uncontrolled study on the effect of plasmaphoresis on beta-cell function found that plasma levels of CP were higher for up to four years after diagnosis, again as compared with earlier controls.[16]

In a study of the possible effects of immunosuppression by azathioprine alone, every other patient presenting with newly diagnosed IDDM was treated with this agent.[17] Of 13 patients who received azathioprine, the drug was discontinued in 5 patients within six months of entry either because of an adverse effect or at the patient's request. Nevertheless seven azathioprine-treated patients achieved NIR remission, six of which were maintained through a 12-month period. Only one patient in the control group was in NIR remission at one year. Moreover basal and glucagon-stimulated CP levels in the plasma were significantly increased in the azathioprine-treated patients. In a further study, continued treatment with azathioprine was combined with short-term use of high doses of glucocorticoids at entry, and the course of the disease was compared with that of concurrent control subjects.[18] In this study, the plasma CP levels through one year were higher in the experimental group. An index of response, depending on the hemoglobin A1C level, the insulin dosage, and the plasma CP level, favored the experimental treatment.

Open Studies with Cyclosporin

Following studies in the BB rat in which it was shown that cyclosporin can prevent the development of diabetes in this animal model of IDDM,[19] and in light of experience with this agent in organ transplantation, cyclosporin was employed in pilot studies in patients with IDDM of recent onset. The first of these studies examined the effects of cyclosporin, given in an initial daily dose of 10 mg/kg, adjusted to maintain serum trough levels in the 150 to 200 ng/ml (polyclonal radioimmunoassay, RIA) range in patients with clinically diagnosed IDDM of up to 11 months duration.[20] An apparent induction of remission to a NIR state occurred, with control of glycemia maintained in the target range (<7.8 mmol/l

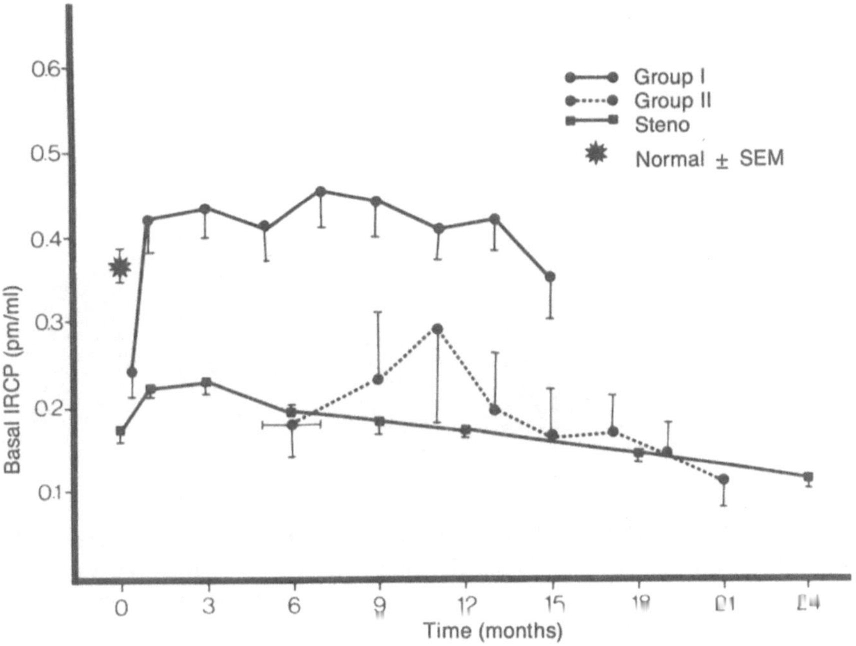

FIGURE 11.1. Mean levels of plasma C-peptide (CP) in a group of conventionally treated patients with IDDM (Steno Hospital, Copenhagen) after an overnight fast and of two groups of patients with recent-onset IDDM who were treated with cyclosporin the first year after diagnosis. Group I patients started the experimental treatment within two months of commencing insulin therapy. Group II patients started cyclosporin therapy later; their mean duration of diabetes was approximately six months at entry. Cyclosporin was discontinued at 12 months in all Group II patients; the drug had no apparent effect in this group. In Group I patients, the mean fasting plasma CP level rose to normal and remained there during administration of the drug for one year or more. (Drawn from data in ref. 34.)

ac meals) in an unexpectedly high proportion of patients. Also glucagon-stimulated CP levels in the plasma increased, and remained in the normal range through 12 months of treatment (Fig. 11.1). Thus the remission rate, of approximately 40% at one year, in patients who entered within eight weeks of starting insulin therapy (Fig. 11.2) was strikingly higher than historically observed in patients undergoing conventional insulin therapy. Moreover the maintenance of stable CP plasma levels through the second half of the year of study in patients with recent-onset diabetes (less than eight weeks insulin therapy) was apparently unprecedented. By contrast patients who entered the study with overt diabetes of more than two months duration showed no obvious clinical response and no significant change in CP plasma levels.

Continued experience in the Canadian open study, in which the great majority of the 81 patients were nine years old or older at the time of clinical diagnosis,

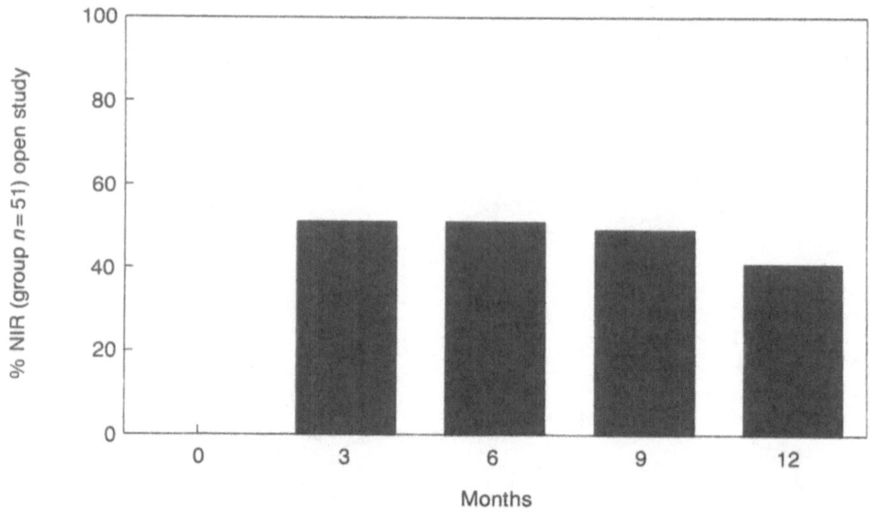

FIGURE 11.2. Percent of NIR patients in whom target control of glycemia (fasting blood glucose less than 7.8 mmol/l, 140 mg/dl) was maintained for at least two weeks. All patients were over nine years of age and had received insulin for less than eight weeks at entry. From ref. 21 with permission.

was consistent with the initial report and led to a number of inferences regarding the characteristics of patients most likely to respond to cyclosporin therapy.[21] Within the group who had been treated with insulin for no more than eight weeks at entry, the patients who had had 3 weeks, or less, of insulin therapy at entry had a significantly higher chance of developing NIR remission. Moreover in these studies, the chance of remission was greater for patients 16 years old or older at entry, compared with younger patients. Even lower rates of clinical remission were observed in an additional group of younger patients, ranging in age from 22 months to 8 years at entry.[22] The clinical outcome did not correlate with the basal or stimulated CP plasma level at entry or with the presence or absence of islet-cell antibodies in the blood. The Canadian open study also yielded the important information that interruption of cyclosporin treatment from 12 to more than 30 months after entry was likely to be followed by clinical relapse within a few weeks. In an open study with cyclosporin in France, similar results in terms of clinical remissions defined by the insulin treatment status were obtained.[23] The anticipated clinical and biochemical adverse effects of cyclosporin administration observed in these open studies did not differ from those documented in other experiences with the drug; they appeared to be fully reversible upon interruption of therapy.

More recently the results of another open study in which cyclosporin was given to children with IDDM of recent onset have been reported.[24] In this French study, the whole blood trough level of cyclosporin was maintained at 150 to 350 ng/ml (polyclonal RIA), a range not associated with a detectable adverse effect on kidney function. The majority of the patients were entered within two months of onset of

TABLE 11.1. Open studies of cyclosporin treatment in children. Frequency of early remission of diabetes, according to selected characteristics at entry.

	All patients (N = 40)	Patients not needing insulin at 4 mos (N = 27)	Patients needing insulin at 4 mos (N = 13)[a]
Duration of polyuria (days)			
< 30	25	20 (80)	5 (20)
30–90	15	7 (47)	8 (53)
Weight loss (% body weight)			
< 5	18	18 (100)	0
5–10	15	8 (53)	7 (47)
> 10	7	1 (14)	6 (86)
Ketoacidosis			
Without	29	24 (83)	5 (17)
With	11	3 (27)	8 (73)
Hemoglobin A_{1c} (%)			
< 10	9	8 (89)	1 (11)
10–13	26	19 (73)	7 (27)
> 13	5	0	5 (100)

[a]Numbers in parentheses indicate percent of patients in remission.
From Bougneres et al.[24] Reprinted with permission from *The New England Journal of Medicine*.

symptoms of diabetes and within one to two weeks of insulin therapy. Noninsulin-receiving remission was recognized when target control of glycemia was maintained without insulin therapy. Among 42 children 7 to 15 years old, 67% were in NIR remission within 4 months, and the remission rate, according to this definition, was 45% at 12 months. The mean duration of symptoms prior to diagnosis was significantly shorter in the patients who entered remission (28.8 versus 48.0 days). The degree of weight loss was lower, the initial hemoglobin A1C level was lower, and the frequency of ketoacidosis was less in patients who entered NIR remission than in patients who did not (Table 11.1). These findings suggested that high rates of clinical remission can be induced even in young children, provided the intervention is made very early in the course of overt diabetes. In this study, the plasma CP response to glucagon stimulation at the time of entry was also significantly higher in those who entered remission later. No differences were found between remitting and nonremitting patients with respect to age, sex, HLA phenotype, presence or absence of antibodies to insulin or islet-cell antigens, or doses and trough levels of cyclosporin.

Randomized Control Trials of the Effects of Cyclosporin in IDDM

The apparent ability in the Canadian and French open studies, of cyclosporin to induce clinical remissions, defined by the insulin therapy status, and the apparent enhancement and preservation of beta-cell function through the treatment period in the Canadian study, indicated the need for randomized control trials with cyclosporin. The objective was to assess these effects in light of their clinical potential and their importance in understanding IDDM. The first randomized, placebo-controlled, double-blind study of an experimental therapy carried out in IDDM[25] demonstrated the need for such trials. This study, which examined the effects of treatment with alpha interferon, was performed in Finland on patients with clinically diagnosed IDDM of recent onset. Alpha interferon had no effect on the course of the disease; at six months, however, 29% of the patients were in clinical remission, defined as the condition in which target glycemic control was maintained without insulin therapy. It was also observed that CP levels in the blood were substantially increased. The clinical remission rate declined through the second half of the year, however, and only 12% of the patients were in NIR remission 12 months later.

The French Randomized Control Trial with Cyclosporin

The first report of a randomized control trial of cyclosporin in IDDM came from a multicenter study group in France.[26] In this trial, 122 patients were randomized for treatment with cyclosporin or a placebo; insulin therapy was continued, but an attempt was made to lower the dose as long as glycemic control was maintained at targeted levels. The patients and the clinicians managing the diabetes were blinded to the treatments. Cyclosporin was given at a pre-determined dose without adjustment on the basis of drug blood levels. The initial daily dose was 7.5 mg/kg; this was increased to 10 mg/kg after three months if the insulin dose exceeded 50% of that required for metabolic control at entry. Remission was considered "complete" when blood glucose levels were maintained at less than 7.8 mmol/l before meals, and less than 11.1 mmol/l after meals, with glyco-hemoglobin levels at less than 7.5% (upper limit of normal, 5.8%) without insulin therapy. Remission was considered "partial" when glycemic control was maintained at targeted levels with a insulin dose of less than .25 u/kg. After six months, cyclosporin was discontinued in patients who showed no evidence of remission. Patients in remission continued to take cyclosporin, but the dose was decreased by 20% monthly, to 5 mg/kg a day by nine months. At the time of reporting, data from 106 patients were available. The complete remission rates at nine months were 24% in the cyclosporin-treated group and 6% in the placebo-treated group ($p < .01$). When the findings were examined relative to the observed cyclosporin blood levels, it was found that a substantial number of patients had maintained levels in the range considered to be subtherapeutic on the basis of experience in organ transplantation (i.e., less than 300 ng/ml whole blood trough levels, polyclonal RIA). The results are summarized in Table 11.2. At six months, the complete remission rate was 37.5% and the complete-plus-partial remission rate was 65.5% in those patients in whom the cyclosporin trough level had been maintained at 300 ng/ml or more through the first three months, compared with 21% and 32%, respectively, in placebo-treated patients. The results of this study demonstrate that cyclosporin can induce remissions of type I diabetes, according to the definitions used in the studies. The rate

TABLE 11.2. Randomized control trial: incidence of remissions according to initial cyclosporin blood trough level.[a]

Type of remission	Placebo	Cyclosporin < 300 (ng/ml)	Cyclosporin ≥ 300 (ng/ml)
Complete			
6 mos	20.8% (11/53)	25.0% (3/12)	37.5% (12/32)
9 mos	6.8% (3/44)	16.7% (2/12)	37.0% (10/27)
Partial or complete			
6 mos	32.1% (17/52)	25.0% (3/12)	65.6% (21/32)
9 mos	15.9% (7/44)	25.0% (3/12)	55.6% (15/27)

[a] Mean of five determinations (days 7 and 15, months 1, 2, and 3), corresponding to cyclosporin dosage of 7.5 mg/kg a day. Patients who had cyclosporin determination on plasma and drop-out patients are not taken into account.
From Feutren et al.[26]

of NIR remission was consistent with that observed in the open studies. Observations on the insulin secretory status and on the later clinical course of patients in remission in this study were not reported. The administration of cyclosporin in so-called therapeutic doses was associated with increased serum creatinine levels, but no major untoward effects were encountered.

The Canadian–European Randomized Control Trial with Cyclosporin

In this multicenter study, 188 patients with clinically diagnosed IDDM were studied within 14 weeks of onset of classic symptoms and with no more than six weeks of treatment with insulin.[27] The patients ranged in age from 9 to 35 years, but the great majority (86%) were 16 years old or older. All had nonfasting ketonuria at diagnosis. The study was designed to examine the effects of cyclosporin on remission, defined in terms of insulin treatment status, and on beta-cell function, defined in terms of plasma CP levels, through a minimum, one-year period. Two definitions of remission were employed. A clinical definition of NIR remission required the maintenance of the targeted control of glycemia (self- and laboratory-determined values equivalent to plasma glucose concentrations up to 7.8 mmol/l before meals) without insulin therapy. A further compound definition of remission required either the NIR state as defined above or the presence of glucagon-stimulated CP levels in the plasma at least 0.6 nmol/l, a value achieved by more than 95% of the normal subjects. The patients and the clinicians managing the administration of insulin were blinded to the treatment. The initial daily dose of cyclosporin was 10 mg/kg; this was adjusted to maintain trough levels of the drug equivalent to 400 to 600 ng/ml of whole blood (polyclonal RIA). The dose of cyclosporin was reduced if serum creatinine levels increased by more than 50% of the entry values or if other unacceptable adverse effects occurred. Placebo was substituted for cyclosporin in patients in the active treatment group who had confirmed glucagon-stimulated CP levels of less than 0.1 pmol/l. Glycemic control was monitored by serial determinations of glycated hemoglobin levels, and the

TABLE 11.3. Cyclosporin-induced remission of IDDM after early intervention.

	6 mos (n = 187)				12 mos (n = 183)			
	%	n	p^a	p^b	%	n	p^a	p^b
NIR or CP ≥ 0.6 nM								
Cyclosporin	51.6	48	.004	.001	33.0	30	.06	.004
Placebo	30.9	29			20.7	19		
NIR								
Cyclosporin	38.7	36	.003	.001	24.2	22	.009	.002
Placebo	19.1	18			9.8	9		

The minor discrepancies between numbers of subjects in the two response groups are accounted for by data falling outside predetermined time windows.
[a] χ^2–Test.
[b] Multiple logistic regression analysis, after adjustment for differences in baseline variables (sex, body mass index, stimulated CP, age ≤ 16 or > 16 years and short- versus long-duration diabetes). From the Canadian–European randomized control trial.[27] Reproduced with permission of the American Diabetes Association, Inc.

insulin secretory status was established by determining glucagon-stimulated CP levels in the plasma at three-month intervals.

Glycated hemoglobin levels did not differ in the two treatment groups through the period of study. A highly significant increase in the rate of remission to the NIR state was observed in the cyclosporin-treated group (Table 11.3). During the first three months of observation, clinical remissions to the NIR state occurred at similar rates in the two treatment groups, but from four months onwards, the rate was higher in the cyclosporin-treated group (Fig. 11.3). By six months, NIR remission had occurred in 38% of cyclosporin-treated patients, compared with 19% of placebo-treated patients; at 12 months, the rates were 24% and 9%, respectively. Multiple logistic regression analysis adjusting for differences in base-line variables showed that these differences were statistically significant. There were also significant differences between remission rates as outlined in the compound definition (Table 11.3). With respect to the effect on beta-cell function, there was a significantly greater increase in the response of plasma CP to stimulation with glucagon by 90 days in the cyclosporin-treated patients (Fig. 11.4). The enhanced beta-cell response continued at approximately the same level for the year for the cyclosporin-treated patients; in the placebo-treated group, however, the enhancement effect disappeared within 180 days, and responses declined below entry values thereafter.

This trial confirms the efficacy of cyclosporin in inducing clinical remissions in IDDM of recent onset and demonstrated that these remissions are associated with enhancement and preservation of endogenous insulin secretion. The results also indicated a correlation between the time of appearance of diabetic symptoms before entry and the response to cyclosporin (Table 11.4). Clinical remission in the cyclosporin-treated group was largely confined to the 40% of patients who entered the study within six weeks of the appearance of symptoms and within two weeks of starting insulin therapy. The rate of NIR remission at one year was 32%

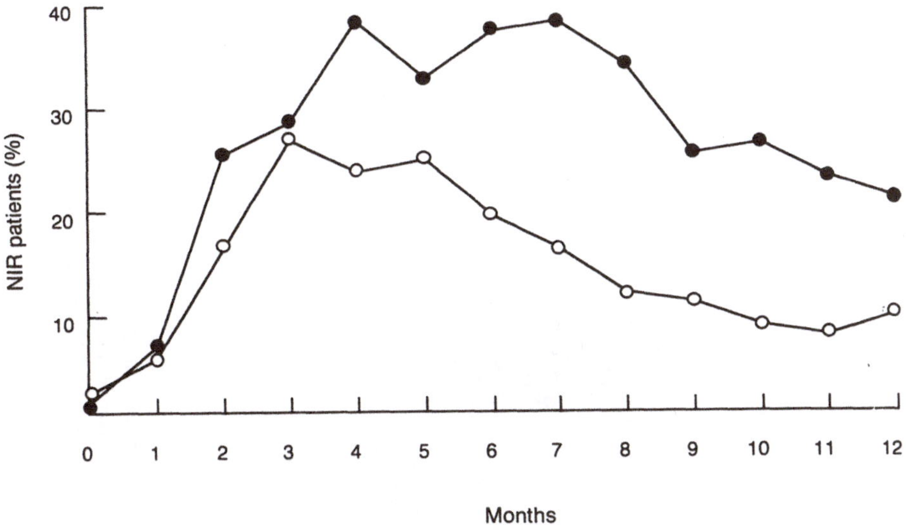

FIGURE 11.3. The NIR remission rates for cyclosporin-treated (●) and placebo-treated (○) groups of patients, at intervals through one year. From Canadian–European Randomized Control Trial.[illegible] Reproduced with permission of the American Diabetes Association, Inc.

in cyclosporin-treated patients with "short-duration" diabetes, as defined, compared with 3% in placebo patients. The other characteristics of patients with short- and longer-duration diabetes in the two treatment groups, including the levels of glucagon-stimulated CP in the plasma, were the same. The anticipated clinical and biochemical adverse effects of cyclosporin in this trial were consistent with those reported in the open studies in which similar doses of cyclosporin were used. Changes in calculated creatinine clearance agreed well with direct measurements of glomerular filtration rate in sets of 49 and 45 patients in the cyclosporin and placebo groups, respectively; thus the mean reduction of the glomerular filtration rate in terms of calculated creatinine clearance was 20%, and by direct measurement, 21%, in the cyclosporin-treated group.

Efficacy and Safety of Immunosuppression in Remission Induction in IDDM

Efficacy of Cyclosporin

The results of these randomized controlled trials with cyclosporin clearly establish the efficacy of this agent in inducing remissions, as defined; the results of the Canadian-European Study further demonstrate that this effect is associated with enhancement and preservation of beta-cell function. It is notable that at one year the rate of NIR remission, with satisfactory glycemic control, was higher in the

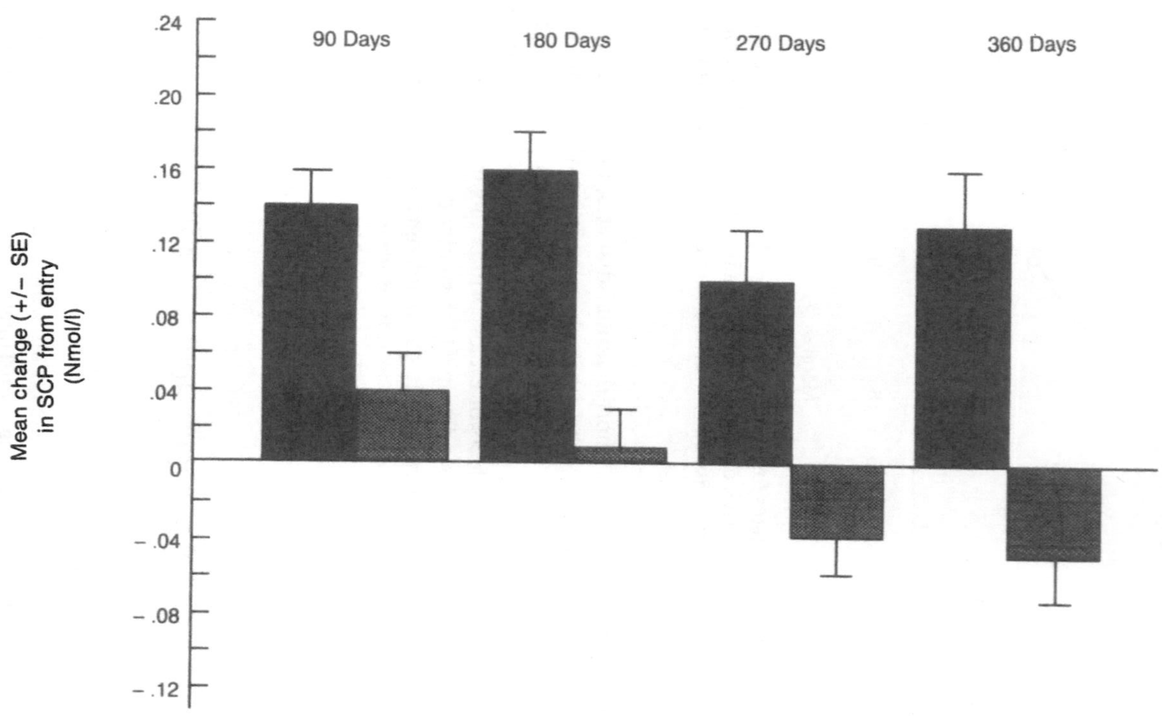

FIGURE 11.4. Mean changes (± SE) from base-line values in stimulated CP (SCP) levels in plasma in cyclosporin-treated patients (solid bars) and placebo-treated patients (hatched bars). Differences in the responses of the two groups were statistically significant at all times ($p < .01$). From Canadian–European Randomized Control Trial.[27] Reproduced with permission of the American Diabetes Association, Inc.

TABLE 11.4. Cyclosporin-induced remission of IDDM in patients with "short-duration" and "long-duration" diabetes. From the Canadian–European randomized control trial.[27]

Outcome	6 mos Short[a]	6 mos Long	12 mos Short[a]	12 mos Long
NIR or CP $\geq$ 0.6 nM				
Cyclosporin	71.1 (38)	38.2 (55)	39.5 (38)	28.3 (53)
Placebo	31.6 (38)	30.4 (56)	16.2 (37)	23.6 (55)
P	.001	.39	.02	.58
P diff[b]	.02		.15	
NIR				
Cyclosporin	55.3 (38)	27.3 (55)	31.6 (38)	18.9 (53)
Placebo	13.2 (38)	23.2 (56)	2.7 (37)	14.5 (55)
P	.001	.62	.001	.55
P diff[b]	.003		.06	

Values are percentages with number of subjects in parentheses. The minor discrepancies between numbers of subjects in the two response groups are accounted for by data falling outside predetermined time windows.

[a] $\leq$ 6 wk symptoms and $\leq$ 2 wk insulin before entry.

[b] Difference in treatment effect between short-duration and long-duration subsets.

Reproduced with permission of the American Diabetes Association.

open studies than in the randomized controlled trials. Although this may represent a study effect in the open condition, it should be pointed out that cyclosporin therapy was intensified, within defined safety limits, at the time of suspected relapse in the pilot study, and that this was not done in the double-blinded trials. The relatively high remission rate at one year in the Canadian open study also may be related to the use of oral hypoglycemic agents by 10% of the patients in NIR remission.

The potential for high rates of clinical remission as a result of cyclosporin treatment was realized at the time of optimum response, approximately six months from entry, when 55% of the patients with short-duration diabetes in the Canadian-European study were in NIR remission, and when 75% of the children who participated in the subsequent French open study, were in similar remission while receiving lower doses of cyclosporin.[24]

Whereas the physiologic significance of these results with respect to an understanding of the disease process is clear, the reversion of remissions in the randomized control study during the six- to twelve-month interval is of concern with respect to the use of this experimental treatment. In this regard, it is important to note that the effect of cyclosporin on the beta-cell response to glucagon was approximately at a plateau through the year of study. It is not clear whether clinical relapse, as indicated by deteriorating glycemic control, despite maintenance of beta-cell responses to nonphysiologic stimulation, can be explained by inadequate immunotherapy, by patient noncompliance with other modalities of treatment, or by the effect of one or more other mechanisms. Preliminary studies suggest that relapse may be related to changes in insulin sensitivity in the cyclosporin-treated patients,[28] as documented during conventional treatment in the early months after IDDM diagnosis.[29] It is clear that continued studies to define and characterize relapse, and its response to a variety of interventions, are necessary if a search for clinically applicable immunomodulatory programs in IDDM is to proceed.

Combinations of Drugs in Immunotherapeutic Regimens

The efficacy of cyclosporin, and the probable efficacy of azathioprine, in inducing IDDM remissions[17,18] suggests that other combinations of treatments that control allograft rejection might also be effective in IDDM. Although high-dose glucocorticoid therapy plus azathioprine therapy[18] may be effective in inducing higher rates of remission, longer-term treatment with steroids has obvious disadvantages in diabetes. The various combinations of immunosuppressive agents that have been useful in controlling rejection of transplanted organs have not yet been effectively assessed in IDDM. An apparently beneficial combination of bromocriptine and cyclosporine to treat autoimmune uveitis[30] and to control allograft rejection is another possibility.[31] Combination therapies that include several compatible, nonspecific antiinflammatory drugs also may prove to be safe and effective.

Use of Cyclosporin in IDDM

Cyclosporin is mainly toxic to the kidneys.[32,33] Selection of cyclosporin doses in the initial open studies and in the randomized control trials was based on the therapeutic range established in allograft rejection control. The immunosuppression regimens first employed in studies in diabetes were similar to the long-term regimen used to prevent rejection of a kidney transplant, during which there was no clear evidence of progressive or irreversible nephrotoxicity.[34] More recently, however, it has been reported that histologic damage, attributable to cyclosporin, in the transplanted kidney is related to the length of time the drug was administered.[35] In view of the importance of this question, the kidneys of 17 patients taking part in the Canadian–European trial were biopsied. The kidneys of 23 patients in studies using similar protocols in France and Canada were also biopsied. Histologic examination of these biopsied specimens showed changes associated with cyclosporin administration, though not specifically attributable to the drug.[27] The lesions included glomerular obsolescence or segmental focal sclerosis, arteriolopathy, tubular atrophy, and interstitial fibrosis, each of which was rated on a five-point scale. The median score for glomerular change was zero, but one biopsy had abnormalities in more than 10% of glomeruli. The commonest finding was interstitial fibrosis, with a median score of "slight," but this was rated "moderate" in eight biopsies. The median score for tubular atrophy was "minimal–slight," but this lesion was rated "moderate" in seven biopsies. This last abnormality was the only one that could be correlated with trough levels of cyclosporin in the blood. The incidence of tubular atrophy showed a median score of "minimal" in biopsies from patients in whom the mean cyclosporin trough levels had been less than the approximate median value for the studies. These findings suggest that an upper limit of 400 ng/ml whole blood for the 12-hour trough levels of cyclosporin (polyclonal RIA) would minimize kidney damage. This expectation is supported by the results from the recent open study with cyclosporin in n=40 children in France, discussed above,[24] in which the cyclosporin trough level in the blood was below 350 ng/ml and in which there were no functional effects on the kidney, and no definite histologic abnormality in the 19 kidney biopsies. In this study, the remission rate was comparable to that observed with higher doses of cyclosporin. Thus it appears that a cyclosporin dose that is not demonstrably nephrotoxic during these relatively short studies can induce high rates of remission and is safe to use in further studies.

Timing of Immunomodulatory Interventions in IDDM

Preventive Immunosuppression in Prodromal IDDM

Phenomena that appear to be autoimmune have been detected over a period of years before overt diabetes occurred in high-risk siblings of patients with IDDM. The presence of these phenomena strengthens the autoimmune hypothesis for the etiology of IDDM, and raises the question whether immunomodulatory intervention in the prodromal phase of the disease should be considered. Given the generally accepted view that, by the time overt diabetes develops, there is an irreversible destruction of beta-cells amounting to 90% or more of the mass of insulin-secreting tissue, it appeared that this might be the only feasible approach to immunotherapy in IDDM. As discussed below, this assessment of the potential for recovery of insulin-secretory capacity in overt IDDM is open to question. In any case, preventive immunomodulatory therapy in prodromal IDDM clearly depends on correctly identifying patients at sufficient risk and appropriately timing the intervention. The high rate of discordance in identical twins,[36] and the highly variable time-course of development of the overt disorder in concordant twins and in haplo-identical siblings of patients with IDDM, emphasizes the need for means of staging the process.[37] Further complicating this issue is the unresolved question of whether progression of the process in prodromal IDDM occurs at a steady rate until an insufficient mass of functioning beta-cells remains or whether the disease goes through remitting and relapsing phases. On the basis of long-term studies of humoral immune autoreactivity, it has been suggested that the process in identical twin- or haplo-identical siblings who develop IDDM progresses steadily, and that subnormal insulin secretory responses develop gradually.[37] It might therefore be argued that immune and metabolic parameters can be monitored to obtain prognostic information. This conclusion may be questioned in light of other studies in which a waxing and waning of the immunologic markers of autoimmune reactivity over long periods of time have been described.[38,39] Furthermore the results from studies of high-risk siblings of patients with IDDM may or may not apply to the great majority of patients with no family history of IDDM who develop the syndrome. Until these questions are answered, preemptive intervention in high-risk siblings of diabetic subjects, or in subjects without a strong family history, who might be identified by screening programs for immune autoreactivity, may not be justifiable.

Secondary Interventional Immunosuppressive Therapy in Overt IDDM

With the development of assays for the insulin-connecting peptide, it was recognized that the beta-cell secretory responses, that are very low at the time of diagnosis of IDDM, commonly improve during the initial few months of conventional insulin therapy and then decline over a longer period.[11,12] These observations are not consistent with a constant-rate model for beta-cell destruction, and demonstrate a capacity for partial functional recovery of the beta-cells at this stage of the disease. The results of the Canadian pilot study further suggested that immunosuppression therapy started at eight weeks or more after the initiation of insulin treatment would have little effect, regardless of the prevailing level of CP in the plasma.[20] In this study, as well as in the Canadian–European randomized controlled trial and the French open study in children, it also appeared that a good clinical outcome depends on the timing of the intervention. In the Canadian–European randomized control trial, the beneficial effect of cyclosporin was largely confined to patients with symptoms of less than six weeks at entry and with no more than two weeks of insulin therapy (short-duration diabetes).[27] This did not appear to depend simply on time of appearance of symptoms before entry, since the "spontaneous" remission rate in the placebo group was lowest in patients with short-duration diabetes. It was therefore postulated that the short period of overt diabetes associated with remission induction is related to an acute acceleration of beta-cell damage, which is particularly marked in patients with a brief clinical history. This interpretation suggests that the plasma CP response in relation to residual beta-cell mass is different for patients with short-duration diabetes than for patients with longer-duration diabetes, and that greater impairment of beta-cell function with a lesser degree of beta-cell destruction may occur in patients who have had diabetic symptoms for less than six weeks. This is consistent with the finding that cyclosporin can induce high rates of remission in children with very short histories of symptomatic disease.[24]

These findings suggest that the beta-cells may be undergoing a specific immunoreactive attack, with an associated inflammation, compounded by the adverse effects of metabolic decompensation. It follows that immune interventions may be more effective earlier in the course of the disease. This might be tested by undertaking such treatments on an emergent basis in patients with newly diagnosed IDDM. Since it is also clear that such patients go through a period of many weeks with supranormal blood glucose levels in the absence of symptoms, it may be possible to plan studies in which such patients are identified in screening programs to detect overt hyperglycemia in definable, high-risk groups.

IDDM in Remission: Endocrine Metabolic Status and Its Possible Implications

Insulin Secretory Capacity in Patients in NIR Remission

Patients in NIR remission maintain normal-range glycated hemoglobin levels, and many of them also have glucagon-stimulated CP levels within the normal range. The NIR patients in the Canadian–European randomized trial had mean values for glucagon-stimulated CP that were approximately 50% of the mean normal level in both cyclosporin- and placebo-treated groups.[27] Moreover determinations during the first year of observation of glucagon-stimulated CP in all patients receiving cyclosporin in the Canadian open study showed that patients in NIR remission had a normal mean glucagon-stimulated plasma CP response that could not be distinguished from that of normal subjects (Fig. 11.5).[21] There was, however, considerable overlap of glucagon-stimulated CP levels between patients who entered NIR remissions and those who did not. This could reflect a normal range of insulin sensitivity among patients with a relatively limited insulin secretory capacity. It is evident, however, that the changes in insulin sensitivity, discussed above,[28,29] may not only play a role in determining which patients enter NIR remission, but may also affect the longer term stability of remissions.

Endocrine Responses to Nutrients in Remission Phase IDDM

There have been few studies of the physiologic responses of the endocrine pancreas in patients under conventional treatment for IDDM during the remission phase, but preliminary reports of studies in patients with cyclosporin-induced remissions have appeared.[40] There is little or no release of immunoreactive insulin or CP into the blood in response to intravenous glucose alone, in patients in NIR remission with normal levels of glycated hemoglobin. Upon ingestion of glucose, a grossly obtunded, but statistically significant increase of blood levels of insulin and CP occurred. The responses of insulin and CP following ingestion of a standardized mixed meal (Sustacal), however, were greater and were associated with only a minor degree of glucose intolerance. Thus these patients showed an obvious glucose intolerance with intravenous or oral glucose, but glucose tolerance after ingestion of the mixed meal was only mildly impaired and was consistent with their maintenance of normal-range glycated hemoglobin levels under physiologic conditions.

It appears that patients in remission have the same abnormal response to glucose as that described in the preclinical phase of the disease in high-risk siblings of diabetic patients. This suggests that beta-cell function persists, that patients in remission revert to a condition similar to that of patients in the preclinical phase, or that the defective insulin response to glucose in remission-phase patients is attributable wholly or in part to the inhibitory effect of cyclosporin on insulin release.[41]

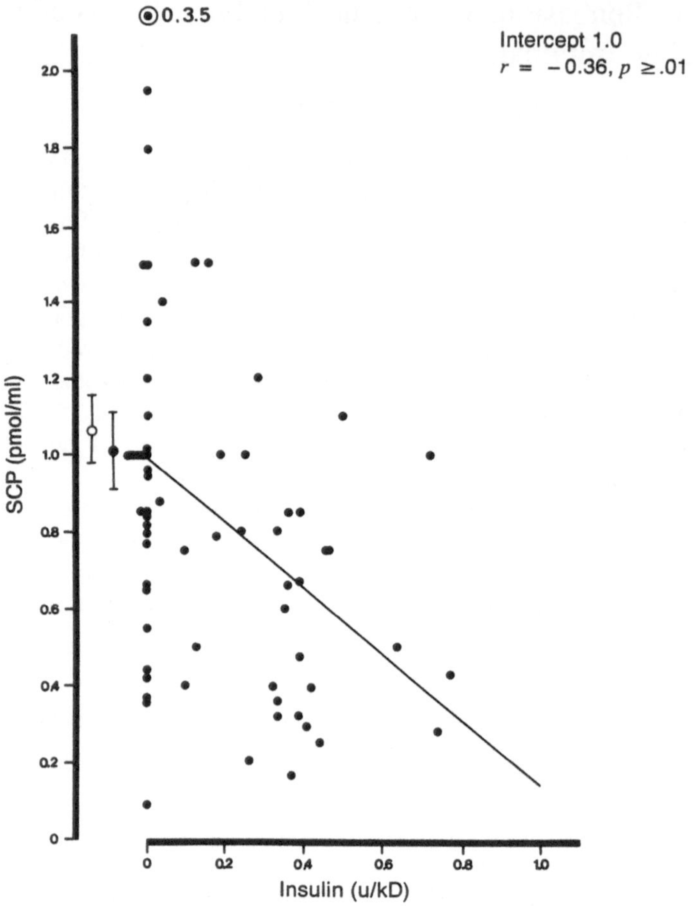

FIGURE 11.5. Stimulated plasma CP levels (SCP) and the concurrent daily insulin dose in individual patients after three months of treatment. The solid symbol with bars indicates the mean ± SEM for all NIR patients; the open symbol with bars indicates the mean ± SEM in 22 normal subjects. From ref. 21, reprinted with permission.

Endocrine Metabolic Effects of Cyclosporin

At present it seems unlikely that any direct, endocrine-metabolic effects of cyclosporin are important in determining the metabolic status of cyclosporin-treated patients with IDDM. Intravenous glucose tolerance and the associated insulin response, for example, are normal in multiple sclerosis patients receiving doses of cyclosporin that overlap those doses employed in studies in diabetes.[42] Studies of erthrocyte binding of insulin have demonstrated no major abnormality in cyclosporin-treated patients with IDDM.[43] Moreover preliminary observations

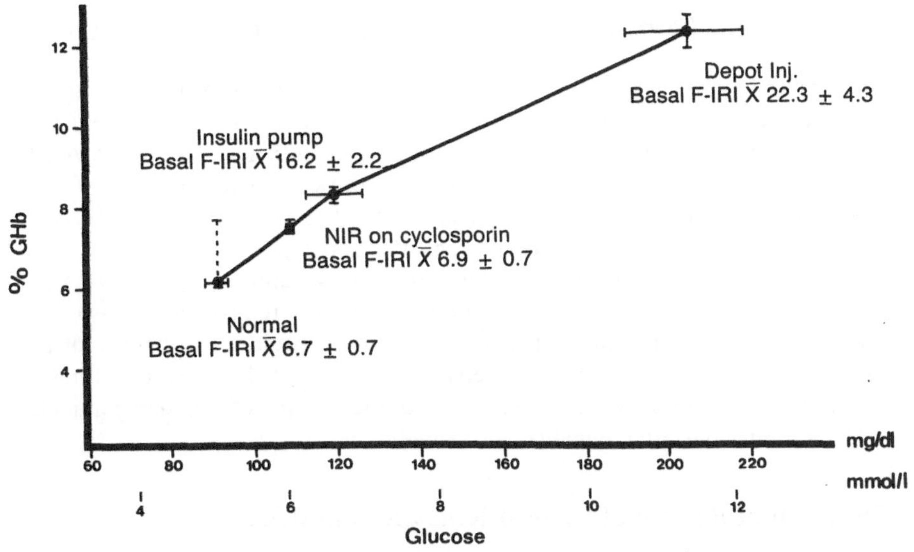

FIGURE 11.6. The blood glycated hemoglobin level (colorimetric method, normal value with SD shown) versus the mean diurnal plasma glucose level in 21 normal subjects, 31 patients in NIR remission of type I IDDM undergoing cyclosporin therapy, patients with CP-negative IDDM undergoing intensified insulin therapy by long-term continuous subcutaneous (SC) insulin infusion, and 23 conventionally treated (depot SC injections of insulin once or twice a day) patients with CP-negative IDDM. The mean basal (postabsorptive) free insulin level (F-IRI, μ/ml ± SEM) for each group is shown (unpublished observations of the authors).

of insulin sensitivity in the cyclosporin-treated, IDDM patients cited above indicate that this function is normal in patients in NIR remission.

In summary it appears that IDDM remissions induced by immunosuppression with cyclosporin are characterized by a continuing abnormality of the beta-cell response to stimulation with glucose, with relatively well-preserved responses to mixed meals. Such responses under everyday conditions are presumably mediated by nonglucose nutrients and/or by enteroinsular mechanisms. Thus these patients can maintain normal-range glycated hemoglobin levels associated with a subnormal but substantial insulin response to meals. In this respect they differ from truly insulin-dependent diabetics who achieve similar control as a result of intensive insulin therapy and who manifest elevated fasting and integrated diurnal plasma levels of free immunoreactive insulin (Fig. 11.6).[44] Although the limited information available indicates that there is no endocrine or metabolic effect of cyclosporin other than that of enhancing beta-cell function in IDDM, much remains to be learned about the physiology of "spontaneous" and treatment-induced remission of IDDM.

Strategies for Continuing Studies of Immunotherapy in IDDM

The results to date suggest that immunosuppression, with cyclosporin as the model agent, can induce remission to an NIR state by six months from entry in approximately 50% of patients selected according to the criteria discussed. A significant rate of relapse to insulin requirement in the second half of a year of such treatment is to be expected under the conditions already tested. If the objective of treatment is induction of NIR remissions, it is apparent that experimental programs designed to explore the usefulness of immunosuppression in IDDM not only must achieve comparable or higher optimal rates of remission, but also must maintain remissions through a period of time that would allow studies of the risk/benefit ratios of conventional versus experimental therapies to be made. These considerations raise the question of whether relapses from therapeutically induced remissions are amenable to treatment.

Treatment of Relapse of Clinical Remissions in IDDM

Although experience with immunosuppression in IDDM has emphasized the importance of very early intervention by showing that clinical remissions of the disease are rarely induced by treatments initiated more than a few weeks after onset of symptoms, it also has been noted in open studies that cyclosporin can induce secondary remissions after relapse occurring after interruption or reduction of dose, many months after diagnosis.[21] In the Canadian open study, exploratory trials of alternative immunotherapies with agents other than cyclosporin, implemented more than one year after entry, resulted in loss of glycemic control, which was recovered upon resumption of protocol treatment with cyclosporin in 60% of the cases. It should also be noted, however, that relapses in this open study were treated with much higher doses of cyclosporin than would now be employed and that the apparent efficacy of cyclosporin under these conditions does not apply to the treatment of relapses occurring in the course of continued administration of cyclosporin according to protocols that have been shown to induce remissions. Further experience in the detection and treatment of immunologic relapse in IDDM is clearly essential to improve IDDM immunotherapy. Although monitoring of glycemic control is a practical means of detecting relapse, it is nonspecific and cannot be used to anticipate a clinical relapse. There is clearly a need for assays that will monitor the immune response in this context. A further problem in interpreting these studies depends on the means and effects of glycemic control with insulin. The importance of this factor is emphasized by evidence that treatment with exogenous insulin can promote recovery of beta-cell function,[45] as suggested by the results of earlier studies of intensive insulin therapy in IDDM of recent onset.[46] Further studies on the effects of insulin alone on beta-cell function in early IDDM are needed. It is reasonable, meanwhile, to adopt strict normoglycemia as the target for metabolic control in our search for treatments to induce remission.

Duration of Immunotherapy in IDDM

The available information strongly suggests that long-term immunomodulatory therapies will be necessary to maintain remissions in IDDM. This is indicated not only by the studies of the short-term experimental immunosuppression in IDDM of recent onset, in which there is a likelihood of relapse on interruption of therapy at one- to four-year intervals from immunosuppression initiation,[21] but also by studies of segmental pancreas transplantation in discordant identical twins cited earlier.[6] In these studies, in the absence of immunosuppressive therapy, beta-cells derived from the long-term IDDM discordant twin were destroyed by an inflammatory process confined to the islets, which was associated with evidence of a resumption of immune autoreactivity in the recipient. It is clear that a situation in which a missing beta-cell antigen is again presented after a long interval is quite different from a situation in which residual beta-cells may survive as a result of adaptation or manipulation of the immune response.

Possible Modulation of Immunotherapy Relative to the Autoimmune Process

A pathologic, but subclinical level of immune autoreactivity that is qualitatively similar to the level of immune reactivity leading to overt disease may occur in the absence of progression to clinical expression, as is recognized in the case of autoimmune disorders affecting other endocrine target cells. In this regard, it is of interest that there are detectable immunologic[38] and endocrine-metabolic abnormalities[47] in nondiabetic siblings among discordant identical twins. These abnormalities may persist over long periods without evidence of progression. Although it has been suggested that discordance in this situation may be the result of variations in the nature or intensity of exogenous provocative stimuli, it is also possible that differences in autoregulation of the immune system, attributable to the developmental generation of immunologic diversity among individuals, may allow one twin to maintain an equilibrium in which the "activated" process does not progress. Self-tolerance might also be reattainable as a result of appropriate immunomodulatory interventions, as suggested by the observation that monoclonal antibodies directed against a helper T-cell population can induce long-term remissions of IDDM in the NOD mouse, in spite of a continuing inflammatory process in the islets of Langerhans.[48] The relevance of this finding to the human disease is suggested by a recent report documenting the presence of analogous T-cells with HLA-restricted cytotoxicity in the peripheral blood of patients with recent-onset IDDM.[49]

It has been pointed out that cell-mediated effector mechanisms of autoimmune disease may fluctuate in expression or intensity. Such fluctuations would not necessarily be accompanied by parallel changes in indices of humoral reactivity. It is therefore important to develop tests of immunologic reactivities that might serve to predict exacerbations or relapses, so that intermittent immunosuppressive therapies might be tested for their ability to maintain beta-cell mass, with or without a background of less intensive immunomodulatory therapy. It is also possible that identification of beta-cell antigens associated with the development of IDDM[50] would facilitate more specific approaches to treatment, by neutralizing or eliminating selected effector cells and "retolerizing" the host to self.

Clincal Significance of Beta-Cell Function Preservation in IDDM

It is important to consider the possibility that a long-term clinical benefit might accrue from preserving a level of beta-cell secretory capacity that is not sufficient to allow withdrawal of insulin therapy. Although the NIR remission to date generally has been the goal of experimental immunotherapy in IDDM in humans, preserving the characteristics of remission-phase IDDM, even with continued administration of insulin, might be of clinical benefit, provided the effective therapy were sufficiently low risk. In the remission phase of IDDM, maintaining virtual normoglycemia with simple regimens of low-dose insulin therapy, with very little risk of hypoglycemia, is possible in many patients. It has been pointed out that blood levels of insulin in patients in NIR remission are in the normal range, and the authors unpublished data indicate that this is also true of patients undergoing the low-dose insulin therapy required for strict glycemic control under these conditions. There is evidence that endogenous insulin secretion contributes to the efficacy of insulin in controlling glycemia even when residual beta-cell function is low.[51] It also appears that modest preservation of beta-cell function is associated with reduced doses of insulin and relatively low levels of glycated hemoglobin in remission-phase IDDM.[52] There is also evidence that patients receiving less than 0.5 units/kg of insulin a day are at low risk for the long-term complications of IDDM and that demonstrable residual beta-cell function occurs in many patients in this category decades after diagnosis.[53,54] It is therefore possible that the preservation of definable levels of CP in the blood might be associated with reduced morbidity and mortality in insulin-treated IDDM.

Summary and Conclusions

The autoimmune hypothesis for the mechanism of progressive beta-cell destruction in IDDM was developed largely on the basis of circumstantial evidence in man and has been supported by the results of interventional studies with immunomodulatory therapies in animal models. The finding that immunosuppression can enhance and preserve beta cell function, at least temporarily, in overt insulin-dependent diabetes of recent onset constitutes virtual proof that beta cell damage is mediated by an immune process in the human disease. These effects of immunosuppression with cyclosporine are associated with an increased rate of remission to a state in which glycemic control is maintained without administration of insulin. Studies in patients in remission have also revealed the

Phase	Stages of Type I Diabetes Mellitus (DM)	Principal Objectives of Therapy	Possible Modes of Therapy
PRE-CLINICAL PHASE	I PRE-PRO-DM (0 - 20 YEARS)	ABROGATE AUTOIMMUNE PROCESS PRESERVE NORMAL METABOLISM PREVENT ACUTE/CHRONIC COMPLICATIONS	(? ENVIRONMENTAL) IMMUNOTHERAPY
	II PRO-DM (1 - 10 YEARS)	CONTROL AUTOIMMUNE PROCESS PRESERVE NORMAL METABOLISM PREVENT ACUTE/CHRONIC COMPLICATIONS	IMMUNOTHERAPY ± ENVIRONMENTAL
REMISSION-PHASE	III CP-POS DM (1 - 5 YEARS)	PRESERVE/ENHANCE RESIDUAL BETA CELL FUNCTION PRESERVE NORMAL METABOLISM PREVENT ACUTE/CHRONIC COMPLICATIONS	IMMUNOTHERAPY ± ENVIRONMENTAL ± EXOGENOUS INSULIN
INSULIN-DEPENDENT PHASE	IV CP-NEG DM	OPTIMIZE METABOLIC CONTROL, AMELIORATE ACUTE AND MANAGE CHRONIC COMPLICATIONS	EXOGENOUS INSULIN, OR/AND PANCREAS/ISLET TRANSPLANT WITH IMMUNOTHERAPY
	V CP-NEG DM WITH ORGAN FAILURE	PRESERVE ORGAN FUNCTION COMPENSATE ORGAN FAILURE	AS FOR STAGE IV ± ORGAN REPLACEMENT WITH IMMUNOTHERAPY

FIGURE 11.7. Possible treatment modalities and their objectives at different stages of IDDM.

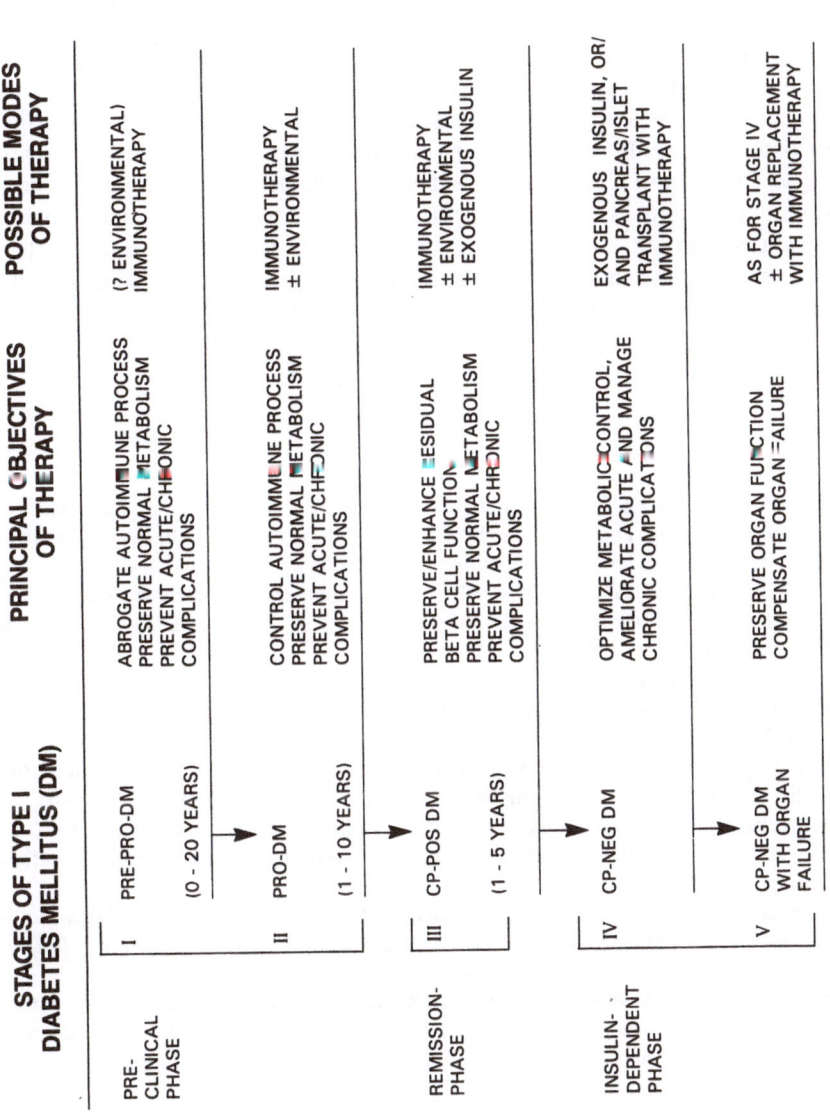

importance of insulin sensitivity as a determinant of the clinical response, indicating the need for further studies of the factors controlling this function.

In view of the present difficulty of recognizing and staging the prodromal phase of IDDM, the search for improved therapies must continue in overt disease of very recent onset. Concurrent development of more powerful and safer modes of immunosuppression may proceed more effectively in the context of control of allograft rejection, and can be expected to suggest new treatments for autoimmune disease. In the future, better means of identifying a patient at high risk for development of IDDM may provide for studies of similar inverventions at earlier stages of the disease. In considering the various treatment modalities that might be employed (Figure 11.7) it is clear that effective control of the immune process is essential in all approaches to management of the active disease, and probably also in attempts to prevent it. All such studies must be viewed as means of arriving at protocols that can ultimately be tested for clinical benefit, taking account of the longterm risks of the therapy as well as of the disease.

References

1. Nerup J, Mandrup-Poulsen T, Molvig J. The HLA–IDDM association: Implications for etiology and pathogenesis of IDDM. *Diabetes Metab Rev* 1987;3:779–802.
2. Bruserud O, Jerwell J, Thorsby E. HLA-DR3 and -DR4 control T-lymphocyte responses to mumps and Coxsackie B4 virus: Studies on patients with Type-1 (insulin-dependent) diabetes and healthy subjects. *Diabetologia* 1985;28:420–426.
3. Irvine WJ, Gray RS, Steel JM. Islet cell antibody as a marker for early stage type I diabetes mellitus, in Irvine WJ (ed): *The Immunology of Diabetes*. Edinburgh: Teviot Publishing, 1980, pp 117–154.
4. Gorsuch AN, Spencer KM, Lister J, Wolf E, Bottazzo GF, Cudworth AG. Can future Type I diabetes be predicted? A study in families of affected children. *Diabetes* 1982; 31:862–866.
5. Boitard C, Debray-Sachs M, Pouplard A, Assan R, Hamburger J. Lymphocytes from diabetic patients suppress insulin release in vitro. *Diabetologia* 1981;21:41–46.
6. Sutherland DER, Sibley R, Chinn P. Twin-to-twin pancreas transplantation (tx): Reversal and reenactment of the pathogenesis of type 1 diabetes. *Clin Res* 1984;32: 561A.
7. Rossini AA, Mordes JP, Like AA. Animal models of insulin-dependent diabetes mellitus, in Andreani D, DiMario U, Federlin KF, Heding LG (eds): *Immunology in Diabetes*. London, Kimpton, 1984, pp 35–44.
8. Like AA, Biron CA, Weringer EJ, Byman K, Sroczynski E, Guberski DL. Prevention of diabetes in Bio-Breeding/Worcester rats with monoclonal antibodies that recognize T-lymphocytes or natural killer cells. *J Exp Med* 1986;174:1145–1159.
9. Godtfredsen CF, Buschard K, Frandsen EK. Reduction of diabetes incidence of BB Wistar rats by early prophylactic insulin treatment of diabetes-prone animals. *Diabetologia* 1985;28:933–935.
10. Makino S, Kanimoto K, Muraoko Y, Mizushima Y, Katagari K, Tochinoy. Breeding of non-obese diabetic strain of mice. *Exp Animals* 1980;29:1–13.
11. Sochett E, Daneman D, Clarson C, Ehrlich R. Factors affecting and patterns of residual insulin secretion during the first year of Type I (insulin-dependent) diabetes mellitus in children. *Diabetologia* 1987;30:453–459.

12. Wallensteen M, Dahlquist G, Persson B, Landin-Olsson M, Lernmark A, Sundkvist G, Thalme B. Factors influencing the magnitude, duration, and rate of fall of B-cell function in Type I (insulin-dependent) diabetic children followed for two years from their clinical diagnosis. *Diabetologia* 1988;31:664–669.
13. Drash AL (ed). *Clinical Care of the Diabetic Child.* Chicago: Year Book Medical Publications, Inc., 1987, pp 33–51.
14. Leslie RDG, Pyke DA, Denman AM. Immunosuppressive therapy in diabetes. *Lancet* 1985;i:516.
15. Elliott RB, Crossley JR, Berryman CC, James AG. Long-term outcomes of children with insulin dependent diabetes mellitus treated ab initio with prednisone. *Pedr Adolesc Endocrinol* 1986;15:345–349.
16. Ludvigsson J, Heding L, Leiden G, Marner B, Lernmark A. Plasmapheresis in the initial treatment of insulin dependent diabetes mellitus in children. *Br Med J* 1983;286: 176–179.
17. Harrison LC, Colman PG, Dean B, Baxter R, Martin FIR. Increase in remission rate in newly diagnosed type I diabetic subjects treated with azathioprine. *Diabetes* 1985; 34:1306–1308.
18. Silverstein J, MacLaren N, et al. Immunosuppression with azathioprine and prednisone in recent onset IDDM. *N Engl J Med* 1988;319:599–604.
19. Stiller CR, Laupacis A, Keown PA, Gardell C, Dupre J, Thibert P, Wall W. Cyclosporin action, pharmacokinetics and effect in the BB rat model. *Metabolism* 1983;32 (Suppl 1):69–72
20. Stiller CR, Dupre J, Gent M, Jenner MR, Keown PA, Laupacis A, Martell R, Rodger NW, von Graffenried B, Wolfe BMJ. Effects of cyclosporine immunosuppression in insulin-dependent diabetes mellitus of recent onset. *Science* 1984;223:1362–1367.
21. Stiller CR, Dupre J, Gent M, Donner A, Mahon J, Jenner MR, Keown PA, Laupacis A, Martell R, Rodger NW, von Graffenried B. Effects of immunosuppression with cyclosporine in an insulin dependent diabetes mellitus of recent onset: The Canadian open study at 44 months. *Transplant Proc* 1988;XX(Suppl 4):184–192.
22. Stiller CR, Dupre J, Gent M, Heinrichs D, Jenner MR, Keown PA, Laupacis A, Martell R, Rodger NW, Wolfe BMJ, Mahon J. Effects of cyclosporine in recent onset juvenile type I diabetes: Impact of age and duration of disease. *J Pediatr* 1987;111: 1069–1072.
23. Assan R, Debray-Sachs M, Laborie C, Chatenoud L, Feutren G, Quiniou-Debrie MC, Thomas G, Bach JF. Metabolic and immunological effects of cyclosporin in recently diagnosed type I diabetes mellitus. *Lancet* 1985;i:67–71.
24. Bougneres PF, Carel JC, Castano L, Boitard C, Gardin JP, Landais P, Hors J, Mihatsch MJ, Paillard M, Chaussain JL, Bach JF. Factors determining very early remission of type I diabetes in children treated with cyclosporin A. *N Engl J Med* 1988;318: 663–671.
25. Koivisto VA, Aro A, Cantell K, Haataja M, Huttunen J, Karonen SL, Mustajoki P, Pelkonen R, Seppala P. Remissions in newly diagnosed type I (insulin-dependent) diabetes: Influence of interferon as an adjunct to insulin therapy. *Diabetologia* 1984;27: 193–197.
26. Feutren G, Papoz L, Assan R, Vialettes B, Karsenty G, Vexiau P, Du Rostu H, Rodier M, Sirmai J, Lallemand A, Bach JF. Cyclosporin increases the rate and length of remissions in insulin-dependent diabetes of recent onset. *Lancet* 1986;ii:119–129.
27. Canadian–European Diabetes Study Group. Cyclosporin-induced remission of IDDM after early intervention: Association of 1 year of cyclosporin treatment with enhanced insulin secretion. *Diabetes* 1988;37:1574–1582.

28. Hramiak I, Finegood D, Dupre J. Insulin sensitivity in cyclosporine treated type I diabetes in remission. *Diabetes* 1988;37(Suppl. 1):16A.
29. Yki-Jarvinen H, Koivisto VA. Natural course of insulin resistance in Type I diabetes. *N Engl J Med* 1986;315:224–230.
30. Nussenblatt RB, Palestine AG, Chan CC, Breen L, Caruso R. Improvement of uveitis and optic nerve disease by cyclosporine in a patient with multiple sclerosis. *Am J Ophthalmol* 1984;97:790–791.
31. Hiestand PC, Gale JM, Mekler P. Soft immunosuppression by inhibition of prolactin release: Synergism with cyclosporin in kidney allograft survival and in the localized graft-versus-host reaction. *Transplant Proc* 1986;28:870–872.
32. Myers BD, Ross J, Newton L, Luetscher J, Perlroth M. Cyclosporine-associated chronic nephropathy. *N Engl J Med* 1984;311:699–705.
33. Mihatsch MJ, Thiel G, Ryffel B. Cyclosporin-associated nephropathy, in Schindler R. (ed): *Cyclosporin in Autoimmune Diseases*. Berlin: Springer-Verlag, 1985, pp 50–58.
34. The Canadian Multicentre Transplant Study Group: A randomized clinical trial of cyclosporine in cadaveric renal transplantation. *N Engl J Med* 1986;314:1219–1225.
35. Ruiz P, Kolbeck PC, Scroggs MW, San Filippo F. Association between cyclosporine therapy and interstitial fibrosis in renal allogrant biopsies. *Transplantation* 1988;45: 91–95.
36. Barnett AH, Eff C, Leslie RDG, Pyke DA. Diabetes in identical twins. A study of 200 pairs. *Diabetologia* 1981;20:87–93.
37. Srikanta S, Ganda OP, Eisenbarth GS, Soeldner JS. Islet-cell antibodies and beta-cell function in monozygotic triplets and twins initially discordant for Type-I diabetes mellitus. *N Engl J Med* 1983;308:322–325.
38. Milward BA, Alviggi L, Hoskins PJ, et al. Immune changes associated with insulin dependent diabetes may remit without causing the disease: A study in identical twins. *Br Med J* 1986;292:793–796.
39. McCulloch DK, Benson E, Johnston C, Raghu P, Klaff L, Palmer JP. Subclinical insulin dependent diabetes may be common among siblings of type I diabetics. *Diabetes* 1986;35:77A.
40. Dupre J, Stiller CR, Jenner M, Mahon J, Keown P, Rodger NW, Wolfe BM. Responses to nutrients in non-insulin requiring (NIR) remission of type I diabetes during administration of cyclosporin. *Diabetes* 1987;36(Suppl I):74A
41. Yale JF, Roy RD, Grose M, Seemayer TA, Murphy GF, Marliss EB. Effects of cyclosporine on glucose tolerance in the rat. *Diabetes* 1985;34:1309–1313.
42. Robertson RP, Franklin G, Nelson L. Intravenous glucose tolerance and pancreatic islet B cell function in patients with multiple sclerosis during two years' treatment with cyclosporine. *Diabetes* 1989;38:58–64.
43. Behme MT, Dupre J, Stiller C. Effect of cyclosporine on insulin binding to erythrocytes in Type I diabetes mellitus of recent onset. *Clin Invest Med* 1988;11:113–122.
44. Bergenstal BM, Dupre J, Lawson PM, Rizza RA, Rubenstein AH for the Kroc Collaborative Study Group. Observations on C-peptide and free insulin in the blood during continuous subcutaneous insulin infusion and conventional insulin therapy. *Diabetes* 1985;34(Suppl 31):31–36.
45. Shah SC, Malone JI, Simpson NE. Suppression of endogenous insulin secretion in new onset IDDM prolongs beta cell function. *Diabetes* 1987;36(Suppl 1):61A.
46. Mirouze J, et al. Les remissions diábetiques en cours d'insulinotherapie conventionelle ou asservie a l'aide d'un pancréas artificiel. *Sem Hop Paris* 1979;55:354–359.

47. Heaton DA, Millward BA, Gray P, Tun Y, Hales CN, Pyke DA, Leslie RDG. Evidence of B cell dysfunction which dose not lead on to diabetes: A study of identical twins of insulin dependent diabetics. *Br Med J* 1987;294:145–146.
48. Shizuru JA, Taylor-Edwards Cariel, Banks BA, Gregory AK, Fathman CG. Immunotherapy of the nonobese diabetic mouse: Treatment with an antibody to T-helper lymphocytes. *Science* 1988;240:659–662.
49. De Berardinis P, James RFL, Wise PH, Londei M, Lake SP, Feldman M. Do CD4-positive cytotoxic T cells damage islet B cells in type I diabetes? *Lancet* 1988; ii:823–824.
50. Baekkeskov S, Landin M, Kristensen JK, et al. Antibodies to a 64,000 M_r human islet cell antigen precede the clinical onset of insulin-dependent diabetes. *J Clin Invest* 1987;79:926–934.
51. Rodger NW, Dupre J, Canny CLB, Brown WF. Continuous subcutaneous insulin infusion in adults: Glycemic advantage is predicted by venous plasma C-peptide concentrations. *Diabetes Care* 1985;8:447–455.
52. Effects of age, duration, and treatment of IDDM on residual B-cell function: Observations during eligibility testing for the DCCT. *J Clin Endocrinol Metab* 1987;65:30–36.
53. Eff CL, Faber O, Deckert T. Persistent insulin secretion assessed by plasma C-peptide estimation in long-term juvenile diabetes with a low insulin requirement. *Diabetologia* 1978;15:169–172.
54. Deckert T. Influence of supervision and endogenous secretion of insulin on the course of IDDM. *Acta Endocrinol (Copenh)* 1980;94(Suppl 238):31–38.

Index

Note: Page numbers in *italics* refer to illustration; page numbers followed by t refer to tables.

C

D

H

I